AF790288

For over 35 years, biological scientists have come to rely on the research protocols and methodologies in the critically acclaimed Methods in Molecular Biology series. The series was the first to introduce the step-by-step protocols approach that has become the standard in all biomedical protocol publishing. Each protocol is provided in readily-reproducible step-by step fashion, opening with an introductory overview, a list of the materials and reagents needed to complete the experiment, and followed by a detailed procedure that is supported with a helpful notes section offering tips and tricks of the trade as well as troubleshooting advice. These hallmark features were introduced by series editor Dr. John Walker and constitute the key ingredient in each and every volume of the Methods in Molecular Biology series. Tested and trusted, comprehensive and reliable, all protocols from the series are indexed in PubMed.

Manuscript Instructions https://drive.google.com/file/d/1K4Az5vpAFVzYadJ3jKx2-QSo7Dqa_SNA/view?usp=sharing

Bispecific Antibodies

Methods and Protocols

Edited by

Masahiro Yasunaga

Division of Developmental Therapeutics, National Cancer Center, Kashiwa City, Japan

Humana Press

Editor
Masahiro Yasunaga
Division of Developmental Therapeutics
National Cancer Center
Kashiwa City, Japan

ISSN 1064-3745 ISSN 1940-6029 (electronic)
Methods in Molecular Biology
ISBN 978-1-0716-5036-3 ISBN 978-1-0716-5037-0 (eBook)
https://doi.org/10.1007/978-1-0716-5037-0

© Springer Science+Business Media, LLC, part of Springer Nature 2026

This work is subject to copyright. All rights are solely and exclusively licensed by the Publisher, whether the whole or part of the material is concerned, specifically the rights of translation, reprinting, reuse of illustrations, recitation, broadcasting, reproduction on microfilms or in any other physical way, and transmission or information storage and retrieval, electronic adaptation, computer software, or by similar or dissimilar methodology now known or hereafter developed.
The use of general descriptive names, registered names, trademarks, service marks, etc. in this publication does not imply, even in the absence of a specific statement, that such names are exempt from the relevant protective laws and regulations and therefore free for general use.
The publisher, the authors and the editors are safe to assume that the advice and information in this book are believed to be true and accurate at the date of publication. Neither the publisher nor the authors or the editors give a warranty, expressed or implied, with respect to the material contained herein or for any errors or omissions that may have been made. The publisher remains neutral with regard to jurisdictional claims in published maps and institutional affiliations.

This Humana imprint is published by the registered company Springer Science+Business Media, LLC, part of Springer Nature.
The registered company address is: 1 New York Plaza, New York, NY 10004, U.S.A.

If disposing of this product, please recycle the paper.

Preface

Bispecific antibodies (BsAbs) have emerged as a transformative modality in modern immunotherapy, offering unprecedented versatility and potency by simultaneously binding two distinct targets. These engineered molecules have shown remarkable clinical efficacy, particularly in hematologic malignancies, and are being actively explored for applications in solid tumors, autoimmunity, and beyond.

This volume, part of the *Methods in Molecular Biology* series, is dedicated to providing researchers and developers with a comprehensive set of protocols and perspectives for the design, production, characterization, functional evaluation, and clinical translation of bispecific antibodies. It is intended for a broad audience of molecular biologists, translational scientists, and biotechnology professionals engaged in antibody discovery and development.

The chapters are organized into the following four major sections:

1. **Antibody Engineering**: This section covers various platforms for generating bispecific antibodies, including knobs-into-holes, Fab-arm exchange, domain rearrangement, FAST-Ig™, ImmTAC, chemical conjugation, and Fc-fusion strategies. These techniques enable precise control over antibody structure and assembly, addressing issues such as chain mispairing, expression efficiency, and manufacturability.
2. **Bioassays**: Practical protocols are provided to evaluate T-cell-dependent bispecific antibodies (TDBs), including immunocytochemistry for visualizing immune synapse formation, T-cell migration assays, and models of T-cell exhaustion. This section also describes in vivo fluorescence imaging in both subcutaneous and brain tumor models, as well as the use of humanized mouse models to evaluate therapeutic activity in immunocompetent contexts.
3. **Quantitative Pharmacology**: This section introduces modeling and simulation approaches that support the clinical development of T-cell-engaging BsAbs. Topics include pharmacokinetic/pharmacodynamic (PK/PD) modeling, mitigation of cytokine release syndrome, and dose-optimization strategies informed by quantitative systems pharmacology.
4. **Clinical Applications**: This section provides a comprehensive overview of the current state and future directions of BsAb development in both hematologic malignancies and solid tumors, highlighting clinical successes, ongoing challenges, and emerging solutions to improve safety and efficacy.

This collection aims to bridge the gap between molecular design and translational success by equipping investigators with not only the technical know-how but also a contextual understanding of each method's relevance to therapeutic development. The diversity of contributors, spanning academia and industry, reflects the collaborative nature of bispecific antibody research.

We hope this volume will serve as a practical and inspiring resource for both new and experienced investigators working in the rapidly evolving field of bispecific antibody development.

Kashiwa City, Japan *Masahiro Yasunaga*

Contents

Contributors

ANNELISE VUIDEPOT • *Immunocore, Abingdon, UK*

ANRI SAITOH • *Hematology Department, Oncology, Medical Affairs, Johnson & Johnson, Tokyo, Japan*

AYA UCHIDA • *Department of Technical Information Sciences, The Jackson Laboratory Japan, Yokohama, Kanagawa, Japan*

BENJAMIN P. OESTRINGER • *Immunocore, Abingdon, UK*

CHI-CHUNG LI • *Genentech, Inc., South San Francisco, CA, USA*

CHIEW YING YEO • *Research Division, Chugai Pharmabody Research Pte Ltd, Singapore, Singapore*

CHIKA OGAMI • *Chugai Pharmaceutical Co. Ltd., Tokyo, Japan*

HARUHIKO KAMADA • *Graduate School of Pharmaceutical Sciences, Kyoto University, Kyoto, Japan; National Institutes of Biomedical Innovation, Health and Nutrition, Osaka, Japan*

HIROKI AKIBA • *Graduate School of Pharmaceutical Sciences, Kyoto University, Kyoto, Japan; National Institutes of Biomedical Innovation, Health and Nutrition, Osaka, Japan*

HIROAKI OHNO • *Graduate School of Pharmaceutical Sciences, Kyoto University, Kyoto, Japan; National Institutes of Biomedical Innovation, Health and Nutrition, Osaka, Japan*

KEUN HO LEE • *iProgen Biotech Inc., Burnaby, BC, Canada*

KOHEI SHIBA • *Refeyn Japan K.K., Kobe, Japan*

KOUHEI TSUMOTO • *School of Engineering and The Institute of Medical Science, The University of Tokyo, Tokyo, Japan; The Institute of Medical Science, The University of Tokyo, Tokyo, Japan*

LEO YEN-CHENG LIN • *iProgen Biotech Inc., Burnaby, BC, Canada*

MASAHIRO YASUNAGA • *Division of Developmental Therapeutics, EPOC, National Cancer Center, Kashiwa, Chiba, Japan*

MAYU OKA • *Division of Developmental Therapeutics, EPOC, National Cancer Center, Kashiwa, Chiba, Japan*

MIKIKO ITSUKAICHI • *Division of Developmental Therapeutics, EPOC, National Cancer Center, Kashiwa, Chiba, Japan*

RIKUTO NAKAMURA • *Division of Developmental Therapeutics, EPOC, National Cancer Center, Kashiwa, Chiba, Japan*

RONG DENG • *Genentech, Inc., South San Francisco, CA, USA*

RYO TSUMURA • *Division of Developmental Therapeutics, EPOC, National Cancer Center, Kashiwa, Chiba, Japan*

RYUTARO ASANO • *Department of Biotechnology and Life Science, Graduate School of Engineering, Tokyo University of Agriculture and Technology, Tokyo, Japan*

SHU FENG • *Research Division, Chugai Pharmabody Research Pte Ltd, Singapore, Singapore*

TAKAHIRO ANZAI • *Division of Developmental Therapeutics, EPOC, National Cancer Center, Kashiwa, Chiba, Japan; Department of Chemistry and Materials Science, National Institute of Technology (KOSEN), Gunma College, Maebashi, Gunma, Japan*

TAKAHIRO ISHIGURO • *Chugai Pharmaceutical Co. Ltd., Tokyo, Japan*
TEPPEI AZUMA • *Division of Developmental Therapeutics, EPOC, National Cancer Center, Kashiwa, Chiba, Japan*
TOMOHIRO FUJII • *Ajinomoto Co. Inc., Kawasaki-Shi, Kanagawa, Japan*
YURI NAKAMURA • *Division of Developmental Therapeutics, EPOC, National Cancer Center, Kashiwa, Chiba, Japan*
YUTAKA MATSUDA • *Ajinomoto Co. Inc., Kawasaki-Shi, Kanagawa, Japan; Ajinomoto Bio-Pharma Services, San Diego, CA, USA*

Part I

Antibody Engineering

Check for updates

Chapter 1

Production of Bispecific Antibody Combining Knobs-into-Holes and Controlled Fab-Arm Exchange Methods

Leo Yen-Cheng Lin and Keun Ho Lee

Abstract

This method is for producing a pure bispecific antibody starting from modifying the amino acid sequences of the two monoclonal antibodies to be combined. A mild reduction condition for "splitting the monoclonal antibodies in half" cleaves only the interchain disulfide bonds between two heavy chains, and maintains the cognate light chain-to-heavy chain association or the intact antigen-binding domain (i.e., Fab). The sequence modifications involve the "knobs-into-holes" mutations in the Fc fragment's CH3 dimerization interface for enabling the two different heavy chains to dimerize or the two different Fab's to recombine (i.e., Fab-arm exchange) in the formation of bispecific antibody. In addition to high yield, the pure bispecific antibody can be produced through a fast two-step affinity chromatographic method.

Keywords Knobs-into-holes, Fab-arm exchange, Bispecific, Antibody, Disulfide, 2-Mercaptoethylamine, 2-MEA, Complementary, Hybrid

1 Introduction

In 1961, Nisonoff and Rivers [1] reported that bi-specific antigen binding fragments, (Fab')2, can be produced by heterodimerization of antigen binding fragments produced from two different antibodies (i.e., pepsin digestion to release the bivalent (Fab)2 fragment, chemical reduction to split (Fab)2 to monovalent Fab, and followed by oxidation and recombination with a different Fab); however, such a bi-specific Fab fragment isn't popular because its production difficulties (i.e., labor-intensive purification procedures, extensive optimization and chemical manipulations), low yield, and loss of activity (i.e., lack of Fc domain for effector function) [2]. Since then the design and production of bi-specific antibodies have been continuously evolving. In 1983, Milstein and Cuello [3] demonstrated that bi-specific antibody can be made in hybrid hybridoma (i.e., fusion of two different antibody-producing cell lines); however, co-expression of two different antibodies simultaneously from 4 antibody structural genes (i.e., two different heavy chains and two

Masahiro Yasunaga (ed.), *Bispecific Antibodies*, Methods in Molecular Biology, vol. 2997,
https://doi.org/10.1007/978-1-0716-5037-0_1,
© The Author(s), under exclusive license to Springer Science+Business Media, LLC, part of Springer Nature 2026

different light chains) can result in 10 combinations (i.e., two parent and eight hybrids) of immunoglobulin structures (Fig. 1), and just only one hybrid is the desirable bispecific antibody. To improve the production of bispecific antibody, technologies have been invented to (1) favor heterodimizeration of the two heavy chains from different mAbs, and to (2) minimize light-chain mispairing (i.e., opposed to retaining the specific association of light chains to its cognate heavy chain in the original mAb).

Ridgway et al. (1996) [4] demonstrated that heterodimerization of two different heavy chains can be "enforced" (i.e., formation of bi-specific hybrid antibody) by engineering the "Knobs-into-holes"

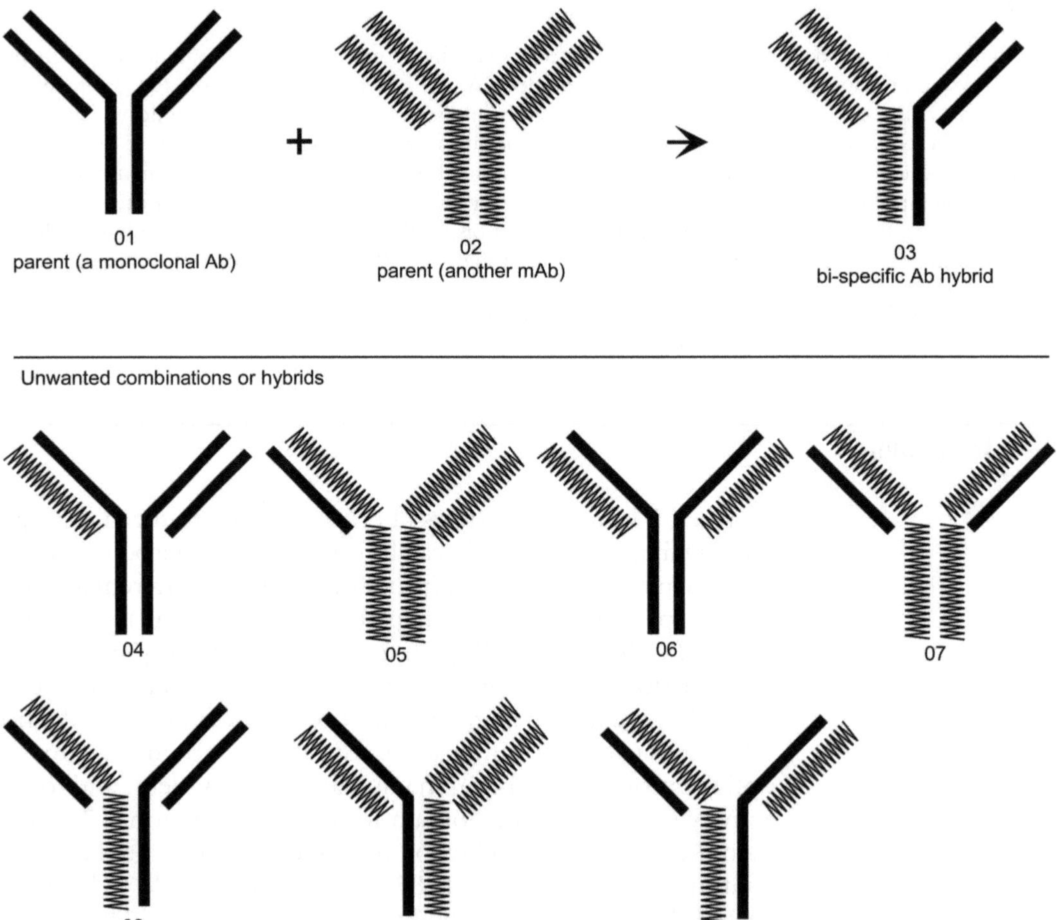

Fig. 1 Schematic drawing for the ten possible antibody species from recombining two monoclonal antibodies (i.e., assuming a random and equal probability for association from the completely dissociated heavy and light chain polypeptides). The desirable bispecific antibody (No. 03 in the figure) is the hybrid that inherits one intact Fab (antigen binding fragments) from each parent monoclonal antibodies, and the bispecific antibody can simultaneously bind to or bridge between two antigens. The other seven hybrid combinations all contain the mis-paired light and heavy chains (i.e., non-functional Fab)

mutations at the dimerization contact interface between CH3 domains. The second technical issue in making the bi-specific antibody is the mis-pairing of light chain (see Fig. 1, the seven unwanted hybrids in, combination 04-10). Mispairings happen when the contact interface between the variable heavy (VH) and the variable light (VL) and between the constant heavy (CH1) and the constant light (CL) domains in both arms of an immunoglobulin are very similar. Merchant et al. (1998) [5] averted the light chain mis-pairing issue by identifying a common light chain sequence used by different scFv antibodies that are specific against different antigens, and used that common light chain together with the two heavy chains from two different antibodies for producing bispecific antibody. Schaefer et al. [6] devised CrossMab technology to avoid mispairing of light chain to other heavy chain. Briefly, Schaefer et al. [6] re-engineered the structures of light chain (VL-CL domains) and its cognate binding heavy chain (VH-CH1 domains) by swapping their modular domain structural gene sequences. As a result, only the engineered light chain can pair/associate with the engineered heavy chain, but will no longer match with an original heavy chain [6]. Since the sequences of the swapped domains are intercrossing between light chain and heavy chain, this interchange is also termed domain crossover. The interchange of sequence (or CrossMab) is generic and applicable to any two antibodies. However, it is assumed that the domain-interchanged light chain and heavy chain should be structurally similar to that of original antibody and should retain excellent antigen binding specificity and affinity as the Fab of original antibody. In 2013, Gramer et al. [7] developed a controlled Fab-arm exchange (cFAE) method to generate a bispecific antibody from two separately purified antibodies, both of which contain complementary mutations in the dimerization CH3 domain interface for facilitating heterodimerization of heavy chain. The procedure can be robust with >95% conversion to the desirable bispecific antibody product and with high yield of recovery. Here, we describe the controlled Fab-arm exchange procedures for making a stable bi-specific antibody at high purity and high recovery that involves knobs-into-hole mutations in the heavy chain CH3 domain to enforce heterodimization of two heavy chains, and also utilizes the controlled Fab-arm exchange to enforce correct light chain pairing to its cognate heavy chains.

2 Materials

2.1 Analysis and Manipulation of Antibody Sequences

1. A computer with internet access for retrieving the protein's DNA and/or amino acid sequences from online sequence databases (such as GenBank, NCBI-BLAST)
2. A computer program, such as Clone Manager, for reading and editing the DNA and the translated protein sequences.

2.2 Plasmid DNA Amplification

1. *E. coli.* DH5α competent cells.
2. Ampicillin (stock of 100 mg/ml in water and then filter-sterilized) (*see* **Note 1**).
3. LB media with 100 μg/ml Ampicillin (*see* **Note 1**).
4. LB agar plates with 100 μg/ml Ampicillin (*see* **Note 1**).
5. 37°C shaking incubator.
6. 37°C incubator oven.
7. Plasmid DNA purification kit.
8. High speed centrifuge.

2.3 In Vitro Cell Culture

1. Cell line—CHO suspension cells.
2. Cell culture medium and supplements—F17 media with 4 mM L-Glutamine and 0.10% (v/v) Kolliphor
3. Temperature-controlled water bath.
4. Laminar flow cabinet.
5. Temperature-controlled CO_2 humidified incubator (5% CO_2, 90% humidity).
6. Tissue culture plates (6-, 12-, or 24-well plates).
7. Erlenmeyer baffled cell culture flasks (100 ml, 250 ml, 500 ml, 1 L).
8. Variable-speed platform shaker 100–110 rpm.
9. Transfection reagent—LPEI-Max (stock of 3 mg/ml in deionized water), total of 12 μg of LPEI-Max in 8 ml, final concentration of 1.5 μg per ml,
10. Swing-Bucket centrifuge.
11. 0.45 μm syringe filters
12. −80°C freezer.
13. Liquid nitrogen storage.
14. Dialysis buffer PBSE: 20 mM K/Na/HPO4 (pH 7.0), 150 mM NaCl, and 1 mM EDTA

2.4 Purification of the Monoclonal Antibodies

1. High speed centrifuge.
2. Protein A affinity chromatography column.
3. PBS: 20 mM K/Na/HPO4, pH 7.0, and 150 mM NaCl).
4. Protein A high salt wash buffer: 20 mM mM K/Na/HPO4, pH 7.0, 500 mM NaCl.

5. Protein A elution buffer: 100 mM Citrate, pH 3.0.
6. Protein A neutralization buffer: 1M Tris-HCl, pH 9.0.
7. Dialysis tubing (3.5 kD, 8 kD, or 12–14 kD molecular weight cut-off).
8. Centrifugal concentrator (10 kD, or 30 kD, or 50 kD molecular weight cut-off).
9. PBS-EDTA: 20 mM K/Na/HPO4, pH 7.0, 150 mM NaCl, 1 mM EDTA.
10. Swing-Bucket centrifuge.
11. SDS-PAGE electrophoresis.
12. UV/vis spectrophotometer.

2.5 Controlled Fab-Arm Exchange Reaction

1. Ni-column binding buffer: 20 mM Tris-HCl, 20 mM imidazole, 300 mM NaCl, pH 8.0
2. Ni-column buffer B: 20 mM Tris-HCl, 500 mM imidazole, 300 mM NaCl, pH 8.0
3. 2-Mercaptoethylamine-HCl (2-MEA, directly dissolve in the protein mixture solution, and use at final concentration of 50 mM in the reaction).
4. Temperature-controlled shaking incubator (37°C).
5. PD-10 Desalting column chromatography (e.g., PD10, or Sephadex G25).
6. PBS: 20 mM K/Na/HPO4, pH 7.0, and 150 mM NaCl.
7. Bradford Reagent.

2.6 Bispecific Antibody Purification Step 1: Ni-Affinity Column Chromatography

1. Ni-affinity chromatography column.
2. Ni-affinity column binding buffer: 20 mM Tris-HCl, 20 mM imidazole, 300 mM NaCl, pH 8.0.
3. Ni-affinity column buffer A: 20 mM Tris-HCl, 300 mM NaCl, pH 8.0.
4. Ni-affinity column buffer B: 20 mM Tris-HCl, 500 mM imidazole, 300 mM NaCl, pH 8.0.

2.7 Bispecific Antibody Purification Step 2: Strep-Tactin Column Chromatography

1. Strep-Tactin chromatography column.
2. Strep-Tactin column binding buffer: 20 mM K/Na/HPO4, pH 8.0, 300 mM NaCl.
3. Strep-Tactin column elution buffer: 20 mM K/Na/HPO4, pH 8.0, 300 mM NaCl, 2.5 mM desthiobiotin.

3 Methods

3.1 Design the Sequences of Bispecific Antibody (e.g., from Trastuzumab X Pertuzumab) on Computer

1. Retrieve the structural sequences of the two monoclonal antibodies to be combined.
   (For example, Pertuzumab is Accession DB06366 in DrugBank Online (https://go.drugbank.com), or is Accession ACK19003 (for heavy chain) and Accession ACK19002 (for light chain) in the NCBI Protein database (i.e., BLASTP with Patented Protein Sequences database). Trastuzumab is Accession DB00072 in DrugBank Online, or is Accession WDS83891 (for heavy chain) and Accession WDS83892 (for light chain) in NCBI Protein database.
2. Identify the CH3 domain in the heavy chain sequences and apply knobs-into-holes mutations
   Use Fig. 2 as a reference for CH3 domain sequence. The "knobs" (Lys409 to Arg, i.e., K409R) mutation in Trastuzumab can fit into the "holes" (Phe405 to Leu, i.e., F405L) mutation in Pertuzumab. Figure 2 also provides some alternative "knobs-into-holes" mutation points that are known to enhance production of bi-specific antibodies (*see* **Note 2**).
3. Add two different purification tags (i.e., His-tag and STREP2-tag), one for each monoclonal antibody. Add a (Gly)4-Ser

```
human IgG1 heavy chain Fc domain:

221 DKTHTCPPCP APELLGGPSV FLFPPKPKDT LMISRTPEVT CVVVDVSHED 270
                    CH2 domain

271 PEVKFNWYVD GVEVHNAKTK PREEQYNSTY RVVSVLTVLH QDWLNGKEYK 320

                                                      Δ
321 CKVSNKALPA PIEKTISKAK GQPREPQVYT LPPSRDELTK NQVSLTCLVK 370
                   CH2 domain  CH3 domain                    366

                                              ● Δ ●
371 GFYPSDIAVE WESNGQPENN YKTTPPVLDS DGSFFLYSKL TVDKSRWQQG 420
                                              405 407 409

421 NVFSCSVMHE ALHNHYTQKS LSLSPGK 447
                          CH3 domain
```

Fig. 2 Amino acid sequence (Kabat numbering scheme) of human IgG1 heavy chain Fc fragment. The CH2 domain (single-underlined) dimer interface is mediated by glycosylations, and the CH3 domain (double-underlined) dimer interface involves side chain-to-side chain association. The working "Knobs-into-holes" mutations in the CH3 domains were highlighted. The F405L mutation (i.e., hole mutation) is complementary to the K407R (i.e., knobs mutation). Alternatively, the T366Y "knobs" mutation is complementary to the Y407T "hole" mutation [5]. For more "knobs-into-holes" pairs of mutations, useful for bispecific antibody design, refer to Table 1 in Merchant et al. [5]

pentapeptide spacer (His)6-tag (i.e., GGGGSHHHHHH) to the C-terminus of light chain in Trastuzumab. For the other, add (Gly)4-Ser STREP2-tag (i.e., GGGGSWSHPQFEK) to the C-terminus of light chain in Pertuzumab. For clarity and better understanding, see diagrams in Fig. 3 to help visualize the process. Figure 3 illustrates the step-by-step transition or generation of a bi-specific antibody, with one antibody carrying a His-tag and the other carrying a STEP2-tag (*see* **Note 3**).

4. Add a secretory signal peptide sequence at the N-terminus of the amino acid sequence (i.e., one secretory signal peptide sequence for each chain) (*see* **Note 4**).
5. Submit (i.e., through email or the website) both the heavy and light chain sequences to a gene-synthesis service provider for codon optimization for production in mammalian cell (i.e., CHO), and also instruct the gene synthesizer to produce the structural gene directly in the pTT5 expression vector, a good vector for both bacterial cloning as well as protein expression in mammalian cell culture (*see* **Note 5**).

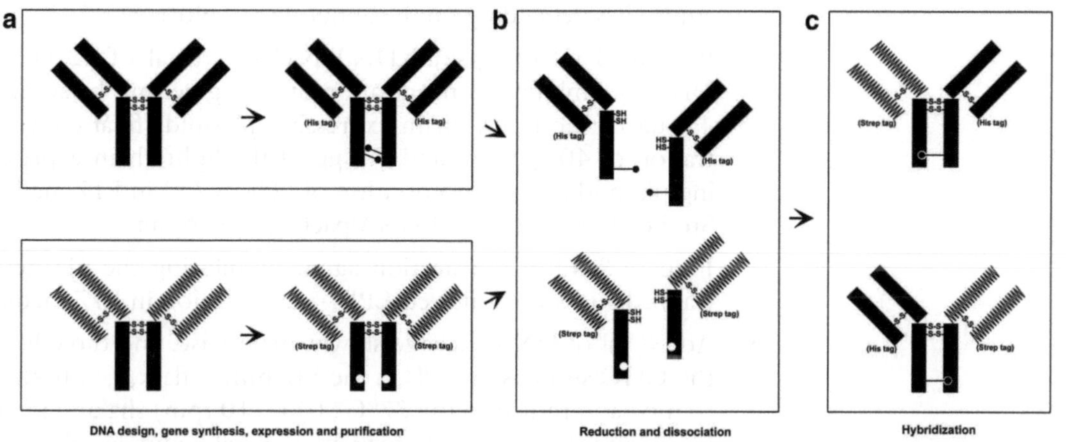

Fig. 3 Schematic drawing for the workflow, a summary—Design and production of bispecific antibody (hybrid) from two monoclonal antibodies (parent). (**a**) (DNA cloning/gene synthesis) Engineering two different purification tags to the two monoclonal antibodies (i.e., to the C-terminus of light chain, a His-tag or a STREP2-tag) for isolation of the bispecific antibody hybrid. Design of "knobs-into-holes" mutation in the CH3 domain of two monoclonal antibodies for high yield production of bispecific antibody hybrid. Briefly, in the controlled Fab-arm exchange (cFAE) method, the two antibodies with "knobs-into-holes" mutations are separately expressed and purified. (**b**) The two purified antibodies are subjected to controlled reduction with 2-Mercaptoethyl amine (2-MEA) to break the hinge disulfide bond in vitro. Meanwhile, the two heavy chains with the "knobs" mutations (so as the other reduced antibody with "holes" mutations) dissociates due to unfavorable dimerization interface in their CH3 domain, splitting the immunoglobulin in half (i.e., dissociation of Heavy chain:Light chain (HL) homodimers). (**c**) Heterodimerization of two different heavy chains guided by pairing of "knobs-into-holes" mutations, and formation of interchain disulfide bonds in the Fc domain. This allows the HL monomer with "knobs" mutations to form a stable complex with the other HL monomer with "holes" mutation, reconstituting a stable bispecific antibody upon reoxidation

6. Once all four plasmids (one for heavy chain and one for light chain in Trastuzumab, and one for heavy chain and one for light chain in Pertuzumab) arrive, transform separately into *E. coli* DH5α, and plate on LB Agar plate. Incubate overnight in a stationary (non-shaking) 37°C incubator.
7. Inoculate from single colonies in fresh LB media, and incubate overnight in a shaking (~200 rpm) 37°C degree C incubator.
8. Purify all four plasmid DNAs separately using commercial plasmid DNA purification kit.

3.2 Transient Transfection and Protein Expression

1. Seed CHO suspension cells (doubling time ~18 h) in 300 ml of F17 media in a 1-L size Erlenmeyer flask, and grow in a shaking (100–110 rpm) 37°C incubator supplemented with 5% carbon dioxide.
2. On the day cell density reaches a >95% viable count (with Trypan Blue) of 1.5 to 2.2 million cells per ml, add 4 ml of F17 media, total volume of 304 ml, and ready for DNA transfection operation. This starting culture size (i.e., 0.30 L) is sufficient for the production of on average 5 mg of monoclonal antibodies; however, it can vary between 1 and 10 mg of monoclonal antibodies depending on the antibody (*see* **Note 6**).
3. Prepare 8 ml of plasmid DNA package (total of 12 μg of DNA in 8 ml, final concentration of 1.5 μg per ml) by adding 4.8 μg of the heavy chain-expressing plasmid (final concentration of 40%, w/w) and 7.2 μg of the light chain-expressing plasmid (final concentration of 60%, w/w) in F17 media. Sterile filter the plasmid DNA package (*see* **Note 7**).
4. Prepare 8 ml of transfection agent by diluting the pH-neutralized and sterile-filtered LPEI-Max solution in F17 media.
5. Add 8 ml of DNA package slowly, drop-wise, and directly to the CHO suspension cells in the Erlenmyer flask, gently swirl to mix, and return to the 37°C (100–110 rpm) shaking incubator, and shake incubate for 5 min. Total volume of 312 ml.
6. Then, add 8 ml of transfection agent slowly and drop-wise. This is Day 0. Allow transfection to take place in the following 24 h in the 37°C (100–110 rpm) shaking incubator. Total volume of 320 ml.
7. After 24 h of transfection incubation (Day 1), add 8 ml of the sterile-filtered 40% (w/v in F17 media) tryptone N1 (1% final concentration). Total volume of 328 ml. Transfer the flask to a lower temperature (32°C) shaking (100–110 rpm) incubator for protein expression to begin.
8. Seventy-two hours after transfection (Day 3), add 0.328 ml (1-in-1000 of culture volume, v/v) of Anti-Clumping Agent,

and continue shaking incubation (32 degree C, 100–110 rpm) (*see* **Note 8**).

9. On Day 5 or 6 (120 to 144 h after transfection), perform a cell count with Trypan Blue. When the viable cell count is 60% or lower, stop the culture, and harvest the clear supernatant for protein purification. Centrifuge the cell culture supernatant to sediment the CHO cells in suspension.
10. Pour the cell culture supernatant into sterile centrifuge bottles, and centrifuge the supernatant at 4000 rpm for 40 min.
11. Perform syringe filtration (0.45 μm syringe filter) to further remove the particulate matter.
12. Proceed to purification of monoclonal antibodies, or store the harvested culture supernatant for long term at −80°C until purification.

3.3 Protein A Purification of Monoclonal Antibodies

1. Prepare 2 ml of Protein A resin in a gravity flow column, and equilibrate the column with PBS.
2. Load the culture supernatant containing the secreted monoclonal antibody onto Protein A by gravity, and collect flow through (*see* **Note 9**).
3. Wash the antibody-bound Protein A column by passing through at least 10 column volumes (or 20 ml) of PBS, and collect flow through.
4. Elute the Protein A-bound antibodies with 2 ml of Protein A elution buffer in a fraction, and immediately neutralize it by adding Protein A neutralization buffer.
5. Repeat Step 4 (i.e., elution and neutralization) to collect a total of six elution fractions.
6. Identify the fractions with protein with Bradford Assay Reagent. Add 10 μl of the neutralized Tris-Citrate elution to 200 μl of Bradford Assay Reagent, and the fractions with the highest protein content will turn Bradford Assay Reagent from brown to deep blue color.
7. Combine the fractions with high protein content.
8. Pour into a dialysis tubing (12–14 kD MWCO), and dialyze overnight in 2 l of buffer PBSE (*see* **Note 10**).
9. Harvest the dialyzed protein from dialysis tubing.
10. Use a 30 kD MWCO centrifugal concentrator to concentrate the protein (spin at 3000 × *g* in a benchtop refrigerated swing-bucket centrifuge) until the protein concentration is at least 5 mg/ml.

3.4 Controlled Fab-Arm Exchange

1. Dilute the purified mAbs to 5 mg/ml with dialysis buffer PBSE.
2. Weigh 5.7 mg or (0.0057 g) of 2-Mercaptoethylamine-HCl into a 1.5 ml-microcentrifuge tube, and then gently add 0.5 ml (2.5 mg) of Trastuzumab-(His tag) and 0.5 ml (2.5 mg) of Pertuzumab-(STREP tag). Mix to dissolve the 2-Mercaptoethylamine-HCl, and incubate the mixture at 37°C incubator overnight (*see* **Notes 11, 12, 13, and 14**).
3. In the next morning, change the incubation temperature to 22°C (i.e., room temperature), and continue incubate for an hour. Meanwhile, equilibrate a PD-10 desalting column (Sephadex G25 resin) with PBS at room temperature.
4. Load the 1 ml antibody reduction mixture onto the PD-10, PBS as the running buffer, collect 0.70 ml per elution fraction and collect ten fractions (*see* **Note 15**).
5. Identify the fractions with protein with Bradford Assay Reagent. Add 10 μl of the eluted fraction to 200 μl of Bradford Assay Reagent, and the fractions with the highest protein content will turn Bradford Assay Reagent from brown to deep blue color.
6. Combine the fractions with high protein content (i.e., test positive in Bradford Assay).
7. Use a 30 kD MWCO centrifugal concentrator to concentrate the protein (i.e., from ~4 ml to ~0.20 ml), and then dilute 60-fold with 12 ml of PBS to completely remove the 2-MEA and EDTA. Concentrate to ~0.20 ml again.

3.5 Two-Step Bispecific Antibody Purification: Ni Affinity Column Chromatography

1. Prepare a 50 ml Ni-column binding buffer by mixing 2 ml of Ni-column buffer B and 48 ml of buffer A (20 mM Tris-HCl, 300 mM NaCl, pH 8.0) (*see* **Note 16**).
2. Dilute the Fab-arm exchanged antibody mixture (~5 mg in 0.20 ml) ~60-fold with ~12 ml of Ni-column binding buffer, and then load onto a Ni-affinity column (1CV of 1.2 ml) pre-equilibrated with the Ni-column binding buffer (*see* **Note 17**).
3. Apply at least 12 ml (ten column volumes) of Ni-column binding buffer to wash the protein-bound Ni-NTA column.
4. Prepare a step-wise gradient (i.e., increasing concentrations) of imidazole concentrations by mixing Ni-column buffer A and B, (i.e., 20 mM Tris-HCl, 300 mM NaCl, pH 8.0, and an increasing concentrations of imidazole at 40, 60, 80, 150, 200, 250, 300, 400, 500 mM). Prepare 5 ml for each concentration.
5. Apply the imidazole concentration gradient (one at a time, and in the order of increasing imidazole concentrations) to elute and then collect the eluted Ni-column-bound antibodies.
6. Identify the fractions with protein with Bradford Assay Reagent. Add 10 μl of the eluted fraction to 200 μl of Bradford

Assay Reagent, and the fractions with the highest protein content will turn Bradford Assay Reagent from brown to deep blue color. Combine the fractions with proteins, and ready for purification on a STREP-Tactin affinity column.

3.6 Two-Step Bispecific Antibody Purification: STREP-Tactin Affinity Column Chromatography

1. Equilibrate a STREP-Tactin affinity column (1 CV = 1 ml) with STREP-Tactin column binding buffer (*see* **Note 18**).
2. Load the pool of protein collected from Ni-column elutions onto the STREP-Tactin affinity column (1 CV = 1 ml) (*see* **Note 19**).
3. Apply at least 10 ml (ten column volumes) of the Strep-Tactin column binding buffer to wash the protein-bound column.
4. Elute the bound protein with STREP-Tactin column elution buffer, and collect elutions in fractions (*see* **Note 20**).
5. Identify the fractions with protein with Bradford Assay Reagent. Add 10 μl of the eluted fraction to 200 μl of Bradford Assay Reagent, and the fractions with the highest protein content will turn Bradford Assay Reagent from brown to deep blue color. Combine the fractions with proteins, and use a 30 kD MWCO centrifugal concentrator to concentrate and then do buffer exchange with PBS to remove the desthiobiotin. The final product is the wanted hybrid (i.e., bispecific antibody).

4 Notes

1. The plasmid DNA pTT5 expresses an Ampicillin resistance marker, or beta-lactamase.
2. The human IgG1 antibody CH3 domain begins at amino acid sequence Pro343-Arg344-Glu345-Pro346-Gln347-Val348-Tyr349-Thr350, and ends at Ser441-Leu442-Ser443-Leu444-Ser445-Pro446-Gly447-Lys448. Simply locate these two stretches of amino acid sequences to mark the N-terminus and the C-terminus of CH3 domain. Then, assign the amino acid sequence number to the amino acid residues in sequence as it would be convenient to precisely locate the residues for knobs-to-holes mutations.
3. Using this sequence design, there will be one His-tag at the Trastuzumab Fab arm and one STREP2-tag at the Pertuzumab Fab arm in the bispecific antibody product (see Fig. 3), and only the bispecific antibody product bearing both His-tag and STREP2 tag will be purified in the two-step affinity chromatographic purification (i.e., Ni- and STREP-TACTIN affinity column chromatography. Monoclonal antibodies are typically very stable and soluble; however, its good physical properties can change dramatically with changes in structures (i.e.,

changes in amino acid sequence, addition of purification tag, point mutations, etc. ...). Without knowing whichever the combination of purification tag and antibody is compatible (i.e., or incompatible), it is highly adviced to make several different versions of fusion genes, and try them out (i.e., His-tag on mAb#1, His-tag on mAb#2, STREP2-tag on mAb#1, STREP2-tag on mAb#2, G4S-His-tag on mAb, and (G4S)2-His-tag on mAb, etc. ...). The reason for creating these redundancy of genes is due to the following concerns—(1) Tag may not be exposed to space such that it doesn't bind to affinity chromatography column, (2) Tag may inhibit expression of protein, and (3) Tag may decrease the solubility of protein. Therefore, to maximize work efficiency and to increase success, simply switch to a different construct whenever stability or solubility of a certain construct design becomes an issue (i.e., resulting in no expression and no production). Sometimes, a small change in sequence remedies all the problems (i.e., solubility and stability) downstream, and it is easier to work with a clone that is stable and soluble, rather than trying to develop a special buffer and/or condition (i.e., stabilizer, pH, ionic strength, detergent) to maintain an unstable protein.

4. Antibodies are extracellular proteins, and antibody heavy and light chains are secreted to the extracellular space.

5. pTT5 is a good vector for expressing the recombinant monoclonal antibodies. Important features/components of pTT5 includes (1) the constitutive CMV promoter that enhances expression of protein of interest, (2) the PMB1 ori replicon that helps DNA replication and maintains the level of pTT5 plasmid and hence its stability in the mammalian cells, (3) the β globin poly A signal terminator that helps ending the transcription correctly, focusing the mRNA generation at the structural gene of interest. For resistance marker selection, pTT5 contains an ampicillin resistance gene (β-lactamase), so that only the bacteria with pTT5 plasmid can multiply in the ampicillin-enriched media. pTT5 doesn't carry other drug resistance gene for selection in mammalian cells, and is particularly beneficial for transient transfection purpose as it allows the host cell to focus on the expression and production for the gene of interest (i.e., recombinant antibody). For that, pTT5 is particularly suitable for transient transfection and recombinant protein expression in mammalian cells.

6. With the methods and materials described, at best 30 mg of monoclonal antibodies can be produced in a liter of F17 culture media. The production yield of 5 mg antibody per liter of F17 media is considered low yield. Proportionally scale-up the starting culture volume and number of seeding cells to increase the production of monoclonal antibody.

7. Excess number of light chain generating plasmid is used typically to generate a recombinant monoclonal antibody. If the production yield of heavy chain-to-light chain in the expressing cell is not a 1-to-1 molar ratio (as this can be visualized on SDS-PAGE), adjust the molar ratio of the heavy chain-to-light chain plasmid accordingly to compensate.
8. This is a transient transfection method of expressing the recombinant antibody in CHO. However, this culture is good for approximately one week only, and not suitable for subsequent splitting (i.e., or continuous culturing) because (1) the cell viability is already low after one week and (2) there is no drug selection to minimize contamination and to exclude non-expressing CHO, leading to greater heterogeneity in the subsequent CHO populations. The advantage of transient transfection method is the simple, quick, and economical generation of the antibody-producing cells in the condition that is also ready for overexpression of recombinant monoclonal antibodies.
9. The binding capacity of Protein A resin is about 20 mg of antibody per ml of Protein A. Adjust the amount of Protein A resin, or adjust the volume of culture supernatant accordingly.
10. Dialysis dilutes out the citrate and Tris, so that the purified antibody is primarily in 20 mM K/Na/HPO4 (pH 7.2), 150 mM NaCl, and 1 mM EDTA.
11. Strong reducing agents, such as DTT or 2-mercaptoethanol, can completely reduce all the disulfide bonds in an IgG, and can generate individual heavy and light chains and can adversely destroys antigen-binding function. 2-Mercaptoethylamine-HCl is a mild reducing agent used for selectively reducing the hinge-region disulfide bonds of IgG heavy chains to sulfhydryls without breaking other disulfide bonds, effectively producing two functional half-antibodies for hybridization. The CH3 homodimerization interface is destabilized due to the unfavorable combination of "Knobs-vs-Knobs" mutations (so as the "holes-vs-holes" mutations) facilitating the heterodimerization (knobs-into-holes) between two different heavy chains.
12. Total volume is 1 ml, the total protein concentration is 5 mg/ml (or 2.5 mg/ml of mAb1-(His tag) and 2.5 mg/ml of mAb2-(STREP tag)), and the final concentration of 2-Mercaptoethylamine-HCl is 50mM. Incubate the mixture at 37°C to initiate the Fab-Arm exchange process between mAb1-(His tag) and mAb2-(STREP tag) overnight.
13. An option to improve the hybridization is to use a higher temperature setting for the overnight incubation, such as 42°C, 47°C, or 52°C. Screen different temperature. Avoid too high a temperature as some antibody can become precipitated.

14. Equally important, another option to improve the hybridization is to include a denaturing agent or detergent (i.e., Urea, guanidine hydrochloride, CHAPS, TWEEN20 at different concentrations) to help splitting the antibody in half, so that two different Fab's can recombine to bispecific antibodies.
15. Use PD-10 to buffer exchange the antibody product to separate 2-MEA and EDTA facilitating the formation of disulfide bonds between the heterodimer heavy chains.
16. The bispecific antibody bear one His-tag and one STREP-tag, and only the heterodimerized (i.e., bispecific) antibodies will get captured and purified in a two-step purification procedure using Ni-affinity and followed by a STREP-Tactin affinity binding chromatographies.
17. The Ni-column flow-through contains primarily the parent antibody that doesn't have the His-tag (i.e., a parent antibody with only the Strep tag).
18. Regarding the binding buffer for STREP-Tactin affinity chromatography—(1) Avoid denaturing agents, chaotropic salts (i.e., guanidine hydrochloride, urea) as it can denature/disrupt the affinity column and/or its interaction with the STREP2 tag, (2) Ensure that pH is >6.5, (3) Avoid and remove (i.e., through dialysis, buffer exchange, or desalting column chromatography) biotin from both the recombinant monoclonal antibodies and all purification buffers as biotin irreversibly blocks the binding sites on STREP-Tactin column
19. Strep-Tactin Superflow Plus column (1 CV of 1 ml, binding capacity of up to 9 mg/ml, Qiagen Catalog No. 30004). The Strep-Tactin column flow-through contains primarily the parent antibody that doesn't have the STREP-tag (i.e., a parent antibody with only His tag).
20. In rare cases that antibody doesn't get eluted, 10 mM biotin can be used. Biotin can displace and elute the bound antibodies off the binding sites on STREP-Tactin column; however, biotin can irreversibly inactivate STREP-Tactin and its reusability.

References

1. Nisonoff A, Rivers MM (1961) Recombination of a mixture of univalent antibody fragments of different specificity. Arch Biochem Biophys 93:460–462. https://doi.org/10.1016/0003-9861(61)90296-x
2. Hammerling U, Aoki T, de Harven E et al (1968) Use of hybrid antibody with anti-gamma-G and anti-ferritin specificities in locating cell surface antigens by electron microscopy. J Exp Med 128(6):1461–1473. https://doi.org/10.1084/jem.128.6.1461
3. Milstein C, Cuello AC (1983) Hybrid hybridomas and their use in immunohistochemistry. Nature 305(5934):537–540. https://doi.org/10.1038/305537a0
4. Ridgway JB, Presta LG, Carter P (1996) 'Knobs-into-holes' engineering of antibody CH3 domains for heavy chain heterodimerization.

Protein Eng 9(7):617–621. https://doi.org/10.1093/protein/9.7.617
5. Merchant AM, Zhu Z, Yuan JQ et al (1998) An efficient route to human bispecific IgG. Nat Biotechnol 16(7):677–681. https://doi.org/10.1038/nbt0798-677
6. Schaefer W, Regula JT, Bahner M et al (2011) Immunoglobulin domain crossover as a generic approach for the production of bispecific IgG antibodies. Proc Natl Acad Sci USA 108(27):11187–11192. https://doi.org/10.1073/pnas.1019002108
7. Gramer MJ, van den Bremer ET, van Kampen MD et al (2913) Production of stable bispecific IgG1 by controlled Fab-arm exchange: scalability from bench to large-scale manufacturing by application of standard approaches. MAbs 5(6):962–973. https://doi.org/10.4161/mabs.26233

Check for updates

Chapter 2

Multimodal Physicochemical Characterization of Cross-Linking Activities of Biparatopic Antibodies

Hiroki Akiba, Kohei Shiba, Kouhei Tsumoto, Hiroaki Ohno and Haruhiko Kamada

Abstract

A biparatopic antibody (BpAb) is a bispecific antibody that recognizes two different epitopes of the same antigen. Many BpAbs have been developed to control the structures of immunocomplexes different from that with the original antibodies. Various binding modes are enabled, including the cross-linking pattern of binding and 1:1-binding, which influence the biological activities of BpAbs. In this chapter, we describe the analytical methods for characterizing cross-linking activities. Taking a development campaign for bivalent asymmetric BpAbs targeting TNFR2 as an example, combined analysis is described using size-exclusion chromatography–multi-angle light scattering, mass photometry, and surface plasmon resonance.

Keywords Biparatopic antibody, Biepitopic antibody, Immune complex, Interaction, Light scattering, Size-exclusion chromatography, Mass photometry, Surface plasmon resonance

1 Introduction

A biparatopic antibody (BpAb) is a type of bispecific antibody consisting of two variable regions that bind to different epitopes of the same antigen molecule [1]. This design enables adopting various binding modes [2]. The BpAbs can have symmetric or asymmetric structures (Fig. 1a). Symmetric structures, typically containing four or more variable regions, are often adopted to enable the formation of large cross-linked immunocomplexes [3–6], whereas asymmetric structures, typically mimicking immunoglobulin G (IgG) structures, are used both for cross-linking and for specific immunocomplex formation, such as antibody:antigen = 1:1 or 2:2 (Fig. 1b) [7–9]. BpAb development campaigns are often aimed at selecting optimal BpAb structures and epitope combinations for the size and structure of immunocomplexes that are most advantageous. Cross-linking pattern of binding is generally efficient for generating cellular events such as internalization or signaling. Limitation into specific

Masahiro Yasunaga (ed.), *Bispecific Antibodies*, Methods in Molecular Biology, vol. 2997,
https://doi.org/10.1007/978-1-0716-5037-0_2,
© The Author(s), under exclusive license to Springer Science+Business Media, LLC, part of Springer Nature 2026

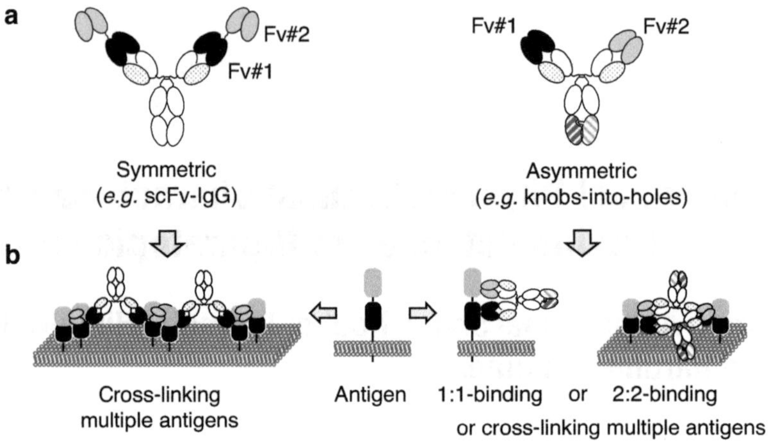

Fig. 1 Different immunocomplex formation by symmetric and asymmetric biparatopic antibodies (BpAbs). (**a**) Symmetric and asymmetric design. (**b**) Possible immunocomplex formation

immunocomplex such as 1:1 may be advantageous for controlling signaling at a specific level. Although the goal of these campaigns differs according to the target, they commonly require appropriate analytical methods to evaluate cross-linked binding [1, 2]. Physicochemical characterization methods can be used in these analyses to provide both quantitative and qualitative information.

Analytical methods for cross-linking can be categorized into two types: solution systems, such as size-exclusion chromatography with multi-angle light scattering (SEC-MALS), analytical ultracentrifugation, and mass photometry (MP); and immobilized systems, such as surface plasmon resonance (SPR), biolayer interferometry, and electro-switchable nanolevers (Fig. 2a). As two-dimensional movement of the receptors is allowed on the cell membrane, characterization through the two extreme systems with free movement in three dimensions in solution systems and restricted movement on the surface in immobilized systems provide a variety of information for cross-linking activities.

In this chapter, we describe the multimodal characterization of immunocomplex formation using SEC-MALS, MP, and SPR among such analytical methods. Descriptions of each instrument required are published in earlier studies [10–12]. We mainly focus on the characterization of BpAbs. This common strategy for sample preparation is also applicable to other analytical systems.

Both SEC-MALS and MP are used to determine the sizes of polymer particles in solution. SEC-MALS is a liquid chromatography system composed of a size-exclusion column for separation and MALS detector that detects scattered light depending on the average size of the particles at elution [12, 13]. BpAb–antigen complexes of

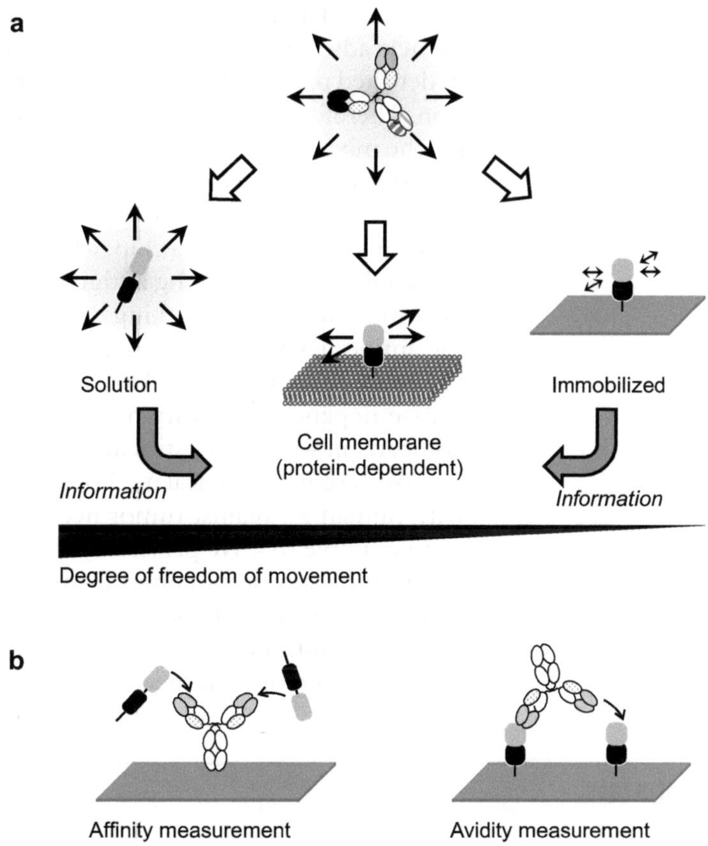

Fig. 2 Systems for analyzing cross-linking activities. (**a**) Solution and immobilized systems as two extreme models for analyzing cross-linking activities. (**b**) Affinity and avidity measurement modes in immobilized systems including surface plasmon resonance

various sizes are separated size-dependently in SEC, and the average molecular weights of the fractions containing immunocomplexes are analyzed. MP is an interferometric scattering microscopy-based method used for mass analysis at the single-particle level [10, 14]. Signal detection depends on nonspecific adsorption of the particles onto the coverslip, which generates static light scattering. Owing to the different mechanisms of SEC-MALS an MP, the optimal concentration range for the analyses differs (approximately 100-fold higher concentration is required for SEC-MALS). Additionally, different types of information are obtained. Using SEC-MALS, all particles with measurable concentrations are detected, enabling determination of the overall distribution profiles of the immunocomplexes [3, 5, 7, 8]. In MP, all particles within the detection range are recorded. The particle mass is analyzed at the single-particle level, and precise characterization of the mass of each immunocomplex is possible even when a wide distribution is present [8]. However, the detection sensitivity depends on the physicochemical characteristics of the particle adsorbed onto the glass surface; thus, the absolute number of detected particles may not precisely match the distribution in solution. Therefore, these two methods are complementary, particularly for the mixture of immunocomplexes of multiple sizes.

SPR can be used for two binding-mode analyses: affinity and avidity (Fig. 2b). The former can be analyzed by immobilizing or capturing antibodies on a chip, whereas the latter can be analyzed by immobilizing or capturing antigens on a chip. Avidity SPR provides information on cross-linking activities [8, 15]. For BpAbs, both the antibodies and antigens have multiple binding regions, making the analysis complicated. Although the optimal analytical method is case-dependent, avidity measurements of BpAbs are useful for comparison with conventional antibodies.

We describe analytical methods using asymmetric bivalent BpAbs with human Fc against tumor necrosis factor receptor 2 (TNFR2) as an example [8]. We previously observed that strong agonists and antagonists can be obtained from a variety of asymmetric BpAbs whose biological activities are closely associated with cross-linking activities. Multimodal physicochemical characterization aids in the understanding of their mechanisms of action.

2 Materials

2.1 Preparation of Recombinant Antibody/Antigen Proteins

1. Recombinant proteins designed suitable for SEC-MALS and MP (either purchased or produced in-house). Optimization of protein size is recommended. A large difference in the sizes of the antigen and antibody enables easier analysis of the complex (*see* **Note 1**). For TNFR2, we designed monomeric TNFR2 without a tag (sTNFR2, ~35 kDa) to correspond to the size of asymmetric IgG-like BpAbs (150 kDa). When an expression/purification tag is used, a tag that preserves the native oligomeric state of the recombinant antigen is required (*see* **Note 2**).
2. Recombinant proteins designed suitable for SPR (either purchased or produced in-house). Fusion of an affinity tag to the antigen is recommended for allowing regeneration of the sensor chip (*see* **Note 3**). C-terminal fusion of maltose-binding protein (MBP) is an established example as we conducted for TNFR2.
3. SEC columns with a separation range suitable for purification and/or analysis of the recombinant proteins.
4. Fast protein liquid chromatography system used for SEC.

2.2 SEC-MALS

1. High-performance liquid chromatography (HPLC) with UV-Vis, refractive index, and MALS detectors.
2. SEC columns with a separation range suitable for analysis.
3. Phosphate-buffered saline (PBS): 8.1 mM Na_2HPO_4, 1.5 mM KH_2PO_4, 137 mM NaCl, and 2.7 mM KCl as the running buffer, passed through a 0.22 μm filter, and degassed. Other buffers containing 100–300 mM NaCl or KCl can be used.
4. Bovine serum albumin (BSA) as a control (at least 0.1 mg).
5. The antigen and antibody of interest (at least 0.1 mg per the antigen–antibody combination).
6. Analysis software suitable for the analytical system.

2.3 MP

1. Instrument for MP.
2. Microscope coverslips with the thickness of #1.5H (170 ± 5 μm). Wash the coverslips successively with ethanol and ultrapure water, and then sonicated for 5 min. Pretreated coverslips are also available.
3. Sample cassettes of 3 mm diameter.
4. PBS as the measurement buffer passed through a 0.22 μm filter. Other buffers containing 100–500 mM NaCl or KCl can be used.
5. The protein standard: 100 nM BSA (67 kDa), human IgG (150 kDa), catalase (240 kDa), and thyroglobulin (660 kDa) in an equimolar mixture. Other combinations of standards can be used. Ensure that the measurement range of the samples does not exceed the standardized range.
6. Antigen and antibody of interest, at least 5 pmol (750 ng for a 150 kDa protein) per the antigen–antibody combination.
7. Analysis software.

2.4 SPR

1. Instrument for SPR. Biolayer interferometry or similar systems with surface modification can be used.
2. Running buffer: 10 mM HEPES-NaOH, 150 mM NaCl, 3 mM EDTA, and 0.05% Surfactant P20 (HBS-EP+). Other buffers containing 100–300 mM NaCl and nonionic detergents may be used.
3. Series S Sensor Chip CM5 and Amine Coupling Kit (Cytiva, Marlborough, MA, USA).
4. 10 mM sodium acetate buffer, pH 5.5.
5. Antigen and antibody of interest, at least 5 pmol (750 ng for a 150 kDa protein) per the antigen–antibody combination.

6. Anti-MBP monoclonal antibody (#E8032S; New England Biolabs, Ipswich, MA, USA). If the antigen is tagged differently, a secondary antibody should be screened prior to analysis.
7. 3 M $MgCl_2$ or other regeneration solution suitable for the tag antibody.

3 Methods

3.1 Preparation of Recombinant Antibody/Antigen Proteins

1. Prepare purified antibody/antigen proteins in suitable expression systems (not described here).
2. Confirm that the proteins used for analysis (both antibodies and antigens) are monodisperse. They should be >95% pure according to SEC. Final purification via SEC is recommended prior to use.
3. Remove aggregates by filtration or centrifugation at >15 000 × *g* for 15 min before use.

3.2 SEC-MALS Measurement

1. Equilibrate the system and column with running buffer for at least 4 h before measurement. An overnight run is recommended. Equilibration of five to ten column volumes is required for the baseline to settle.
2. Run the HPLC at 0.5 mL/min.
3. Run the BSA sample to confirm the machine conditions (2–10 mg/mL, 50 μL). Two main peaks appear with molar mass values of ~60 000 and ~120 000, corresponding to the monomer and dimer of BSA, respectively.
4. Run the antigen and antibody separately (at least 0.2 mg/mL). Both samples should elute as a single peak and exhibit monodispersity.
5. Analyze the average molecular weight and confirm that the prepared material matches the protein design (*see* **Notes 4–6**).
6. Mix the antigen and antibody at the determined molar ratio (*see* **Notes 7 and 8**) in 55 μL solution. Allow the samples to stand for 15–30 min (at least 10 min) at an appropriate temperature (typically at 25°C) to enable immunocomplex formation. The final total protein concentration should be at least 0.2 mg/mL. For BpAbs with cross-linking activities, it is recommended to measure a mixture of samples in two or more molar ratio patterns. For TNFR2, an 8:1 mixture of sTNFR2 (35 kDa, 16 μM) and BpAb (150 kDa, 2 μM) was measured (total protein concentration of ~0.8 mg/mL), and then a 1:1 mixture at 2 μM was measured.

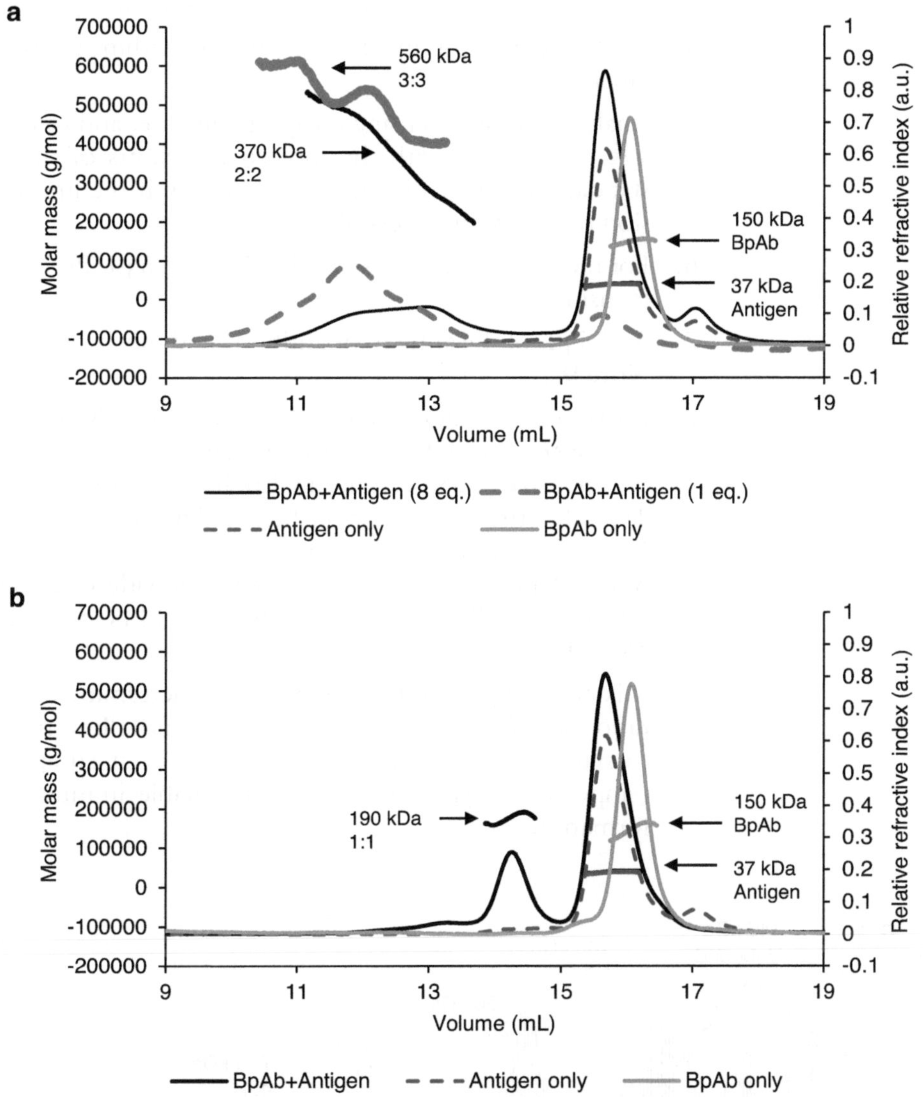

Fig. 3 Size-exclusion chromatography–multi-angle light scattering (SEC-MALS) charts of the immunocomplexes formed by asymmetric biparatopic antibodies (BpAbs). (**a**) BpAb with cross-linking activity (Bp96-92). (**b**) BpAb without cross-linking activity (Bp109-92). The charts were drawn using data published in [8]

7. Run the mixture (50 μL) and analyze the size of immunocomplex (*see* **Note 6**) (Fig. 3).

3.3 MP Measurement

1. Turn on the instrument 1 h before the measurement. Clean the lens, drop immersion oil, and settle the coverslip attached with a sample cassette.
2. Fill one well with 10 μL PBS. Conduct focusing. Confirm that homogeneous background signals are observed.

3. Add 2 μL of the protein standard and mix by pipetting.
4. Start the measurement immediately within 5–20 s after pipetting.
5. Analyze the signals using analyzer software, and calibrate the size distribution with the molecular weights of the protein standard samples. Re-run the calibration if the linear regression of size is unsuccessful.
6. For measurement, fill a second well with 15 μL PBS. Conduct focusing.
7. Add 5 μL of the antigen or antibody and mix by pipetting (*see* **Note 9**).
8. Measure and analyze the sample (same as steps 5–6). Optimize the concentration for the measurement, as sensitivity depends on the size, charge, and hydrophobicity of the samples. Typically, the concentration before dilution is 20–200 nM. For TNFR2, the concentrations before and after dilution were 50 and 12.5 nM, respectively. The video was recorded for 60 s. A bell-shaped distribution should be confirmed (*see* **Note 10**).
9. Mix the antigen and antibody at the determined molecular ratio (*see* **Note 7**) in 10 μL solution and allow the sample to stand for 15–30 min (at least 10 min) at an appropriate temperature (typically at 25°C) to enable immunocomplex formation.

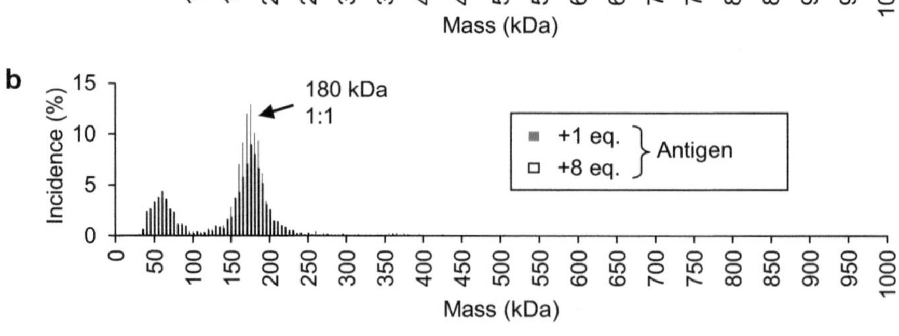

Fig. 4 Histograms of the particles detected in mass photometry (MP). (**a**) Biparatopic antibody (BpAb) with cross-linking activity (Bp96-92). (**b**) BpAb without cross-linking activity (Bp109-92). The charts were drawn using data published in [8]

10. Measure and analyze the sample, as described in step 8. As in SEC-MALS, two or more conditions with different molar ratios may be required. This concentration may be optimized for distribution analysis (*see* **Note 11**).
11. Analyze mass distribution using the manufacturer-provided software. The cumulative plot may be redrawn by exporting events in CSV format (Fig. 4).

3.4 SPR Measurement

1. Place the sensor chip in the instrument and run in the running buffer.
2. Immobilize the anti-tag antibody via amine coupling (*see* **Note 12**). Dilute the anti-MBP antibody with 10 mM sodium acetate, pH 4.5, by 20-fold to 50 μg/mL (*see* **Note 13**) and immobilize the antibody on the CM5 chip using an amine coupling kit. Flow the antibody at 5 μL/min for 5 min. The immobilized level of the anti-tag antibody reached approximately 6000 RU.
3. Test the antigen capture condition. Dilute the tagged antigen at 20 ng/mL, set the flow rate at 30 μL/min, and determine the capture rate in the unit of RU/min. This parameter is used to determine the capture conditions. For TNFR2, the TNFR2-MBP was captured at 320 RU/min under these conditions.
4. Test the regeneration condition. Typical conditions are 3 M $MgCl_2$, 10 mM sodium citrate (pH 3.0) or 10 mM glycine hydrochloride (pH 2.0), run at 30 μL/min for 30 s. To capture TNFR2-MBP using anti-MBP, two successive injections of 3 M $MgCl_2$, run at 30 μL/min for 30 s, enable regeneration of the sensor chip.
5. Confirm regeneration by repeating steps 3–4.
6. Set up the capture conditions at different levels. Using the capture rate determined in step 3, determine the conditions required to enable three or more capture levels. For TNFR2, the condition described in Table 1 was adopted for capture the steps run at 30 μL/min.

Table 1
Conditions for antigen capture at different levels

Response of captured TNFR2-MBP (RU)	Concentration (ng/mL)	Time (s)
12	5	8
45	5	30
120	20	20
300	20	60

7. Perform the run (*see* **Note 14**). Optimization of the concentration range is typically necessary depending on the association and dissociation kinetics of the antigen–antibody combinations. For TNFR2, single-cycle kinetics of 0.4–6.4 nM by two-fold dilution series with contact and dissociation times of 180 s and 1200 s was performed at four capture levels.
8. Qualitatively analyze the SPR data (*see* **Note 15**). Extract the raw data and draw a baseline for the captured antigen, and divide the response value at each time point by the capture

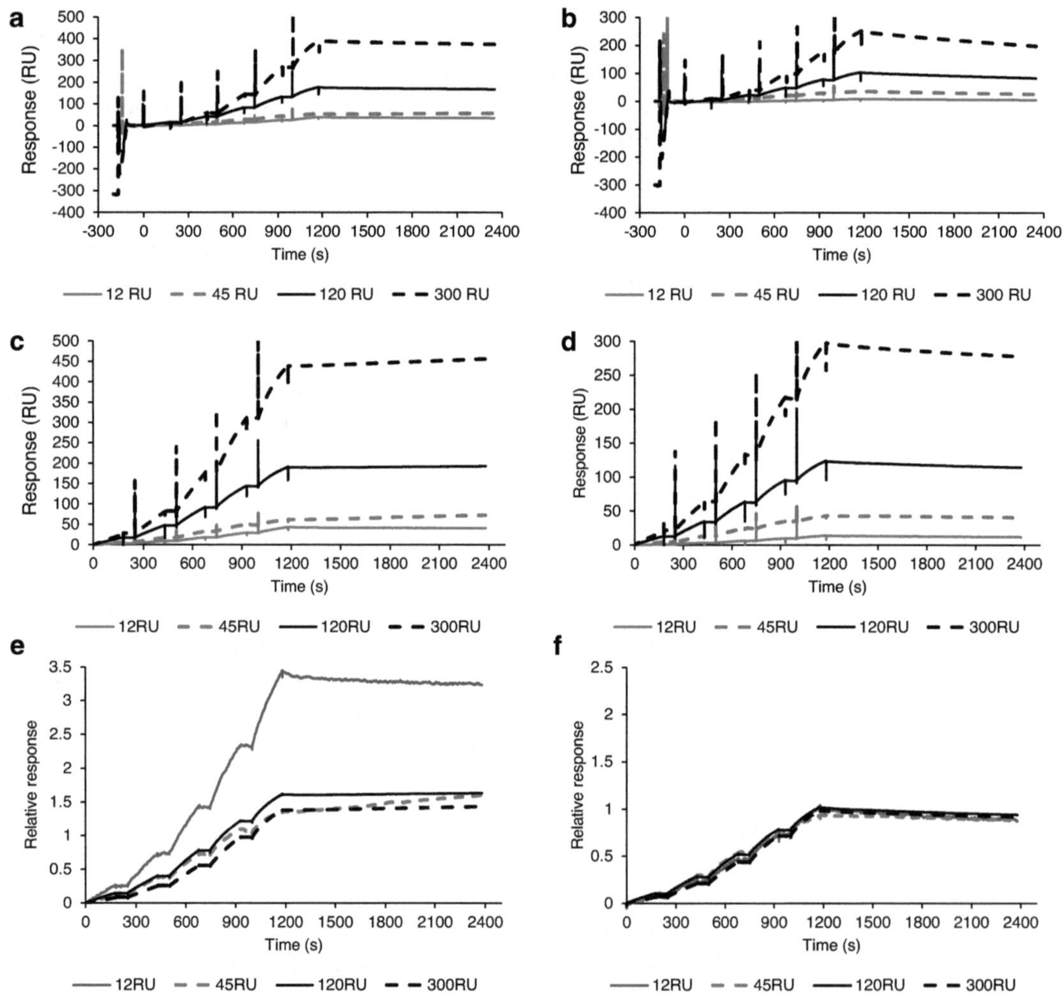

Fig. 5 Surface plasmon resonance (SPR) charts for analyzing cross-linking activities. (**a and b**) Raw sensorgrams subtracted only with the reference sensorgram. (**c and d**) Sensorgrams subtracted with the blank run (antigen-captured). (**e and f**) Relative sensorgrams with the response values divided by the capture level. (**a, c, and e**) Biparatopic antibody (BpAb) with cross-linking activity (Bp96-92). (**b, d, and f**) BpAb without cross-linking activity (Bp109-92). The charts were drawn using data published in [8]

level. The values represent the relative number of antibody molecules bound to each antigen.

9. The analyzed sensorgrams measured for different capture levels are plotted in a single chart (Fig. 5) (*see* **Note 16**).

4 Notes

1. As size analyses are conducted in label-free systems, a small gap in size between the antigen and antibody makes data analysis difficult. The ratio of the molecular weights of the antibody and antigen should be >3 or <1/3 for analysis. Conventional IgG (150 kDa) corresponds to an antigen with a molecular weight of <50 or >450 kDa.
2. Avoid using Fc or GST tags, which produce dimeric proteins, for size analysis (SEC-MALS and MP), unless the native state of the antigen is a dimer.
3. For tag selection in affinity capture, ensure that the tag and anti-tag antibody are high affinity without apparent dissociation during a cycle of SPR measurement (typically 30–60 min). Cross-linking between antigen molecules results in multivalent binding of the tagged antigen to the immobilized anti-tag antibody; thus, irregular dissociation pattern can be observed when the anti-tag antibody does not meet these criteria.
4. Concentration determination by refractive index is recommended because antibodies and antigens often show different values of the ratio of extinction coefficient to molecular weight. In addition, antibodies are typically glycoproteins, and calculating the molecular weight from the amino acid sequence is difficult. For TNFR2, the antigen was also highly glycosylated, and the same issue occurred.
5. Refractive index increment (*dn/dc*) values affect the refractive index and light-scattering intensities. Thus, this value is important in determining the absolute molecular weight. In immunocomplex analysis, the number of antibodies and antigens in multimeric complexes is of interest; and thus, the molecular weight relative to the monomer, not the absolute weight, is required. Thus, the *dn/dc* value may be determined manually to match other observations. For TNFR2 case, $dn/dc = 0.175$ was adopted because the molecular weights of the antibodies (145 kDa) matched the calculated values or those detected in the MP, and that of TNFR2 (37 kDa) matched those determined in sodium dodecyl sulfate-polyacrylamide gel electrophoresis analysis.

6. Various equations are available for molecular weight determination from the angle-dependent scattering intensity; in our case of TNFR2, the molecular weight did not vary widely using the Zimm, Debye, and Berry plots because the concentration was low and the size of the particles was far smaller than the wavelength of the incident light. Carefully select optimal equations depending on the objective.
7. To avoid misinterpretation, either the antibody or the antigen should be excess enough for the initial measurement. For the comparison with conventional antibodies, conditions with excess antigens are recommended. For conventional IgG antibodies, the number of epitopes and paratopes per molecule is 1 and 2; thus, an excess of antigen above twice the concentration of the antibody is required. The optimal conditions differ according to the project objective.
8. Concentration of the mixture may also affect size distribution. Although we do not usually observe qualitative differences (presence or absence of cross-linked binding pattern) by different concentrations, samples prepared in two different concentrations may be analyzed to avoid misinterpretation if other complementary techniques such as MP are not used.
9. The optimal dilution ratio depends on the analyte because of the different detection sensitivities among samples. The sensitivity depends on the surface charge and hydrophobicity of the particles and is not controllable as the particles are detected when they adsorb onto the glass surface. The concentration of samples in the stand in the mixture should be standardized, and a dilution factor between 1/2 and 1/20 should be chosen.
10. Sensitivity is low when the molecular weight is under 40 kDa, and small samples do not show a bell-shaped distribution. In this case, mass determination is not possible.
11. Images of large particles are processed preferentially compared with those of small samples, and careful optimization of the concentration during measurement is recommended. Typically, a single 60–120 s measurement provides enough detected particles. Replicate measurements rather than further extend the measurement if an additional number is required.
12. Direct immobilization of the antigen is also applicable if the regeneration conditions can be established or high-throughput systems with discardable sensor chip are used.
13. Optimization of the pH and concentration may be required for other anti-tag antibodies.

14. Single-cycle kinetics with multiple injections of increasing concentrations is recommended to emphasize the capture level-dependent behavior.
15. As avidity binding involves multiple binding modes, the parameters obtained by fitting a single model, such as 1:1 binding kinetics, are not always theoretically supported. Therefore, quantitative characterization may not be appropriate for comparisons among BpAbs.
16. Because the binding of analytes to sensor chip at contact steps depends both on association and dissociation rates, high response or quick binding in the relative sensorgram does not represent stronger binding or faster association.

Acknowledgments

We thank Reiko Satoh for technical assistance. This work was partially supported by Japan Agency for Medical Research and Development (AMED) grant numbers JP22ak0101099, JP24ama121042, Japan Society for the Promotion of Science (JSPS) grant number JP21K06453, Kyoto University Foundation, and Takeda Science Foundation.

Conflict of Interest K.S. is the CEO of Refeyn Japan, K.K. and the country manager of Refeyn LTD, which manufactures the instruments of mass photometry. Other authors do not declare any conflict of interest.

References

1. Niquille DL, Fitzgerald KM, Gera N (2024) Biparatopic antibodies: therapeutic applications and prospects. mAbs 16(1):2310890. https://doi.org/10.1080/19420862.2024.2310890
2. Akiba H, Tsumoto K (2020) Development and activities, including immunocomplex formation, of biparatopic antibodies and alternative scaffold proteins. Transl Regul Sci 2(1):1–6. https://doi.org/10.33611/trs.2_1
3. Li JY, Perry SR, Muniz-Medina V, Wang X, Wetzel LK, Rebelatto MC, Hinrichs MJM, Bezabeh BZ, Fleming RL, Dimasi N, Feng H, Toader D, Yuan AQ, Xu L, Lin J, Gao C, Wu H, Dixit R, Osbourn JK, Coats SR (2016) A biparatopic HER2-targeting antibody-drug conjugate induces tumor regression in primary models refractory to or ineligible for HER2-targeted therapy. Cancer Cell 29(1):117–129. https://doi.org/10.1016/j.ccell.2015.12.008
4. Kasturirangan S, Rainey GJ, Xu L, Wang X, Portnoff A, Chen T, Fazenbaker C, Zhong H, Bee J, Zeng Z, Jenne C, Wu H, Gao C (2017) Targeted Fcγ receptor (FcγR)-mediated clearance by a biparatopic bispecific antibody. J Biol Chem 292(10):4361–4370. https://doi.org/10.1074/jbc.M116.770628
5. Kast F, Schwill M, Stuber JC, Pfundstein S, Nagy-Davidescu G, Rodriguez JMM, Seehusen F, Richter CP, Honegger A, Hartmann KP, Weber TG, Kroener F, Ernst P, Piehler J, Pluckthun A (2021) Engineering an anti-HER2 biparatopic antibody with a multimodal mechanism of action. Nat Commun 12(1):3790. https://doi.org/10.1038/s41467-021-23948-6
6. Yang Y, Yeh SH, Madireddi S, Matochko WL, Gu C, Pacheco Sanchez P, Ultsch M, De Leon Boenig G, Harris SF, Leonard B, Scales SJ, Zhu JW, Christensen E, Hang JQ, Brezski RJ, Marsters

S, Ashkenazi A, Sukumaran S, Chiu H, Cubas R, Kim JM, Lazar GA (2019) Tetravalent biepitopic targeting enables intrinsic antibody agonism of tumor necrosis factor receptor superfamily members. MAbs 11(6):996–1011. https://doi.org/10.1080/19420862.2019.1625662
7. DaSilva JO, Yang K, Perez Bay AE, Andreev J, Ngoi P, Pyles E, Franklin MC, Dudgeon D, Rafique A, Dore A, Delfino FJ, Potocky TB, Babb R, Chen G, MacDonald D, Olson WC, Thurston G, Daly C (2020) A biparatopic antibody that modulates MET trafficking exhibits enhanced efficacy compared with parental antibodies in MET-driven tumor models. Clin Cancer Res 26(6):1408–1419. https://doi.org/10.1158/1078-0432.CCR-19-2428
8. Akiba H, Fujita J, Ise T, Nishiyama K, Miyata T, Kato T, Namba K, Ohno H, Kamada H, Nagata S, Tsumoto K (2023) Development of a 1:1-binding biparatopic anti-TNFR2 antagonist by reducing signaling activity through epitope selection. Commun Biol 6(1):987. https://doi.org/10.1038/s42003-023-05326-8
9. Akiba H, Ise T, Satoh R, Abe Y, Tsumoto K, Ohno H, Kamada H, Nagata S (2024) Generation of antagonistic biparatopic anti-CD30 antibody from an agonistic antibody by precise epitope determination and utilization of structural characteristics of CD30 molecule. bioRxiv 2024:606455. https://doi.org/10.1101/2024.08.02.606455
10. Wu D, Piszczek G (2021) Standard protocol for mass photometry experiments. Eur Biophys J 50(3–4):403–409. https://doi.org/10.1007/s00249-021-01513-9
11. Z Nikolovska-Coleska (2015) Studying protein-protein interactions using surface plasmon resonance. In: Meyerkord CL, Fu H (eds) Protein-protein interactions: methods and applications. Springer New York, New York, NY, 109–138 https://doi.org/10.1007/978-1-4939-2425-7_7
12. Some D, Amartely H, Tsadok A, Lebendiker M (2019) Characterization of proteins by size-exclusion chromatography coupled to multi-angle light scattering (SEC-MALS). J Vis Exp 148:e59615. https://doi.org/10.3791/59615
13. Matson JB, Steele AQ, Mase JD, Schulz MD (2024) Polymer characterization by size-exclusion chromatography with multi-angle light scattering (SEC-MALS): a tutorial review. Polym Chem 15(3):127–142. https://doi.org/10.1039/d3py01181j
14. Sonn-Segev A, Belacic K, Bodrug T, Young G, VanderLinden RT, Schulman BA, Schimpf J, Friedrich T, Dip PV, Schwartz TU, Bauer B, Peters JM, Struwe WB, Benesch JLP, Brown NG, Haselbach D, Kukura P (2020) Quantifying the heterogeneity of macromolecular machines by mass photometry. Nat Commun 11(1):1772. https://doi.org/10.1038/s41467-020-15642-w
15. De Gasparo R, Pedotti M, Simonelli L, Nickl P, Muecksch F, Cassaniti I, Percivalle E, Lorenzi JCC, Mazzola F, Magrì D, Michalcikova T, Haviernik J, Honig V, Mrazkova B, Polakova N, Fortova A, Tureckova J, Iatsiuk V, Di Girolamo S, Palus M, Zudova D, Bednar P, Bukova I, Bianchini F, Mehn D, Nencka R, Strakova P, Pavlis O, Rozman J, Gioria S, Sammartino JC, Giardina F, Gaiarsa S, Pan-Hammarström Q, Barnes CO, Bjorkman PJ, Calzolai L, Piralla A, Baldanti F, Nussenzweig MC, Bieniasz PD, Hatziioannou T, Prochazka J, Sedlacek R, Robbiani DF, Ruzek D, Varani L (2021) Bispecific IgG neutralizes SARS-CoV-2 variants and prevents escape in mice. Nature 593(7859):424–428. https://doi.org/10.1038/s41586-021-03461-y

Check for updates

Chapter 3

Generation of Bispecific Antibodies Using FAST-Ig™

Chiew Ying Yeo and Shu Feng

Abstract

Robust generation of bispecific antibodies (BsAbs) in single-cell expression systems is essential for basic research and industrial manufacturing. However, the low yield of BsAbs and complicated downstream processing caused by the random assembly of heavy chains (HCs) and light chains (LCs) have limited the use of BsAbs in clinical and preclinical studies. Major technological improvements have been established to address the issues of HC homodimerization and noncognate HC and LC pairing. Nevertheless, current existing methods need to determine common light chains, extensively engineer each Fv portion or perform labor-intensive biochemical processing making downstream processing highly challenging. Here, we describe an engineering technology for preferential cognate HC/LC and HC/HC pairing called four-chain assembly by electrostatic steering technology-immunoglobulin (FAST-Ig™). When co-expressed in single host cells, parental monoclonal antibodies incorporating these interfaces can preferentially assemble into BsAbs. The remaining trace amount of mispairing antibodies can be easily removed via ion exchange chromatography. In this chapter, we provide a detailed protocol for the systematic generation of IgG-like BsAb designs based on the FAST-Ig™ technology in combination with knobs-into-holes technology to obtain highly purified BsAbs. Additionally, we present a detailed protocol for the analysis and characterization of target BsAbs.

Keywords Bispecific antibodies, FAST-Ig™, Ion exchange chromatography, Single-cell expression, Knobs-into-holes

1 Introduction

Bispecific antibodies (BsAbs) are emerging as next-generation antibody therapeutics because of their ability to simultaneously bind two different antigens or two different epitopes of the same antigen. This unique dual-binding property enables them to exhibit unique mechanisms of action that cannot be achieved by a combination of two individual monospecific antibodies. The binding versatility of engineered BsAbs has great therapeutic potential across a wide range of diseases in both the oncology and non-oncology fields [1]. To date, 14 BsAbs have been approved by the Food and Drug Administration (FDA), and more than 200 BsAbs are currently in clinical development stages [2].

Masahiro Yasunaga (ed.), *Bispecific Antibodies*, Methods in Molecular Biology, vol. 2997,
https://doi.org/10.1007/978-1-0716-5037-0_3,
© The Author(s), under exclusive license to Springer Science+Business Media, LLC, part of Springer Nature 2026

Despite the various symmetric and asymmetric BsAb formats developed in this fast-growing field [3], immunoglobulin (IgG)-based BsAbs are still the preferred format because of the presence of the Fc portions for longer serum half-life, desired immune effector function, and easier purification by protein A affinity chromatography [4, 5]. Additionally, BsAbs with simple IgG formats often possess lower immunogenicity risk and fewer issues with their physicochemical properties [5, 6]. However, a great challenge in efficient production of BsAbs with respect to quantity and quality of BsAbs in single-cell expression systems has limited their rapid advancement. When four distinct polypeptide chains are co-expressed in single host cell, two heavy chains (HCs), and two light chains (LCs) randomly assemble; this results in a correct BsAb accompanied by nine mispairing species. These byproducts result from the undesired formation of HC homodimers, noncognate HC and LC pairing, unbalanced expression of the different chains and incomplete assembly [7]. In addition to causing low target BsAb yields, these byproducts display similar properties to target BsAbs; therefore, the downstream processing becomes problematic and complicated.

Through decades of development, many technological improvements have been established to overcome mispairing issues and ensure high-yield production of BsAbs using conventional expression and purification techniques. To promote HC heterodimerization, strategies such as knobs-into-holes (KiH) [8, 9] and charge pair mutations [10, 11], which rely on CH3 engineering of the antibody, have been established. In addition to HC heterodimerization, numerous strategies have been developed to ensure correct HC/LC pairing. One effective approach is to use a common LC to pair with both HCs; this eliminates the possibility of HC/LC mispairing [9]. However, the identification of a common LC that can maintain desirable functional properties can be challenging and time-consuming [12]. Other approaches focus on engineering the HC/LC interface to suppress HC/LC mispairing during the co-expression of HCs and LCs including CrossMab [13], orthogonal Fab pairing [14–16], and interchain disulfide bond engineering [17] have been extensively reviewed elsewhere [7, 18].

Here, we describe an engineering technology called four-chain assembly by electrostatic steering technology-immunoglobulin (FAST-Ig™) for preferential cognate HC/LC and HC/HC pairing [19]. In this chapter, we provide a detailed protocol for generating IgG-like BsAb design based on FAST-Ig™ technology in combination with KiH technology to obtain highly purified BsAbs using single host cells. The subsequent protocol for the analysis and characterization of the target BsAbs are covered in more detail.

2 Materials

2.1 Plasmid Preparation

1. Soft program for plasmid design (e.g., GENETXY).
2. DNA polymerase. We use PrimeSTAR® Max DNA polymerase (Takara) as an example.
3. DNA assembly cloning kit. We use In-Fusion HD Cloning Kit (Clontech) as an example.

2.2 Antibody Expression and Purification by Protein A Affinity Chromatography

1. Mammalian cells for recombinant protein expression. We use Expi293F cells (Thermo Fisher Scientific) as an example.
2. Reagents for mammalian cells expression and transfection
3. Protein A for antibody purification. Ready-to-use pre-packed columns or pre-coupling magnetic beads with protein A can be used. We use AbCapcher Mag2 (ProteNova) as an example.
4. 0.22 μm filter unit (e.g., Merck Milipore).

2.3 Sodium Dodecyl Sulfate-Polyacrylamide Gel Electrophoresis (SDS-PAGE) Analysis

1. SDS-PAGE gel electrophoresis equipment, running buffer, and molecular weight marker (e.g., Bio-Rad).
2. SDS-PAGE gel. We use 4–20% and 12% Mini-PROTEAN® TGX™ Precast gel (Bio-Rad) as example.
3. Sample buffer (reducing and non-reducing). We use sample buffers (2ME−) and (2ME+) (x4) (Wako) as example.
4. Coomassie blue staining solution.
5. Gel Imaging system (e.g., Bio-Rad).

2.4 Capillary Isoelectric Focusing (cIEF) Analysis

1. Maurice and Compass for iCE software (Protein Simple) or equivalent.
2. cIEF Method Development Kit and cIEF Cartridge (Protein Simple) or equivalent.

2.5 Size-Exclusion Chromatography (SEC) Analysis

1. ACQUITY UPLC H-Class HPLC system and Empower Software (Waters) or equivalent.
2. ACQUITY UPLC Protein BEH SEC Column, 200 Å, 1.7 μm, 4.6 mm × 150 mm (Waters) or equivalent.
3. Running buffer: 50 mM sodium phosphate, 300 mM sodium chloride, pH 7.0.

2.6 Thermal Shift Assay (TSA) Analysis

1. Real-time PCR system for TSA. We use QuantStudio 12K Flex Real-Time PCR system and Protein Thermal Shift Software (Thermo Fisher Scientific) as an example.
2. SYPRO Orange Protein Gel Stain (5000×)
3. Phosphate buffer for SYPRO Orange dilution: 500 mM NaH-2PO4·2H2O, pH 7.5.

2.7 Antigen-Binding Activity Analysis

1. Biacore™ T200 surface plasmon resonance (SPR) system (Cytiva).
2. Recombinant protein A/G.
3. CM4 sensor chip (Cytiva).
4. Biacore™ T200 Evaluation Software (Cytiva).

3 Methods

Depending on the application of BsAbs, the optimal BsAb design with desired functionalities should consider several factors; these include the number of valences, distance of paratopes, presence or absence of Fc regions, and Fc-mediated effector function [2]. This protocol is intended for application in preclinical research, which requires quick generation and assessment of BsAbs. Here, we provide details on the simple IgG-based BsAb design based on FAST-Ig™ technology, construction of expression plasmids, expression and purification of BsAbs, and characterization of target BsAbs (*see* **Note 1**).

3.1 Design of FAST-Ig Variants

1. Each BsAb consists of two different HCs and two different LCs. The BsAb design using the FAST-Ig™ design principle is shown in Fig. 1. This prototype BsAb contains an anti-A arm that is paired with a kappa light chain and an anti-B arm that is paired with a lambda light chain (*see* **Note 2**).
2. The HC and LC of the anti-A arm are represented as H1 and L1, respectively. The HC and LC of the anti-B arm are

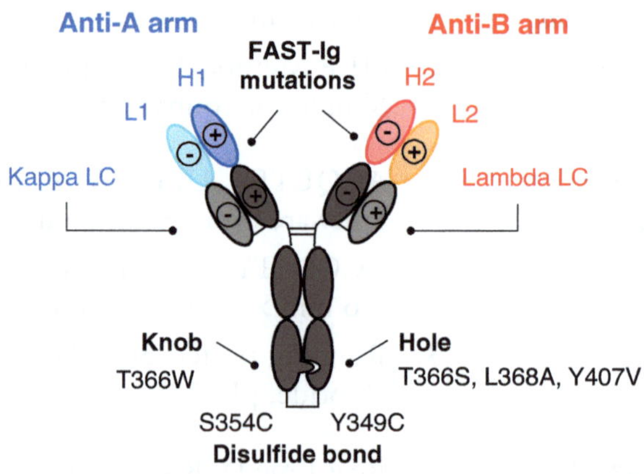

Fig. 1 Schematic illustration of human bispecific antibodies (BsAbs) generated using the FAST-Ig™ design principle. The HC and LC of the anti-A arm with a kappa LC are represented as H1 and L1, respectively. The HC and LC of the anti-B arm with a lambda LC are represented as H2 and L2, respectively. Disulfide-linked knobs-into-holes mutations are introduced to ensure preferential HC heterodimerization. The Fc numbering is defined by EU numbering [24]

represented as H2 and L2, respectively. The target BsAb is represented as H1L1H2L2.

3. FAST-Ig mutations are introduced at the VH/VL and CH1/CL interfaces in all four chains to promote cognate HC/LC pairing. There are multiple versions of FAST-Ig that can be used in human IgG BsAb. Specific charged pair mutations on the CH1/CL interface of each FAST-Ig variant are shown in Table 1.
4. To further improve the correct assembly of BsAb, additional charged pair mutations on the VH/VL interface, as shown in Table 2, can be combined with FAST-Ig mutations on the CH1/CL interface (*see* **Notes 3 and 4**).
5. Disulfide-linked knobs-into-holes mutations (dKiH) are introduced on two heavy chains to promote HC heterodimerization (*see* **Note 5**). The dKiH used in this protocol consists of knob mutations (S354C and T366W) in H1 and hole mutations (Y349C, T366S, L368A, and Y407V) in H2.

3.2 Construction of Expression Plasmids

The generation of BsAbs requires the co-expression of four expression plasmids in single mammalian cells. Thus, HC and LC expression plasmids are separately constructed.

1. Design mutation primers for the gene assembly method with the desired FAST-Ig and KiH mutations using GENETXYN (*see* **Note 6**).
2. Perform polymerase chain reaction (PCR) reaction with PrimeSTAR® Max DNA polymerase according to manufacturer's recommended PCR conditions.
3. Assemble PCR fragments containing the desired FAST-Ig and dKiH mutations into mammalian expression vectors that contain prototype wild-type HC or LC sequences using In-Fusion HD Cloning Kit.
4. Verify the nucleotide sequences of the newly prepared expression vectors via DNA sequencing.

3.3 Expression and Purification of Bispecific Antibodies

BsAbs with a human full-length IgG format can be expressed by transient transfection of Expi293 suspension cells and purified from cell culture supernatants with protein A affinity chromatography using the same procedure as for IgG samples (*see* **Note 7**).

1. Co-transfect the optimal plasmid ratios of both heavy and light chain expression plasmids into 30 mL of Expi293F cells using the Expi293 transfection reagents according to the manufacturer's instructions (*see* **Note 8**).
2. Harvest the cell culture supernatants containing the BsAbs using centrifugation and then filter through a 0.22 μm filter

Table 1
Charged pair residues of FAST-Ig variants on the CH1/CL interface

FAST-Ig variants	Anti-A arm		Anti-B arm	
	CH1 on H1	CL (Kappa LC) on L1	CH1 on H2	CL (Lambda LC) on L2
C1	K147/Q175K	S131E/Q160E	K147E/Q175E	T131K/E160K
C2	K147E/Q175E	S131K/Q160K	K147/Q175K	T131E/E160
C3	K147/Q175K	S131E/T180E	K147E/Q175E	T131K/S180K
C4	K147E/Q175E	S131K/T180K	K147/Q175K	T131E/S180E
C19	K147/Q175K	S131E/Q160E/T180E	K147E/Q175E	T131K/E160K/S180K
C20	K147E/Q175E	S131K/Q160K/T180K	K147/Q175K	T131E/E160/S180E
C21	K147/Q175K/K213	E123/S131E/Q160E/T180E	K147E/Q175E/K213E	E123K/T131K/E160K/S180K
C22	K147E/Q175E/K213E	E123K/S131K/Q160K/T180K	K147/Q175K/K213	E123/T131E/E160/S180E

The mutations are described with their positions based on EU numbering [24] for HC CH1 domain and Kabat numbering [25] for the LC constant region

Table 2
Charged pair residues of FAST-Ig variants on the VH/VL interface

FAST-Ig variants	Anti-A arm		Anti-B arm	
	VH on H1	VL on L1	VH on H2	VL on L2
V1	Q39K	Q38E	Q39E	Q38K
V2	Q39E	Q38K	Q39K	Q38E

The mutations are described with their positions based on Kabat numbering for both VH and VL [25].

unit to remove the remaining cellular debris and any other particles derived from the cell cultures.

3. Purify the BsAbs from the clarified cell culture supernatant using Ab-Capcher MAG2 according to the manufacturer's instructions.
4. Measure the absorbance of the purified antibody and background correction at 280 nm and 320 nm, respectively, using a spectrophotometer. Calculate antibody concentration using the molar absorbance coefficient, also known as the extinction coefficient; this value is predicted directly from a protein sequence (*see* **Note 9**).

3.4 Characterization of Bispecific Antibodies

For quick selection of optimal FAST-Ig mutations for preclinical screening use, basic analysis methods that can be performed easily with smaller sample amount requirements need to be prioritized. The initial characterization of the BsAb purity and correct full-size assembly can be easily determined by reducing and nonreducing sodium dodecyl sulfate-polyacrylamide gel electrophoresis (SDS-PAGE) analysis. The yield of target BsAb can be assessed using capillary isoelectric focusing (cIEF) and expression analysis (*see* **Note 10**). Other analysis methods, such as surface plasmon resonance (SPR), thermal stability assay (TSA) and size exclusion chromatography (SEC) can be performed to evaluate the antigen binding activity, thermal stability, and aggregation level, respectively (*see* **Note 11**).

3.4.1 Sodium Dodecyl Sulfate-Polyacrylamide Gel Electrophoresis (SDS-PAGE)

The purity and molecular weight of antibodies can be easily determined by reducing and nonreducing SDS-PAGE analysis.

1. Mix 1 ug of sample with Sample Buffer Solution (2ME−) (×4) for the nonreducing gel and Sample Buffer Solution (2ME+) (×4) for the reducing gel.
2. Heat the mixture at 70°C for 10 min and subsequently load onto a 4%–20% Mini-PROTEAN® TGX™ Precast gel for non-reducing conditions and a 12% Mini-PROTEAN® TGX™

Precast gel for reducing conditions. Include a molecular weight marker and standard antibody as a positive control.

3. Run electrophoresis at 200 V for 35 min.
4. Stain the gel with Coomassie blue staining solution to visualize the protein bands. Image the stained gel using a gel imaging system after the final destaining step with Milli-Q water.

3.4.2 Expression Level Analysis

Analyze the expression level of each FAST-Ig variants by comparing the concentration of purified antibody in the Methods section (Sect. 3.3, step 4). To accurately estimate the BsAb titer, the results from the assessment of BsAb purity from the cIEF analysis (refer to the Methods section, Sect. 3.4, No. 3) and SEC analysis (refer to Methods section, Sect. 3.4, No. 4) need to be considered since the protein concentration information contains not only properly formed BsAb but also all possible undesired species.

3.4.3 Capillary Isoelectric Focusing (cIEF) Analysis

The purity and yield of the target BsAb are analyzed via cIEF using the Maurice system. Mispairs of H1L1H2L1 and H1L2H2L2 can be used as reference samples to separate the BsAb chromatograms at the overlapping regions (*see* **Note 12**).

1. Determine the theoretical isoelectric point (pI) values of the BsAbs and their mispairing species using Genetyxn (*see* **Note 13**).
2. Dilute 8.5 μL of 0.5 mg/mL of antibody with 41.5 μL of cIEF master mix solution containing methyl cellulose, urea, Pharmalytes, arginine, iminodiacetic acid, and pI markers (*see* **Notes 14 and 15**).
3. Run samples at 1500 volts for 1 min, followed by 3000 volts for 6 min.
4. Analyze the pI values and quantify the peak areas of the BsAb and their mispairing species using Compass for iCE software.
5. Calculate the area ratios between H1L1H2L1, H1L2H2L2, and H1L1H2L2(H1L2H2L1). Determine the BsAb yield using a previously reported calculation method [20].

3.4.4 Size Exclusion Chromatography (SEC) Analysis

The aggregation tendency of BsAbs is determined via analytical SEC using a high-performance liquid chromatography system. Aggregation can potentially arise from different steps of downstream processing, such as the protein A elution step using low pH buffer, fast neutralization of the protein A eluate and virus inactivation steps. Impurities are often associated with high immunogenicity risk. Therefore, aggregation tendency needs to be assessed by quantifying high molecular weight (HMW) and low molecular weight (LMW) species in the early drug discovery process.

1. Equilibrate an ACQUITY UPLC Protein BEH SEC Column with a mobile phase containing 50 mM sodium phosphate, 300 mM sodium chloride, pH 7.0 and preset the column temperature at 30°C.
2. Inject 10 μL of 0.1 mg/mL of antibody into column at a flow rate of 0.3 mL/min.
3. Apply isocratic mobile phase for 10 min and detect UV absorbance at 220 nm.
4. Analyze the SEC profiles of FAST-Ig variants using the Empower software.

3.4.5 Thermal Shift Assay (TSA) Analysis

The thermal stability of antibodies is determined by measuring the melting temperature (Tm) using the TSA method (*see* **Note 16**). Poor thermal stability can cause protein misfolding, which results in decreased efficacy and increased risk of immunogenicity. FAST-Ig mutations that decrease antibody stability cause the protein to unfold more rapidly with increasing temperature, leading to faster incorporation of fluorescent dye into exposed hydrophobic regions. Therefore, FAST-Ig mutations that significantly reduce the BsAb stability need to be excluded.

1. Perform 50-fold dilution of SYPRO orange protein gel stain (5000×) with 500 mM Phosphate buffer, pH 7.5.
2. Mix 18 μL of 0.1 mg/mL samples with 2 μL of Sypro orange protein gel stain (100×).
3. Run melt assay on a real-time PCR. Increase temperature from 30°C to 99°C with a ramp rate of 0.05°C/s.
4. Analyze the melting curves with Protein Thermal Shift software. The Tm is obtained from the top of the peak in the derivative plot. The first Tm values are used to compare the stability among the FAST-Ig variants and the non-FAST-Ig antibody is used as a reference.

3.4.6 Antigen Binding Activity Analysis

The binding specificities, kinetics, and affinities of various FAST-Ig variants are assessed using the T200 surface plasmon resonance (SPR) system (*see* **Note 17**). Characterizing antigen binding activity is critical to ensure that BsAbs has demonstrate the desired selectivity and specificity in vivo. Compared with the parent without FAST-Ig mutations, the FAST-Ig mutations that cause a significant reduction in binding activity should be excluded.

1. Capture purified FAST-Ig variants on a CM4 sensor chip with immobilized recombinant protein A/G.
2. Inject an appropriate concentration of target antigen as an analyte.
3. Fit the sensorgrams with a 1:1 binding model using Biacore™ T200 evaluation software.

4. Analyze the binding kinetic profile of FAST-Ig variants and compare with the parent antibody without the FAST-Ig mutations (*see* 1 **Notes 18 and 19**).

4 Notes

1. FAST-Ig™ technology is currently under patent protection [10, 21, 22]. The use of any FAST-Ig mutations described in this chapter must be performed under an official agreement.
2. The application of FAST-Ig is versatile. It can be applied to different subclasses of human and mouse IgGs [19, 23]. Additionally, FAST-Ig is also applicable for BsAbs paired with both kappa LC or one kappa and one lambda LC.
3. Different antibodies may exhibit different expression levels when mammalian expression systems are used. The effect of different FAST-Ig mutations on individual Fab arms is difficult to predict. Therefore, to obtain the optimal assembly of the target BsAb, we recommend attempting different FAST-Ig mutation pairings for every new BsAb, even if the same mutation positions are selected.
4. The same position of FAST-Ig mutations can be used for kappa and lambda LCs. When designing the FAST-Ig mutations, the original amino acids on the FAST-Ig positions are different from those at S131, Q160, and T180 on the kappa LC and those at T131, E160, and S180 on the lambda LC.
5. FAST-Ig can be used for both preclinical screening and clinical candidate generation. For preclinical screening, efficient generation of BsAbs with sufficient yield and purity is needed. Therefore, we recommend the use of KiH or dKiH variants on both HCs and FAST-Ig mutations in both the VH/VL and CH1/CL interfaces for preclinical screening studies since they have greater pairing potential. In addition to the use of dKiH ver11 in this protocol, other previously reported KiH and dKiH mutations can be introduced to improve HC heterodimerization [9]. However, for the generation of clinical candidate, we recommend screening both KiH and charge mutation pairing on CH3 domains. For example, if the antibody belongs to IgG4 subclass or if stability under acidic conditions is a concern, ART-S3 (E356K/K439E) is preferable [19]. Additionally, screening of different FAST-Ig mutations for clinical use is needed to identify the best FAST-Ig mutations for generating high-quality BsAbs in terms of yield, purity, efficacy, developability and safety. Generally, FAST-Ig

with fewer mutations is preferable for generating a clinical candidate to minimize the risk of immunogenicity.

6. Several free online supporting tools are available for primer design. The plasmids can also be prepared by ordering synthetic gene fragments and cloning them into an expression vector. The gene assembly cloning method is selected in this protocol because of its fast-cloning steps and higher success rate than those of restriction-based cloning methods. Alternatively, the plasmids can be directly ordered from several certified DNA synthesis companies; they can provide codon optimization services to optimize antibody expression depending on the host cells used.
7. Protein A eluate obtained from protein A affinity chromatography can be further purified with Superdex200™ size exclusion chromatography to remove any aggregates and ion exchange chromatography to remove any mispairing species. The BsAb usually possesses a high isoelectric point (pI), which enables cation-exchange chromatography as the first polishing step after affinity chromatography in large-scale production. The mispairing species with different in pIs can be easily separated from the target BsAb and removed during the purification step.
8. In general, providing more LC expression plasmids than HC expression plasmids can improve antibody expression levels. A 3:1 ratio of LC:HC in the transfections provides a better yield than a 1:1 ratio. For BsAb generation, performing plasmid ratio optimization for all four antibody chains is recommended. Plasmid ratio optimization not only improves the expression of target BsAb but also reduces the number of mispairing species. This is a useful step when there is imbalanced expression of both arms observed even without FAST-Ig mutations.
9. Direct spectrophotometric determination of protein concentrations at 280 nm provides a quick and accurate method for protein quantification. However, the expression level may also be affected by the purification efficiency and recovery. If accurate expression levels of antibodies are needed for comparison, the antibody concentration can be directly determined from the clarified culture supernatant with more sensitive assays, such as Octet or ELISA using a human IgG detection system.
10. The integrity of BsAbs and their mispairing species can be further determined using mass spectrometry. Antibody mass heterogeneity derived from post-translational modifications may complicate the data interpretation. Peptide-N-glycosidase F treatment should be performed to remove N-glycans prior

to the mass spectrometry analysis. Another source of mass heterogeneity is derived from proteolysis of the C-terminal lysine (K447); this source can be circumvented by deleting this lysine residue. Additionally, the deconvoluted mass spectra can be used to estimate the BsAb yield using a previously reported calculation method [20].

11. For the selection of the optimal FAST-Ig variant for clinical use, more intensive analyses in terms of immunogenicity, pharmacokinetics, heat acceleration stability and acid stability need to be performed to ensure the efficacy, safety and developability of BsAb therapeutics.
12. Mispairs of H1L1H2L1 and H1L2H2L2 can be generated by removing one of the LCs while maintaining the total amount of DNA during the transfection process. For quick selection of optimal FAST-Ig mutations for preclinical screening, the minute amount of mispaired impurities derived from HCs can be disregarded. When detailed analysis of any mispaired species is needed for clinical candidate selection, all possible mispaired impurities can be generated and used as reference samples in cIEF or in analytical ion exchangechromatography. Any low expression or unpreparable HC/LC mispaired species may imply the effectiveness of FAST-Ig mutations on pairing potency.
13. Several free online supporting tools are available for the pI calculation. Based on our experience, the calculated theoretical pI of antibodies generally correlates well with the actual pI in our cIEF analysis. To achieve sufficient separation via ion exchange chromatography, the pI difference between the mispair and target BsAb is preferred to be greater than 0.5. When the pI difference is less than 0.5, overlapping peaks may be obtained which hinders the subsequent polishing step with ion exchange chromatography.
14. Ampholytes 3–10 and Maurice cIEF pI markers 4.05 and 9.99 can be used for a standard broad-range separation method, as they are applicable to most BsAbs and their mispaired species. The standard antibody should always be included as a control in the run.
15. To further improve the resolution, a narrower range of ampholytes in the pH range of interest can be chosen. Two pI markers from the kit that closely bracket the peaks in the sample need to be selected. Additionally, these markers need to be within the pH range of narrow ampholytes for accurate pI determination.
16. Although the TSA assay provides a simple and rapid method to determine protein stability by monitoring changes in Tm, the TSA profile of Fab and Fc are often overlap; thus, only

the first Tm is useful for comparison. Hence, when more detailed TSA profiles are needed, such as in the comparison of the effects of KiH and ART-S3 mutations on CH3 stability, we recommend the use of differential scanning calorimetry (DSC) for analysis.

17. With respect to membrane-bound antigens, the antigen-binding activity of FAST-Ig variants can be analyzed by SPR using the extracellular domain of antigens or by fluorescence-activated cell sorting (FACS) binding analysis using cell line expressing membrane-bound antigens.

18. Since BsAbs can bind two different targets with distinct mechanisms of action from their parent monospecific antibodies, the intended mechanism of action as therapeutics needs to be considered in the design of assays for evaluation of their dual-target specificities.

19. When the antigen and BsAb interaction involves avidity binding, the binding kinetic values derived from SPR are not accurate. For example, in the case of a biparatopic antibody where both arms can bind the same antigen at nonoverlapping epitopes, monovalent antibodies with and without FAST-Ig mutations of both arms can be used to separately compare the binding affinity.

References

1. Labrijn AF, Janmaat ML, Reichert JM, Parren PW (2019) Bispecific antibodies: a mechanistic review of the pipeline. Nat Rev Drug Discov 18(8):585–608
2. Klein C, Brinkmann U, Reichert JM, Kontermann RE (2024) The present and future of bispecific antibodies for cancer therapy. Nat Rev Drug Discov 23(4):301–319
3. Brinkmann U, Kontermann RE (2017) The making of bispecific antibodies. MAbs 9(2):182–212
4. Husain B, Ellerman D (2018) Expanding the boundaries of biotherapeutics with bispecific antibodies. BioDrugs 32(5):441–464
5. Tsai W-TK, Li Y, Yin Z, Tran P, Phung Q, Zhou Z, Peng K, Qin D, Tam S, Spiess C (2024) Nonclinical immunogenicity risk assessment for knobs-into-holes bispecific IgG1 antibodies. Mabs 16(1):2362789
6. Bivi N, Moore T, Rodgers G, Denning H, Shockley T, Swearingen CA, Gelfanova V, Calderon B, Peterson DA, Hodsdon ME (2019) Investigation of pre-existing reactivity to biotherapeutics can uncover potential immunogenic epitopes and predict immunogenicity risk. MAbs 11(5):861–869
7. Li Y (2019) A brief introduction of IgG-like bispecific antibody purification: methods for removing product-related impurities. Protein Expr Purif 155:112–119
8. Ridgway JB, Presta LG, Carter P (1996) 'Knobs-into-holes' engineering of antibody CH3 domains for heavy chain heterodimerization. Protein Eng. Des Sel 9(7):617–621
9. Merchant AM, Zhu Z, Yuan JQ, Goddard A, Adams CW, Presta LG, Carter P (1998) An efficient route to human bispecific IgG. Nat Biotechnol 16(7):677–681
10. Igawa T, Tsunoda H (2006) Methods for producing polypeptides by regulating polypeptide association. WO2006106905, 12 October 2006
11. Gunasekaran K, Pentony M, Shen M, Garrett L, Forte C, Woodward A, Ng SB, Born T, Retter M, Manchulenko K (2010) Enhancing antibody Fc heterodimer formation through electrostatic steering effects: applications to bispecific molecules and monovalent IgG. JBiol Chem 285(25):19637–19646
12. Sampei Z, Igawa T, Soeda T, Okuyama-Nishida Y, Moriyama C, Wakabayashi T, Tanaka E, Muto A, Kojima T, Kitazawa T (2013) Identification

and multidimensional optimization of an asymmetric bispecific IgG antibody mimicking the function of factor VIII cofactor activity. PLoS One 8(2):e57479
13. Schaefer W, Regula JT, Bähner M, Schanzer J, Croasdale R, Dürr H, Gassner C, Georges G, Kettenberger H, Imhof-Jung S (2011) Immunoglobulin domain crossover as a generic approach for the production of bispecific IgG antibodies. Proc Natl Acad Sci 108(27):11187–11192
14. Lewis SM, Wu X, Pustilnik A, Sereno A, Huang F, Rick HL, Guntas G, Leaver-Fay A, Smith EM, Ho C (2014) Generation of bispecific IgG antibodies by structure-based design of an orthogonal Fab interface. Nat Biotechnol 32(2):191–198
15. Dillon M, Yin Y, Zhou J, McCarty L, Ellerman D, Slaga D, Junttila TT, Han G, Sandoval W, Ovacik MA (2017) Efficient production of bispecific IgG of different isotypes and species of origin in single mammalian cells. MAbs 9(2):213–230
16. Liu Z, Leng EC, Gunasekaran K, Pentony M, Shen M, Howard M, Stoops J, Manchulenko K, Razinkov V, Liu H (2015) A novel antibody engineering strategy for making monovalent bispecific heterodimeric IgG antibodies by electrostatic steering mechanism. J Biol Chem 290(12):7535–7562
17. Mazor Y, Oganesyan V, Yang C, Hansen A, Wang J, Liu H, Sachsenmeier K, Carlson M, Gadre DV, Borrok MJ (2015) Improving target cell specificity using a novel monovalent bispecific IgG design. MAbs 7(2):377–389
18. Madsen AV, Pedersen LE, Kristensen P, Goletz S (2024) Design and engineering of bispecific antibodies: insights and practical considerations. Front Bioeng Biotechnol 12:1352014
19. Koga H, Yamano T, Betancur J, Nagatomo S, Ikeda Y, Yamaguchi K, Nabuchi Y, Sato K, Teranishi-Ikawa Y, Sato M (2023) Efficient production of bispecific antibody by FAST-IgTM and its application to NXT007 for the treatment of hemophilia A. MAbs 15(1):2222441
20. Yin Y, Han G, Zhou J, Dillon M, McCarty L, Gavino L, Ellerman D, Spiess C, Sandoval W, Carter PJ (2016) Precise quantification of mixtures of bispecific IgG produced in single host cells by liquid chromatography-Orbitrap high-resolution mass spectrometry. MAbs 8(8):1467–1476
21. Kuramochi T, Kawazoe M, Hironiwa N, Igawa T (2013) Antigen-binding molecule having regulated conjugation between heavy-chain and light-chain. WO2013065708, 10 May 2013
22. Teranishi Y, Kato K, Koga H, Igawa T, Yamaguchi K, Soeda T (2019) Multispecific antigen-binding molecules having blood coagulation factor VIII (FVIII) cofactor function-substituting activity and pharmaceutical formulations containing such a molecule as an active ingredient. WO2019065795, 2019
23. Koga H, Kuroi H, Hirano R, Hirayama H, Nabuchi Y, Kuramochi T (2024) Rapid generation of murine bispecific antibodies using FAST-IgTM for preclinical screening of HER2/CD3 T-cell engagers. Antibodies 13(1):3
24. Edelman GM, Cunningham BA, Gall WE, Gottlieb PD, Rutishauser U, Waxdal MJ (1969) The covalent structure of an entire γG immunoglobulin molecule. Proc Natl Acad Sci 63(1):78–85
25. Kabat EA, Te Wu T, Perry HM, Gottesman KS, Foeller C (1991) Sequences of proteins of immunological interest, 5th edn. US Department of Health and Human Services, Public Health Service, National Institutes of Health, Bethesda, MD

Chapter 4

ImmTAC: A Novel Platform of T-Cell Receptor-Based Soluble Bispecifics

Benjamin P. Oestringer and Annelise Vuidepot

Abstract

Immune mobilizing monoclonal T-cell receptors Against Cancer (ImmTAC®) molecules are innovative immunotherapies designed to harness the body's immune system to fight cancer. They consist of a soluble affinity-enhanced T-cell receptor (TCR) as the targeting arm, covalently linked to an antibody fragment specific to CD3 as the effector arm, enabling T-cell redirection and cancer cell killing. TCRs can target the intracellular antigenic landscape via recognition of peptides presented by surface localized Human Leukocyte Antigen (HLA) proteins, unlocking a choice of targets that are specific to the tumor. The ImmTAC platform has been validated in the clinic with the approval of tebentafusp in 2022 for the treatment of metastatic uveal melanoma. It is the first TCR therapeutic and first soluble bispecific targeting a solid tumor to be approved for commercial use. Here, we describe the methods to produce soluble, affinity-enhanced TCRs and high-affinity ImmTAC molecules.

Keywords TCR (T-cell receptor), mTCR (monoclonal TCR), ImmTAC (Immune mobilizing Monoclonal TCRs Against Cancer), HLA (human leukocyte antigen), Phage display, Affinity enhancement, IBs (inclusion bodies), Protein expression, Protein refold, Protein purification, SPR (surface plasmon resonance), Biacore

1 Introduction

Since the 1950s, scientists have been working to develop targeted treatments for oncology. Chemotherapy, a breakthrough at its inception, fails to discriminate normal, healthy cells from tumor cells, resulting in severe unwanted side effects. Immunotherapies such as immune checkpoint inhibitors [1], and targeted therapies like T-cell therapies [2] and Antibody-Drug Conjugates [3, 4] have emerged to treat cancers with reduced adverse effects.

Among the various natural ligands of antibodies are aberrantly expressed cell surface proteins. Although cell surface proteins constitute only a small fraction of the entire cellular protein repertoire [5, 6], they are an important class of drug targets in immune therapy [7].

Masahiro Yasunaga (ed.), *Bispecific Antibodies*, Methods in Molecular Biology, vol. 2997,
https://doi.org/10.1007/978-1-0716-5037-0_4,
© The Author(s), under exclusive license to Springer Science+Business Media, LLC, part of Springer Nature 2026

However, specifically targeting tumors using cell surface proteins while sparing normal cells is challenging because most tumor associated antigens are also expressed on normal cells albeit at lower levels. One of the major clinical challenges of CAR-T and antibody therapies is "on-target off-tumor" toxic side effects [7]. In contrast, TCRs (T-cell receptors), can target about 90% of the cellular protein repertoire by recognizing peptide-Human Leukocyte Antigen (pHLA) complexes presented on the cell surface including peptides derived from large number of intracellular tumor antigens. Therefore, Immune mobilizing monoclonal TCRs Against Cancer (ImmTAC) molecules [8] offer a greater flexibility to select targets that are tumor-specific and to minimize off-target activity and associated toxicity.

However, converting a membrane-bound TCR into a soluble ImmTAC therapeutic comes with a set of challenges. TCRs are membrane-bound proteins with a low affinity for their target (K_D in the μM range) and have no function when disconnected from the T-cell surface. The building of the ImmTAC [8] platform is based on three key inventions: the engineering to produce soluble monoclonal TCRs (mTCRs) [9], technology based on phage display to increase affinity by a million-fold [8], and the addition of an effector function to redirect T-cell killing [8].

This chapter describes the materials and methods required to produce mTCR and ImmTAC molecules.

2 Materials

All solutions are prepared using ultrapure water (18 MΩ–cm at 25°C) and analytical grade reagents or higher. All reagents and buffers are filtered using 0.22 μm filters and/or autoclaved and stored at room temperature unless indicated otherwise. Bacterial work is performed using aseptic techniques in standard laminar flow hoods. All plasmids and proteins are listed as GenBank or FASTA formats in the notes section. All plasmid maps and GenBank formats are created with SnapGene version 7 (Dotmatics).

2.1 Construct Design and Transformation

1. LB media: 10 g/L NaCl, 5 g/L yeast extract, 10 g/L tryptone, pH 7.
2. LB agar: 15 g/L Bacto-agar, 10 g/L NaCl, 5 g/L yeast extract, 10 g/L tryptone.
3. LB media and agar with selection antibiotic, e.g., 100 μg/mL ampicillin (or carbenicillin), 50 μg/mL kanamycin, and/or 34 μg/mL chloramphenicol.
4. *E. coli* competent cells: DH5α or equivalent cloning strain using standard heat shock transformation protocols, and Rosetta™(DE3)pLysS or BL21(DE3)pLysS (Merck Millipore; Merck User Protocol TB009) or equivalent strains for protein expression.

5. SOC media: 0.5 g/L NaCl, 0.18 g/L KCl, 5 g/L yeast extract, 20 g/L tryptone, 0.96 g/L $MgCl_2$, 1.2 g/L $MgSO_4$, 0.4% glucose. Alternatively, use 2.4 g/L $MgSO_4$ instead of adding 0.96 g/L $MgCl_2$. Usually, its delivery is included with competent cells.
6. DNA plasmid miniprep kit, e.g., QIAprep Spin Miniprep Kit (Qiagen) or equivalent.
7. Sequence of TCR and/or pHLA of interest. TCR clone 1G4 and its ligand HLA-A*02-NY-ESO-1-SLL are used as examples (*see* **Note 1**).
8. Expression plasmid under control of T7 promotor: pGMT7 [10, 11] (Fig. 1) or similar such as pET23d (Merck Millipore) (*see* **Note 2**).
9. Oligonucleotides for cloning.
10. Restriction endonucleases (NdeI and HindIII).
11. T4 DNA ligase.
12. DNA synthesis and/or site-directed mutagenesis methods.
13. Water bath.
14. Shaker incubator.
15. Static incubator.
16. DNA gel electrophoresis chambers.
17. Tris-acetate-EDTA (TAE) running buffer.
18. Agarose for preparative electrophoresis, such as TopVision agarose tablets (ThermoFisher) or equivalent.
19. DNA gel stain, e.g., SYBR™ Safe (ThermoFisher) or equivalent.

2.2 Affinity Enhancement: Library Building

1. Phagemid used for phage display: In-house designed pEX746 [12], based on a pUC19 phagemid vector, pLitmus28i (NEB) or equivalent (*see* **Note 3**).
2. Sequence of UCHT1 [13, 14] (*see* **Note 4**).
3. Polymerase chain reaction (PCR) tubes.
4. PCR machine/thermocycler.
5. Agarose for preparative electrophoresis, such as TopVision agarose tablets (ThermoFisher) or equivalent.
6. DNA gel stain, e.g., SYBR™ Safe (ThermoFisher) or equivalent.
7. Nuclease-free H_2O.
8. Tris-acetate-EDTA (TAE) running buffer.
9. DNA gel electrophoresis chambers.
10. Oligonucleotides for block/saturation mutagenesis (NNK). Polyacrylamide gel electrophoresis (PAGE)-purified.
11. DNA column purification kit, e.g., QIAquick Gel extraction kit and QIAquick PCR Purification kit (Qiagen) or equivalent.

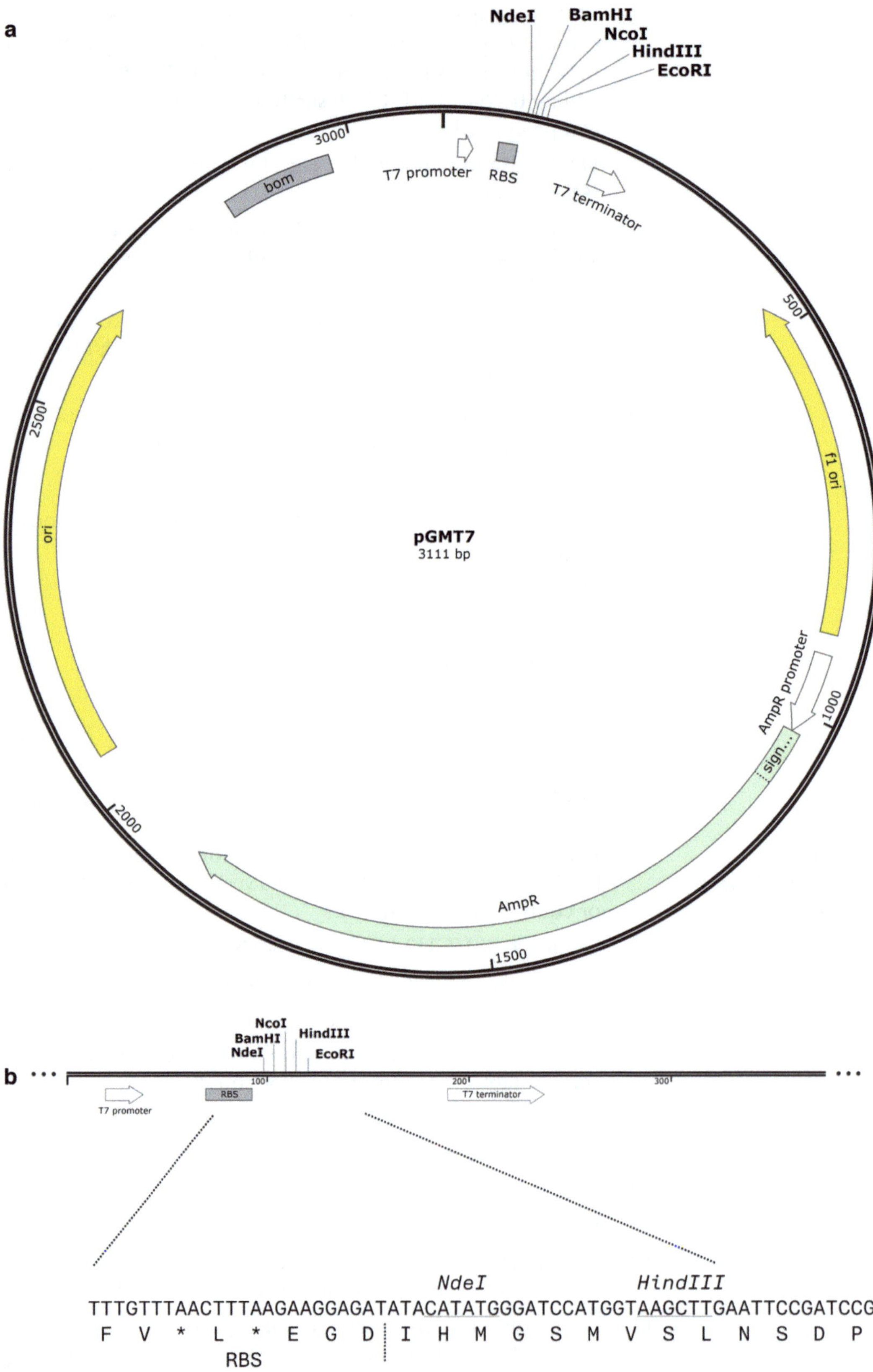
a
NdeI
BamHI
NcoI
HindIII
EcoRI
3000
T7 promoter
RBS
T7 terminator
bom
500
f1 ori
2500
ori
pGMT7
3111 bp
AmpR promoter
1000
sign...
2000
AmpR
1500
b
NcoI
BamHI
NdeI
HindIII
EcoRI
100
200
300
T7 promoter
RBS
T7 terminator
NdeI
HindIII
TTTGTTTAACTTTAAGAAGGAGATATACATATGGGATCCATGGTAAGCTTGAATTCCGATCCG
F V * L * E G D I H M G S M V S L N S D P
RBS

12. Restriction endonucleases (NcoI, NotI, and DpnI).
13. Phusion High-Fidelity DNA polymerase or equivalent.
14. T4 DNA ligase.
15. 50 mL centrifuge tubes.
16. 1.5 mL cryotube or suitable Eppendorf tube for −80°C storage.
17. Taq DNA polymerase or Taq 2× Master Mix or equivalent.

2.3 Affinity Enhancement: Phage Display

1. TG1 electrocompetent *E. coli* cells, genotype: *[F′ traD36 proAB lacIqZ ΔM15] supE thi-1 Δ(lac-proAB) Δ(mcrB-hsdSM)5(rK - mK-)*.
2. Electroporator such as MicroPulser (Bio-Rad) or equivalent.
3. 1.0 cm gap electroporation cuvette, e.g., Gene Pulser/Micropulser (Bio-Rad) or equivalent.
4. Sterile petri dish (90 mm).
5. Sterile square bioassay dish (245 mm × 245 mm).
6. 2xYeast tryptone medium (2xYT): 10 g/L yeast extract, 16 g/L tryptone, 5 g/L NaCl, pH 7.
7. 2xYTag: 2xYT media, 100 μg/mL ampicillin (or carbenicillin), 2% glucose.
8. 2xYTak: 2xYT media, 100 μg/mL ampicillin (or carbenicillin), 50 μg/mL kanamycin.
9. 2xYT glycerol + ampicillin: 2xYT, 20% glycerol, 100 μg/mL ampicillin (or carbenicillin).
10. YTEag plates: 5 g/L yeast extract, 10 g/L tryptone, 8 g/L NaCl, 15 g agar, 100 μg/mL ampicillin (or carbenicillin), 2% glucose.
11. 40% (w/v) glucose (MQ200 grade).
12. Trypsin sensitive helper phage such as KM13 [15] or CM13K (Antibody design labs) (*see* **Note 5**).
13. TBSC buffer: 10 mM Tris pH 7.4, 137 mM NaCl, 1 mM $CaCl_2$.
14. 10 mg/mL trypsin in TBSC buffer.
15. Polyethylene glycol (PEG) precipitation buffer: 20% PEG-8000, 2.5 M NaCl.
16. Sterile Pasteur pipettes.
17. Erlenmeyer flasks.

Fig. 1 The pGMT7 plasmid map. (**a**) The plasmid pGMT7 is a bacterial expression vector enabling expression of proteins using the T7 promotor. It is a derivative of pET-3c. It contains a T7 promotor, a ribosome binding site (RBS), a multiple cloning site and a T7 terminator, as well as an ampicillin resistance gene and a basis of mobility region (bom) from plasmid pBR322, and origins of replications, ori and f1 ori [10, 11]. (**b**) Details of the multiple cloning site show the RBS followed by NdeI and HindIII cleavage sites. Other cloning sites can also be used

18. Blocking buffer, e.g., 1–6% (w/v) skimmed milk powder in PBS (MPBS) (*see* **Note 6**).
19. Spectrophotometer/microvolume spectrophotometer, e.g., NanoDrop (Thermo Fisher) or equivalent.
20. 96-well ELISA plates, e.g., Nunc™ MaxiSorp plates or equivalent.
21. Streptavidin (MQ200 grade).
22. Streptavidin magnetic beads, e.g., Dynabeads™ M-280 Streptavidin (Thermo Fisher) or equivalent, and suitable magnetic racks, e.g., DynaMag™ (Thermo Fisher) or equivalent.
23. Dulbecco's PBS (pH 7.0–7.3, no calcium, no magnesium; referred to as PBS).
24. Tween-20 (Polysorbate 20, MQ300 grade).

2.4 Protein Expression and Inclusion Body (IB) Preparation

1. Tryptone yeast phosphate (TYP) medium: 16 g/L tryptone, 16 g/L yeast extract, 5 g/L NaCl, 2.5 g/L K_2HPO_4 (*see* **Note 7**), 100 μg/mL ampicillin (or carbenicillin), 34 μg/mL chloramphenicol.
2. 1 M Isopropyl β-D-1-thiogalactopyranoside (IPTG).
3. BugBuster® protein extraction reagent (User protocol TB245, Novagen, Merck Millipore) + 20 mg/mL DNAse I + 0.5 M $MgCl_2$ or equivalent bacterial cell lysis methods.
4. Erlenmeyer flasks.
5. Triton wash buffer: 0.5% (w/v) Triton X-100, 50 mM Tris pH 8, 100 mM NaCl, 2–10 mM Ethylenediaminetetraacetic acid (EDTA).
6. Tabletop centrifuge.
7. Spectrophotometer/microvolume spectrophotometer, e.g., NanoDrop (Thermo Fisher) or equivalent.
8. Resuspension buffer: 50 mM Tris, 100 mM NaCl, 10 mM EDTA.
9. 1 M 1,4-dithiothreitol (DTT).
10. Guanidine buffer: 6 M guanidine, 50 mM Tris pH 8.1, 2–10 mM EDTA, 100 mM NaCl. Add 20 mM DTT before use.
11. Sodium dodecyl sulfate–polyacrylamide gel electrophoresis (SDS-PAGE) 4%–20% Tris-Glycine gels or equivalent.
12. Gel electrophoresis equipment for SDS-PAGE with suitable protein markers and gel-loading dye, using Coomassie-based staining methods (e.g., InstantBlue, Expedeon, Sigma-Aldrich or equivalent) or stain-free TGX gels (Bio-Rad).
13. GelDoc Ez Imager (Bio-Rad; ImageLab 6.0) for densitometry analysis or equivalent.
14. Heating block.

2.5 Protein Refold

1. 1 M DTT.
2. Synthetic peptides for HLA refold.
3. Urea (Emprove® Expert ACS grade).
4. Arginine (USP grade).
5. Dimethylsulfoxide (DMSO).
6. mTCR refold buffer: 4–5 M urea, 0.4 M L-arginine, 2 mM EDTA, 100 mM Tris pH 8.1, 1.9 mM cystamine, 6.5 mM cysteamine (2-mercaptoethylamine).
7. ImmTAC refold buffer: 4–5 M urea, 0.4 M L-arginine, 2 mM EDTA, 100 mM Tris pH 8.1, 1 mM cystamine, 10 mM cysteamine (2-mercaptoethylamine).
8. HLA refold buffer: 0.4 M L-arginine, 50–100 mM Tris pH 8.1, 2 mM EDTA, 3.1 mM cystamine, 7.2 mM cysteamine (2-mercaptoethylamine).

2.6 Protein Purification

1. 0.45 μm filters.
2. Anion-exchange columns (AIEX), such as POROS 50HQ™ columns (10/100) (Thermo Fisher) or equivalent.
3. AIEX buffer A: 20 mM Tris pH 8.1.
4. AIEX buffer B: 20 mM Tris pH 8.1 with 1 M NaCl.
5. Cation-exchange columns (CIEX), such as POROS 50HS™ columns 10/100 (Thermo Fisher) or equivalent.
6. CIEX buffer A: 40 mM 2-(N-Morpholino)ethanesulfonic acid hydrate (MES) hydrate pH 5–6.
7. CIEX buffer B: 40 mM MES hydrate pH 5–6 with 1 M NaCl.
8. Size-exclusion chromatography (SEC) columns, such as Superdex® S200 Increase (S200i) 10/300 or Superdex® 75 Increase (S75i) 10/300 columns (Cytiva) or equivalent.
9. Dulbecco's PBS (pH 7.0–7.3, no calcium, no magnesium; referred to as PBS).
10. Spin concentrators (polyethersulfone membrane or equivalent, MWCO: 10 kDa).
11. Dialysis tubing (regenerated cellulose or equivalent, MWCO: 12–14 kDa).
12. BirA500 or Bulk BirA biotin-protein ligase biotinylation kit (Avidity) or equivalent.
13. SEC chromatography system, e.g., ÄKTA pure™ chromatography system (Cytiva) or equivalent.

2.7 Biacore

1. Biacore SPR (surface plasmon resonance) system (Cytiva) with respective evaluation software. Examples include Biacore™ T200 (discontinued), Biacore™ 1 or Biacore™ 8K or equivalent.

2. Surfactant P20 (10% Tween) (Cytiva).
3. 0.4 M 1-ethyl-3-(3-dimethylaminopropyl)-carbodiimide (EDC) (Cytiva).
4. 0.1 M N-hydroxysuccinimide (NHS) (Cytiva).
5. 1 M Ethanolamine-HCl (Cytiva).
6. Dulbecco's PBS (pH 7.0–7.3, no calcium, no magnesium; referred to as PBS).
7. CM5 sensor chip (Cytiva).
8. 50 μM biotin (MQ200 grade) in PBS.

3 Methods

A TCR of interest with wildtype affinity can be converted into a soluble mTCR (monoclonal TCR as it carries invariant engineered constant domains) by addition of an interchain disulfide bond between its TCR alpha constant (TRAC) domain and its TCR beta constant (TRBC) domain. Amino acids are replaced with cysteines at position 48 of TRAC (T48C) and at position 57 of TRBC (S57C), to create an engineered interchain disulfide bond facilitating in vitro refolding [9]. All amino acid numbering is according to IMGT [16]. An unpaired cysteine at position 75 of the native TRBC domain is changed to an alanine (C75A) to prevent incorrect disulfide bonding during refolding [9]. TRAC (T48C) is truncated N-terminal of the natural membrane proximal cysteine. TRBC (S57C, C75A) is similarly truncated [20]. The resulting wildtype mTCR is stable, can be affinity-enhanced using phage display, and produced as a soluble protein [8]. Addition of an anti-CD3 scFv (e.g., UCHT1) to the affinity-enhanced mTCR creates an ImmTAC molecule. HLAs which present peptides (pHLAs) are the natural ligands of TCRs and can be produced as a soluble protein using a similar protocol [17]. The pHLA, mTCR, and ImmTAC molecules are expressed separately in Escherichia coli (*E. coli*) cells, refolded from inclusion bodies (IBs) and purified.

The publicly available NY-ESO-1-specific TCR clone 1G4 is used as an example. The TCR comprises TCR alpha variable (TRAV) domain 21*02, and TCR beta variable (TRBV) domain 6-5*01 (TRAV21*02/TRBV6-5*01). Its ligand is HLA-A*0201-NY-ESO-1-SLL, which consists of HLA-A*0201, beta-2-microglobulin (β2M), and a NY-ESO-1 specific peptide (SLLMWITQC) (PDB: 2BNR) [18, 19] (*see* **Note 1**).

3.1 Protein Expression Construct Design and Transformation

Clone the desired TCRα (TRA, TRAV-TRAC) and TCRβ (TRB, TRBV-TRBC) chains (*see* **Note 8**), β2M and the HLA chains into separate expression plasmids under the control of the T7 promoter, such as pGMT7 [17, 20] (Fig. 1) (*see* **Note 2**).

1. For TCRα and TCRβ constructs, follow steps 2–7.
2. Add NdeI site and ATG at 5′-start (*see* **Note 9a and b**).
3. Add HindIII site at 3′-end (*see* **Note 9c and d**).
4. Add point mutations: TRAC (T48C), TRBC (S57C), and TRBC (C75A) [9, 21, 22].
5. TRAC (T48C) and TRBC (S57C, C75A) are truncated N-terminal of the natural membrane proximal cysteine (*see* **Note 9c and d**) [23].
6. Add stop codon TAA at 3′-end (*see* **Note 9c and d**).
7. Subclone TCR constructs into the bacterial expression vector pGMT7 (Fig. 1) (*see* **Note 9**).
8. For HLA and β2M constructs, follow steps 9–12.
9. Add NdeI site and ATG at 5′-start (*see* **Note 9e and f**).
10. Add HindIII site at 3′-end (*see* **Note 9e and f**).
11. Add stop codon TAA at 3′-end (*see* **Note 9e and f**).
12. Subclone HLA and β2M constructs into the bacterial expression vector pGMT7 (Fig. 1) (*see* **Note 9**).
13. Add a C-terminal $(GGGGS)_1$-linker-AviTag (GLNDIFEAQKIEWHE) to HLAs or TCRs for biotinylation depending on downstream use, e.g., with AviTag for surface plasmon resonance (SPR) experiments (Biacore) [17, 24].
14. Mix ~1 μL of ligation mix with ~50 μL thawed *E. coli* cells (e.g., DH5α or other cloning strain) and incubate on ice for 30 minutes.
15. Heat shock the cells at 42°C for 30 s.
16. Incubate on ice for 5 min.
17. Pipette 950 μL of room temperature SOC media into the mixture.
18. Incubate at 250 rpm at 37°C for 1 h.
19. Plate out 50–100 μL on LB-agar plates containing ampicillin (or carbenicillin).
20. Incubate plates overnight at 37°C.
21. Pick one to three single colonies and grow them separately in 1–5 mL LB-ampicillin (or carbenicillin) media in 50 mL centrifuge tubes and prepare plasmid DNA using a column-based kit, such as QIAprep Spin Miniprep Kit (Qiagen) or equivalent.
22. Sequence using suitable primers to confirm the correct gene sequence in the vector.
23. Following steps 14–20, transform the verified plasmid into an *E. coli* strain suitable for protein expression, e.g., Rosetta(DE3)

pLysS or BL21(DE3)pLysS (*see* **Notes 10 and 11**) or equivalent.

24. Proceed to Sect. 3.2 to affinity-enhance wildtype TCRs. To express, refold and purify pHLA or wildtype affinity TCRs, continue to Sect. 3.3.

3.2 Affinity Enhancement

In general, in vitro affinity enhancement consists of the construction of libraries, followed by screening and selection of clones of interest. Affinity enhancement of TCRs is performed using phage display and degenerate oligonucleotides (NNK, N = A/C/G/T, K = G/T) [8, 23, 25, 26] in block/saturation mutagenesis. Clone TCRs of interest into the phagemid pEX746 to display TCR on phage using the NcoI/NotI restriction sites as shown in Fig. 2 (*see* **Note 3**). TCRα is downstream of the pelB leader sequence and TCRβ is downstream of an internal ribosome binding site followed by the geneIII signal sequence. TCRβ is in-frame with geneIII of M13 bacteriophage that allows the display on phage (Fig. 2) [23].

Building affinity enhancement libraries

1. Design NNK oligonucleotide primers to introduce mutations into the six complementarity-determining region (CDR) segments of TCRα and TCRβ genes using overlap extension PCR, with the NNKs placed centrally into the oligonucleotide and with at least 18 bases at the 5′ and 3′ end of the NNK sequence (Fig. 3).
2. Randomize five consecutive codons of each CDR separately and use overlapping NNK primers spanning the entire length of each CDR to ensure complete coverage to create six libraries; one library for each CDR (Figs. 3 and 4) (*see* **Note 12**). Choose PAGE-purification for NNK oligonucleotides (Table 1).
3. Use mutagenic NNK oligonucleotides as forward primers with YOL22 as reverse primer (e.g., Oligo1_1α in Fig. 3) to create fragment B (PCR2) [23, 25] using Phusion High-Fidelity DNA polymerase or equivalent (Fig. 4a).
4. Use upstream complementary oligonucleotides not covering the NNK region as reverse primers paired with YOL13 (e.g., Oligo1_1α-Rev in Fig. 3) to create fragment A (PCR1) [23, 25] using Phusion High-Fidelity DNA polymerase or equivalent (Fig. 4a).
5. Run the PCR using the following program: initial denaturation at 98°C for 2 min, 25 cycles of amplification consisting of denaturation at 98°C for 16 s, annealing at X°C (X = 45–72°C; depending on primer sequence T_m calculation) for 16 s and extension at 72°C for Y seconds (Y = 15–30 s per kb; depending on fragment length) and a final extension at 72°C

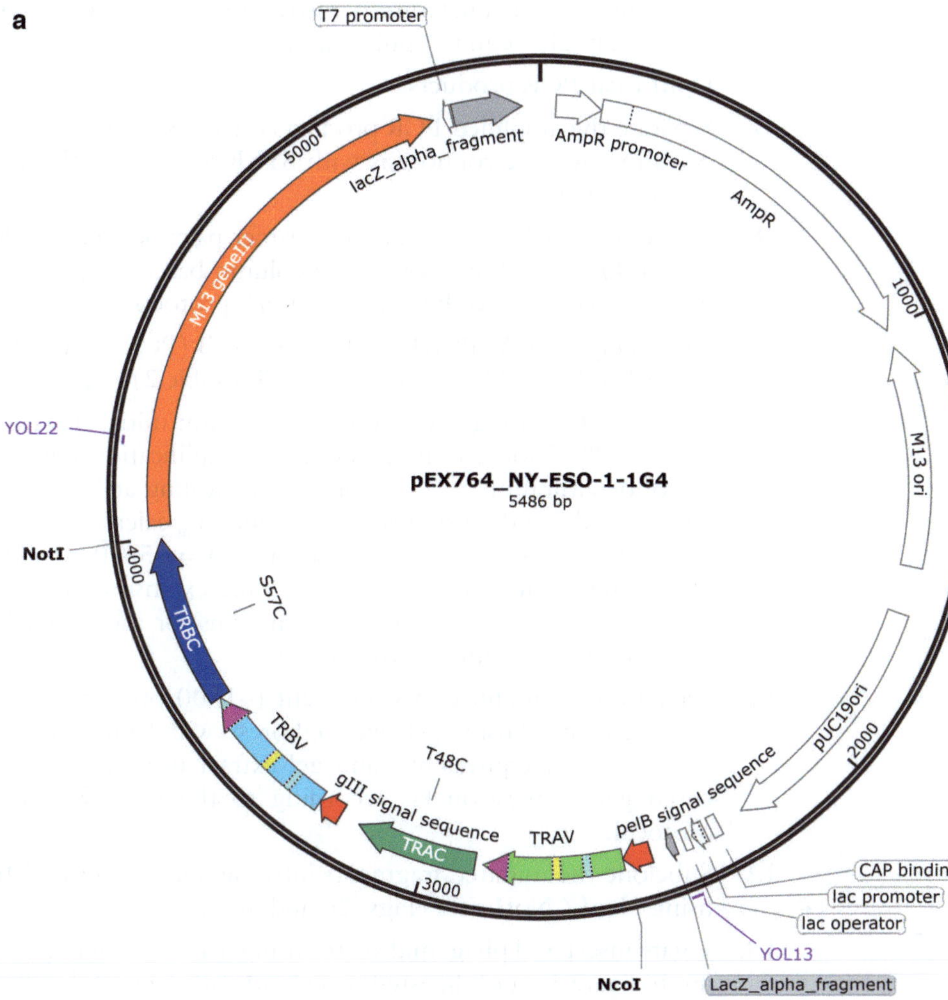

b

```
                                                                                               NcoI
pelB signal sequence and start of TRAV  ATGAAATACCTATTGCCTACGGCAGCCGCTGGATTGTTATTACTCGCGGCCCAGCCGGCCATGGCCCAGGAGGTGACA
                                         M  K  Y  L  L  P  T  A  A  A  G  L  L  L  L  A  A  Q  P  A  M  A¦ Q  E  V  T
                                                          pelB signal sequence                            ¦    TRAV

GeneIII signal sequence and start of TRBV  ATGAAAAAATTATTATTCGCAATTCCTTTAGTTGTTCCTTTCTATTCTCACAGCGCGGGTGTCACTCAGACC
                                         M  K  Y  L  L  F  A  I  P  L  V  V  P  F  Y  S  H  S  A¦ G  V  T  Q  T
                                                          geneIII signal sequence             ¦    TRBV

                                                  NotI
End of TRBC, His-tag and start of geneIII  GGTAGAGCAGACGCGGCCGCATCTAGAATTCACCATCATCACTAGACTGTTGAAAGTTGT
                                         G  R  A  D  A  A  A  S  R  I  H  H  H  H  *  T  V  E  S  C
                                         TRBC                                His-tag  |  geneIII
                                                                                   Amber
```

Fig. 2 The pEX746 phagemid [12]. (**a**) The pEX746 phagemid is shown containing NY-ESO-1-SLL clone 1G4, an example TCR in the monoclonal TCR (mTCR) format (*see* **Note 3**). The pelB signal sequence (red) is followed by the TCR alpha variable (TRAV, light green) and TCR alpha constant (TRAC, dark green) domains. The geneIII signal sequence (red) is followed by the TCR beta variable (TRBV, light blue) and TCR beta constant (TRBC, dark blue) domains. CDR1α and CDR1β are highlighted in cyan, CDR2α and CDR2β are highlighted in yellow, and CDR3α and CDR3β are highlighted in magenta. The restriction sites NcoI and NotI are highlighted in bold. The binding sited for the two oligonucleotide primers YOL13 and YOL22 are indicated in purple. Other

for 5 min. See manufacturer's instructions for more details. Program might require optimization.

6. DpnI treat PCR products.
7. Separate the amplified PCR products on a 1.5% TAE agarose gel using gel electrophoresis with SYBR Safe gel as the dye (or equivalent).
8. Excise bands (size depending on primer pair; between ~100 and 1000 bp) and recover using a column-based gel purification kit according to the manufacturer's protocol.
9. Stitch Fragment A (PCR1) and Fragment B (PCR2) by using YOL13 and YOL22 in an assembly PCR (Table 2) (Fig. 4b, c).
10. Run the PCR using the following program: initial denaturation at 98°C for 2 min, 25 cycles of amplification consisting of denaturation at 98°C for 16 s, annealing at X°C (X = 55–60°C; depending on primer sequence T_m calculation) for 16 s and extension at 72°C for Y seconds (Y = 15–30 s per kb; depending on fragment size) and a final extension at 72°C for 5 min. See manufacturer's instructions for more details. Program might require optimization.
11. Separate the amplified AB-fragment (~1700 bp) on a 1.5% TAE agarose gel using gel electrophoresis with SYBR Safe gel as the dye (or equivalent), and gel extract using a column-based gel purification kit according to the manufacturer's protocol.
12. Subclone the mutated fragments into the phagemid pEX746 using NcoI/ NotI sites (Figs. 2b and 4c–f).
13. Ligate insert and phagemid vector using T4 DNA ligase, aiming to ligate 1 μg of digested phagemid with 1 μg of digested insert.
14. Transfer into PCR tubes and incubate overnight at 16°C in a thermocycler.
15. Purify ligation reactions using a column-based purification kit but elute in nuclease-free H_2O.
16. Transform ~100–200 ng per 30–40 μL of TG1 electrocompetent *E. coli* cells and recover each electroporation using 1 mL recovery medium (*see* **Note 13**).

Fig. 2 (continued) features shown are geneIII of M13 (orange), the lac promotor and operator, an ampicillin resistance gene, M13 and pUC29 origins of replication and a T7 promotor. (**b**) Details of three regions are shown: (i) pelB and the start of TRAV with the restriction site NcoI, (ii) the geneIII signal sequence and the start of TRBV, and (iii) the end of TRBC with its His-tag, start of geneIII and the restriction site NotI, including the amber stop codon upstream of geneIII. Please note that on the nucleotide level, the beta chain (TRBV-TRBC) is one nucleotide out of frame to the alpha chain (TRAV-TRAC)

CDR1α

	D S A I Y N
Phagemid	AACTTGGTTCTCAACTGCAGTTTCACTGATAGCGCTATTTACAACCTCCAGTGGTTTAGGCAGGACCCTGGG
Oligo 1_1α	CTCAACTGCAGTTTCACT**NNKNNKNNKNNKNNK**AACCTCCAGTGGTTTAGG
Oligo 2_1α	AACTGCAGTTTCACTGAT**NNKNNKNNKNNKNNK**CTCCAGTGGTTTAGGCAG

Oligo 1_1α-Rev 5'-AGTGAAACTGCAGTTGAG

Oligo 2_1α-Rev 5'-ATCAGTGAAACTGCAGTT

CDR2α

	I Q S S Q R
Phagemid	GGGAAAGGTCTCACATCTCTGTTGCTTATTCAGTCAAGTCAGAGAGAGCAAACAAGTGGAAGACTTAATGCC
Oligo 1_2α	CTCACATCTCTGTTGCTT**NNKNNKNNKNNKNNK**AGAGAGCAAACAAGTGGA
Oligo 2_2α	ACATCTCTGTTGCTTATT**NNKNNKNNKNNKNNK**GAGCAAACAAGTGGAAGA

Oligo 1_2α-Rev 5'-AAGCAACAGAGATGTGAG

Oligo 2_2α-Rev 5'-AATAAGCAACAGAGATGT

CDR3α

	C A V R P T S G G S Y I P T F
Phagemid	CCTGGTGACTCAGCCACCTACCTCTGTGCTGTGAGGCCCACATCAGGAGGAAGCTACATACCTACATTTGGAAGAGGAACCAGCCTTATTGTTCAT
Oligo 1_3α	TCAGCCACCTACCTCTGT**NNKNNKNNKNNKNNK**TCAGGAGGAAGCTACATAC
Oligo 2_3α	ACCTACCTCTGTGCTGTG**NNKNNKNNKNNKNNK**GGAAGCTACATACCTACA
Oligo 3_3α	CTCTGTGCTGTGAGGCCC**NNKNNKNNKNNKNNK**TACATACCTACATTTGGAAG
Oligo 4_3α	GCTGTGAGGCCCACATCA**NNKNNKNNKNNKNNK**CCTACATTTGGAAGAGGA
Oligo 5_3α	AGGCCCACATCAGGAGGA**NNKNNKNNKNNKNNK**TTTGGAAGAGGAACCAGC

Oligo 1_3α-Rev 5'-ACAGAGGTAGGTGGCTGA

Oligo 2_3α-Rev 5'-CACAGCACAGAGGTAGGT

Oligo 3_3α-Rev 5'-GGGCCTCACAGCACAGAG

Oligo 4_3α-Rev 5'-TGATGTGGGCCTCACAGC

Oligo 5_3α-Rev 5'-TCCTCCTGATGTGGGCCT

CDR1β

	M N H E Y
Phagemid	AGCATGACACTGCAGTGTGCCCAGGATATGAACCATGAATACATGTCCTGGTATCGACAAGACCCAGGCA
Oligo 1_1β	CTGCAGTGTGCCCAGGAT**NNKNNKNNKNNKNNK**ATGTCCTGGTATCGACAAG

Oligo 1_1β-Rev 5'-ATCCTGGGCACACTGCAG

CDR2β

	S V G A G I
Phagemid	GGCATGGGGCTGAGGCTGATTCATTACTCAGTTGGTGCTGGTATCACTGACCAAGGAGAAGTCCCCAATGGC
Oligo 1_2β	CTGAGGCTGATTCATTAC**NNKNNKNNKNNKNNK**ATCACTGACCAAGGAGAAG
Oligo 2_2β	AGGCTGATTCATTACTCA**NNKNNKNNKNNKNNK**ACTGACCAAGGAGAAGTCC

Oligo 1_2β-Rev 5'-GTAATGAATCAGCCTCAG

Oligo 2_2β-Rev 5'-TGAGTAATGAATCAGCCT

CDR3β

	C A S S Y V G N T G E L F F
Phagemid	GCTCCCTCCCAGACATCTGTGTACTTCTGTGCCAGCAGTTACGTCGGGAACACCGGGGAGCTGTTTTTTGGAGAAGGCTCTAGGCTGACCGTACTG
Oligo 1_3β	ACATCTGTGTACTTCTGT**NNKNNKNNKNNKNNK**GGGAACACCGGGGAGCTG
Oligo 2_3β	GTGTACTTCTGTCCAGCC**NNKNNKNNKNNKNNK**ACCGGGGAGCTGTTTTTTG
Oligo 3_3β	TTCTGTGCCAGCAGTTAC**NNKNNKNNKNNKNNK**GAGCTGTTTTTTGGAGAAG
Oligo 4_3β	GCCAGCAGTTACGTCGGG**NNKNNKNNKNNKNNK**TTTTTTGGAGAAGGCTCTAG
Oligo 5_3β	AGCAGTTACGTCGGGAAC**NNKNNKNNKNNKNNK**TTTGGAGAAGGCTCTAGGCTG

Oligo 1_3β-Rev 5'-ACAGAAGTACACAGATGT

Oligo 2_3β-Rev 5'-GCTGGCACAGAAGTACAC

Oligo 3_3β-Rev 5'-GTAACTGCTGGCACAGAA

Oligo 4_3β-Rev 5'-GCCAGCAGTTACGTCGGG

Oligo 5_3β-Rev 5'-GTTCCCGACGTAACTGCT

General

YOL13: 5'-TCACACAGGAAACAGCTATG

YOL22: 5'-CATTTTCAGGGATAGCAAGC

Fig. 3 NNK oligonucleotide design for block/saturation mutagenesis. Using the NY-ESO-1-SLL clone 1G4 TCR as an example, the design for NNK oligonucleotides is shown. Oligos are shown for CDR1α, CDR2α and CDR3α, and CDR1β, CDR2β, and CDR3β, respectively. NNKs are highlighted in bold and anchor residues bordering CDR3s are highlighted in red. Corresponding reverse primers required to create the second fragment are also listed. See Table 1 for primer sequences, Table 2 for PCR set up and Fig. 4 for PCR details

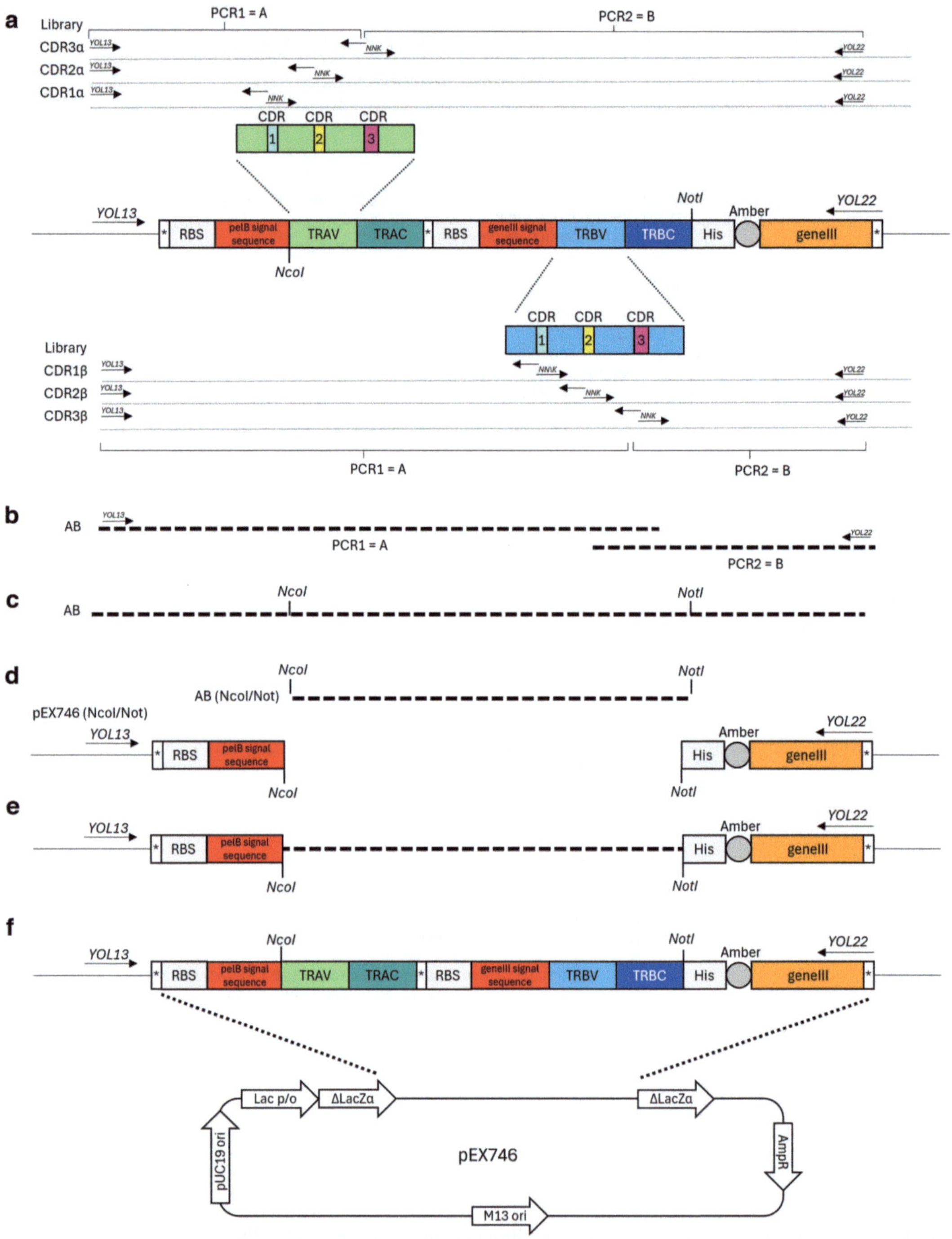

Fig. 4 Library building. (**a**) Schematics of NNK PCR layouts for each CDR are depicted. Arrows indicate fragments A (PCR1) and fragments B (PCR2). CDR1α and CDR1β are highlighted in cyan, CDR2α and CDR2β are highlighted in yellow, and CDR3α and CDR3β are highlighted in magenta. See primer sequence details for example clone NY-ESO-1-SLL-1G4 in Table 1, and general PCR details in Table 2. (**b**) Assembly of fragments A and B for each CDR library is performed using overlap stitch PCRs with primers YOL13 and YOL22 to create fragments AB, which are the final library fragments. The following subcloning steps consist of (**c**) restriction digest using NcoI and NotI and (**d**) ligation of AB (NcoI/NotI) into pEX746 (NcoI/NotI), creating (**e**) the final NNK phagemid libraries. (**f**) The pEX746 phagemid map shows the final layout of the NNK CDR libraries. Plasmid features are described in Fig. 2

Table 1
Oligonucleotide sequences

NNK	Name	Sequence
CDR1α	Oligo 1_1α	5′-CTCAACTGCAGTTTCACTNNKNNKNNKNNKNNKAACCTCCAGTGGTTTAGG
	Oligo 2_1α	5′-AACTGCAGTTTCACTGATNNKNNKNNKNNKNNKCTCCAGTGGTTTAGGCAG
	Oligo 1_1α-Rev	5′-AGTGAAACTGCAGTTGAG
	Oligo 2_1α-Rev	5′-ATCAGTGAAACTGCAGTT
CDR2α	Oligo 1_2α	5′-CTCACATCTCTGTTGCTTNNKNNKNNKNNKNNKAGAGAGCAAACAAGTGGA
	Oligo 2_2α	5′-ACATCTCTGTTGCTTATTNNKNNKNNKNNKNNKGAGCAAACAAGTGGAAGA
	Oligo 1_2α-Rev	5′-AAGCAACAGAGATGTGAG
	Oligo 2_2α-Rev	5′-AATAAGCAACAGAGATGT
CDR3α	Oligo 1_3α	5′-TCAGCCACCTACCTCTGTNNKNNKNNKNNKNNKTCAGGAGGAAGCTACATAC
	Oligo 2_3α	5′-ACCTACCTCTGTGCTGTGNNKNNKNNKNNKNNKGGAAGCTACATACCTACA
	Oligo 3_3α	5′-CTCTGTGCTGTGAGGCCCNNKNNKNNKNNKNNKTACATACCTACATTTGGAAG
	Oligo 4_3α	5′-GCTGTGAGGCCCACATCANNKNNKNNKNNKNNKCCTACATTTGGAAGAGGA
	Oligo 5_3α	5′-AGGCCCACATCAGGAGGANNKNNKNNKNNKNNKTTTGGAAGAGGAACCAGC
	Oligo 1_3α-Rev	5′-ACAGAGGTAGGTGGCTGA
	Oligo 2_3α-Rev	5′-CACAGCACAGAGGTAGGT
	Oligo 3_3α-Rev	5′-GGGCCTCACAGCACAGAG
	Oligo 4_3α-Rev	5′-TGATGTGGGCCTCACAGC
	Oligo 5_3α-Rev	5′-TCCTCCTGATGTGGGCCT
CDR1β	Oligo 1_1β	5′-CTGCAGTGTGCCCAGGATNNKNNKNNKNNKNNKATGTCCTGGTATCGACAAG
	Oligo 1_1β-Rev	5′-ATCCTGGGCACACTGCAG

(continued)

Table 1
continued

NNK	Name	Sequence
CDR2β	Oligo 1_2β	5′-CTGAGGCTGATTCATTACNNKNNKNNKNNKNNKATCACTGACCAAGGAGAAG
	Oligo 2_2β	5′-AGGCTGATTCATTACTCANNKNNKNNKNNKNNKACTGACCAAGGAGAAGTCC
	Oligo 1_2β-Rev	5′-GTAATGAATCAGCCTCAG
	Oligo 2_2β-Rev	5′-TGAGTAATGAATCAGCCT
CDR3β	Oligo 1_3β	5′-ACATCTGTGTACTTCTGTNNKNNKNNKNNKNNKGGGAACACCGGGGAGCTG
	Oligo 2_3β	5′-GTGTACTTCTGTCCAGCCNNKNNKNNKNNKNNKACCGGGGAGCTGTTTTTTG
	Oligo 3_3β	5′-TTCTGTGCCAGCAGTTACNNKNNKNNKNNKNNKGAGCTGTTTTTTGGAGAAG
	Oligo 4_3β	5′-GCCAGCAGTTACGTCGGGNNKNNKNNKNNKNNKTTTTTTGGAGAAGGCTCTAG
	Oligo 5_3β	5′-AGCAGTTACGTCGGGAACNNKNNKNNKNNKNNKTTTGGAGAAGGCTCTAGGCTG
	Oligo 1_3β-Rev	5′-ACAGAAGTACACAGATGT
	Oligo 2_3β-Rev	5′-GCTGGCACAGAAGTACAC
	Oligo 3_3β-Rev	5′-GTAACTGCTGGCACAGAA
	Oligo 4_3β-Rev	5′-GCCAGCAGTTACGTCGGG
	Oligo 5_3β-Rev	5′-GTTCCCGACGTAACTGCT
N/A	YOL13	5′-TCACACAGGAAACAGCTATG
N/A	YOL22	5′-CATTTTCAGGGATAGCAAGC

Table 2

PCR reactions. Primer pairs to create fragments A (PCR1) and B (PCR2) are shown for each alpha and beta CDR library. NNK# indicates in which CDR the NNK mutations are inserted. Template is the phagemid containing the TCR of interest (pEX746-NY-ESO-1-SLL-1G4). Purified fragments A and B are assembled for each CDR library using YOL13 and YOL22 in an overlap stitch PCR to create the full-length library fragments AB. See Fig. 4 for PCR details

Alpha PCRs			
NNK#	**Primers for PCR 1 = A**	**Primers for PCR 2 = B**	**Template**
CDR1α_1	YOL13 + Oligo 1_1α-Rev	Oligo 1_2α + YOL22	
CDR1α_2	YOL13 + Oligo 2_1α-Rev	Oligo 2_2α + YOL22	
CDR2α_1	YOL13 + Oligo 1_2α-Rev	Oligo 1_2α + YOL22	
CDR2α_2	YOL13 + Oligo 2_2α-Rev	Oligo 2_2α + YOL22	
CDR3α_1	YOL13 + Oligo 1_3α-Rev	Oligo 1_3α + YOL22	pEX746-NY-ESO-1-SLL-1G4
CDR3α_2	YOL13 + Oligo 2_3α-Rev	Oligo 2_3α + YOL22	
CDR3α_3	YOL13 + Oligo 3_3α-Rev	Oligo 3_3α + YOL22	
CDR3α_4	YOL13 + Oligo 4_3α-Rev	Oligo 4_3α + YOL22	
CDR3α_5	YOL13 + Oligo 5_3α-Rev	Oligo 5_3α + YOL22	
Beta PCRs			
NNK#	**Primers for PCR 1 = A**	**Primers for PCR 2 = B**	**Template**
CDR1β	YOL13 + Oligo 1_1β-Rev	Oligo 1_1β + YOL22	
CDR2β_1	YOL13 + Oligo 1_2β-Rev	Oligo 1_2β + YOL22	
CDR2β_2	YOL13 + Oligo 2_2β-Rev	Oligo 2_2β + YOL22	
CDR3β_1	YOL13 + Oligo 1_3β-Rev	Oligo 1_3β + YOL22	pEX746-NY-ESO-1-SLL-1G4
CDR3β_2	YOL13 + Oligo 2_3β-Rev	Oligo 2_3β + YOL22	
CDR3β_3	YOL13 + Oligo 3_3β-Rev	Oligo 3_3β + YOL22	
CDR3β_4	YOL13 + Oligo 4_3β-Rev	Oligo 4_3β + YOL22	
CDR3β_5	YOL13 + Oligo 5_3β-Rev	Oligo 5_3β + YOL22	
Full-length assembly PCR (Stitch PCR, PCR AB)			
Library / Alpha	**Primers for AB stitch**		**Template**
CDR1α_1	YOL13 + YOL22		CDR1α_1: PCR1 + PCR2 (A + B)
CDR1α_2	YOL13 + YOL22		CDR1α_2: PCR1 + PCR2 (A + B)
CDR2α_1	YOL13 + YOL22		CDR2α_1: PCR1 + PCR2 (A + B)
CDR2α_2	YOL13 + YOL22		CDR2α_2: PCR1 + PCR2 (A + B)

(continued)

Table 2
continued

Full-length assembly PCR (Stitch PCR, PCR AB)		
Library — Alpha	**Primers for AB stitch**	**Template**
CDR3α_1	YOL13 + YOL22	CDR3α_1: PCR1 + PCR2 (A + B)
CDR3α_2	YOL13 + YOL22	CDR3α_2: PCR1 + PCR2 (A + B)
CDR3α_3	YOL13 + YOL22	CDR3α_3: PCR1 + PCR2 (A + B)
CDR3α_4	YOL13 + YOL22	CDR3α_4: PCR1 + PCR2 (A + B)
CDR3α_5	YOL13 + YOL22	CDR3α_5: PCR1 + PCR2 (A + B)
Library — Beta	**Primers for AB stitch**	**Template**
CDR1β	YOL13 + YOL22	CDR1β: PCR1 + PCR2 (A + B)
CDR2β_1	YOL13 + YOL22	CDR2β_1: PCR1 + PCR2 (A + B)
CDR2β_2	YOL13 + YOL22	CDR2β_2: PCR1 + PCR2 (A + B)
CDR3β_1	YOL13 + YOL22	CDR3β_1: PCR1 + PCR2 (A + B)
CDR3β_2	YOL13 + YOL22	CDR3β_2: PCR1 + PCR2 (A + B)
CDR3β_3	YOL13 + YOL22	CDR3β_3: PCR1 + PCR2 (A + B)
CDR3β_4	YOL13 + YOL22	CDR3β_4: PCR1 + PCR2 (A + B)
CDR3β_5	YOL13 + YOL22	CDR3β_5: PCR1 + PCR2 (A + B)

17. Pool all transformation reactions for the same CDR into a single 50 mL centrifuge tube.
18. Incubate with shaking (250 rpm) at 37°C for 1 h.
19. Estimate total number of library transformants by titrating pooled transformations in serial dilutions down to 10^{-10}.
20. Prepare 1:100 dilutions to 10^{-7}, then continue with ten-fold dilutions to 10^{-10} and plate on petri dishes (90 mm). Expected library sizes range from 10^7–10^9.

21. Plate remaining electroporated TG1 cells onto square YTEag bioassay dishes with 2xYT-agar supplemented with 2% glucose and ampicillin (or carbenicillin).
22. Incubate plates overnight at 30°C.
23. Scrape the colonies on Bioassay dish using 2–5 mL 2xYT glycerol + 100 μg/mL ampicillin (or carbenicillin).
24. Pool and mix thoroughly in 50 mL centrifuge tubes to create the unselected library glycerol stock. Store on ice while scraping other Bioassay dishes.
25. Aliquot ~1 mL into suitable 1.5 mL cryotubes. Freeze on dry ice and store at −80°C.
26. Count the colonies on the small plates and calculate titres by using the formula:

$$\text{Library size cfu/ml} = \frac{\textit{number of colonies} \times \textit{dilution factor}}{\textit{volume plated}}$$

27. Assess library integrity by Sanger sequence analysis focusing on CDR mutations and their diversity. Sequences must be in frame with phagemid backbone.
28. Prepare DNA by performing colony PCR with Taq polymerase on single, random colonies.
29. Set up PCR reaction using one colony, 0.2 μM forward primer (YOL13), 0.2 μM reverse primer (YOL22). Top up the reaction to 25 μL with nuclease-free H_2O.
30. Run the PCR using the following program: initial denaturation at 95°C for 2 min, 30 cycles of amplification consisting of denaturation at 95°C for 16 s, annealing at 56°C for 16 s and extension at 68°C for 2 min 20 s, and a final extension at 68°C for 2 min.

3.2.1 Phage Display Selection

1. Identify high-affinity clones using standard phage display selection methods by using decreasing antigen concentrations per selection round, increasing the selective pressure [23].
2. In selection round one, use 100 nM pHLA. In selection round two, use 10 nM pHLA. In selection round three, use 1 nM pHLA (*see* **Note 14**).
3. To prepare phage for panning, inoculate 250 mL of 2xYTag medium containing 100 μg/mL of ampicillin (or carbenicillin) and 2% glucose with unselected library TG1 glycerol stock in 2 L Erlenmeyer flasks at a starting OD_{600} of 0.05–0.1 (*see* **Note 15**).
4. Grow the TG1 library culture at 37°C with shaking (250 rpm) until at an OD_{600} of 0.5 the mid-exponential phase is reached.

5. Infect 100 mL of TG1 culture at a multiplicity of infection (MOI) of 10–20, using KM13 helper phage and incubate in a 37°C water bath or incubator without shaking for 30 min.
6. Transfer TG1 culture into 50 mL centrifuge tubes and pellet the cells by centrifugation at 3000 × *g* for 10 min at 16°C.
7. Discard supernatant and resuspend the pellet in 250 mL of 2xYTak containing 100 μg/mL ampicillin (or carbenicillin) and 50 μg/mL kanamycin (no glucose) and add to 2 L Erlenmeyer flasks.
8. Grow overnight at 26°C with shaking at 250 rpm.
9. Pellet the cells by centrifugation at 17000 × *g* for 5 min at 4°C and filter supernatant using 0.45 nm filters.
10. Purify phage particles using two rounds of polyethylene glycol precipitations, by adding 1/5 PEG precipitation buffer to 4/5 volume of supernatant (*see* **Note 16**).
11. In selection round one, precipitate 200 mL, in round two and three, 30–50 mL culture is sufficient.
12. Mix well and incubate on ice for 1 h.
13. Pellet phage at 2100–2700 × *g* for 15 min at 4°C.
14. Discard supernatant and remove any remaining PEG by spinning for one minute at 2100–2700 × *g* at 4°C using sterile Pasteur pipettes.
15. Resuspend pelleted selection round one phage in a total volume of 30 mL PBS and resuspend well.
16. Spin to pellet remaining insoluble debris at 17,000 × *g* for 10 min at 4°C.
17. Add 1/5 PEG precipitation buffer to 4/5 volume of the resuspended phage in PBS.
18. Mix well and incubate on ice for 45 min.
19. Pellet phage at 2100–2700 × *g* for 15 min at 4°C.
20. Discard supernatant and remove any remaining PEG by spinning for one minute at 2100–2700 × *g* at 4°C using sterile Pasteur pipettes.
21. Resuspend phage particles in 1–2 mL of PBS [23, 27] and store at 4°C until use, preferably the next day.
22. Determine phage titres using serial plate dilutions or estimate using the following formula:

$$\frac{phage}{ml} = OD260 \times dilution \times 22.14 \times 10^{10} \cdot [27]$$

23. To perform a selection round, block 0.8–1.0 mL phage in blocking buffer (e.g., 1%–6% (w/v) MPBS, using 1–5 × 10^{12} phages/mL as input (*see* **Note 6**).

24. Wash Dynabeads M-280 according to the manufacturer's instructions using PBS and block in blocking buffer for 1 h.
25. Incubate 0.8–1.0 mL of phage with 100 nM biotinylated pHLA and incubate for 1 h.
26. Pull down the pHLA bound phage using streptavidin-magnetic beads.
27. Wash the beads three times with 3% milk and three times with PBS-Tween 0.1%.
28. Elute bound phage by incubating with 0.5–1.0 mL of 0.1 mg/mL trypsin in TBSC buffer for 10 min at 37°C (*see* **Note 5**).
29. Infect 10 mL of TG1 cells growing at log phase (OD_{600} of 0.4–0.5) with the eluted phage for 30 min at 37°C with no shaking.
30. Prepare culture dilutions on petri dishes (90 mm). Expected range of output titres is between 10^3 and 10^7.
31. Pellet remaining culture at 2100–2700 × *g* for 15 min at 16°C.
32. Resuspend in 1 mL 2xYT and plate out on sterile square YTEag bioassay dishes with 2xYT-agar supplemented with 2% glucose and ampicillin (or carbenicillin).
33. Incubate plates overnight at 30°C.
34. Scrape large plates using 2–5 mL 2xYT glycerol + 100 μg/mL ampicillin (or carbenicillin).
35. Pool and add to 50 mL centrifuge tubes. Mix thoroughly to create the selection round one glycerol stocks.
36. Aliquot ~1 mL into suitable 1.5 mL cryotubes. Freeze on dry ice and store at −80°C.
37. These glycerol stocks will be used for selection round two. Repeat steps 3–36 but adjust volume of 2xYT media in step 3 to 100 mL 2xYT medium containing 100 μg/mL of ampicillin (or carbenicillin) and 2% glucose.
38. After three to four rounds of selection (one selection = one repeat of steps 3–36), screen single ampicillin/carbenicillin-resistant colonies in a standard competition/inhibition ELISA phage using 96-well ELISA plates (e.g., Nunc™ MaxiSorp plates or equivalent) coated with streptavidin and biotinylated pHLA of interest [23, 27].
39. Assess clone convergence by Sanger sequence analysis focusing on CDR mutations, by performing colony PCR with Taq polymerase with primers YOL13/YOL22 (see Sect. 3.2, points 28–30).

40. Fuse identified affinity-enhanced mTCRs to an anti-CD3 scFv (e.g., UCHT1) [14] at the N-terminus of the TCRβ chain to generate ImmTAC molecules [8, 9, 28] (*see* **Notes 4 and 9g and h**).

3.3 Protein Expression and Inclusion Body Preparation

The TCRα and TCRβ chains, β2M and the Human leukocyte antigen (HLA) chains are expressed separately as inclusion bodies [9, 20].

1. Inoculate 1 L of TYP medium in Erlenmeyer flasks, with single ampicillin/carbenicillin-resistant colonies for each construct, either in the *E. coli* strain BL21(DE3)pLysS or Rosetta™(DE3) pLysS or equivalent (*see* **Note 17**).
2. Incubate at 37°C with shaking at 250 rpm.
3. Induce in mid-log phase (OD_{600} of 0.4–0.5) with 0.5–1.0 mM IPTG (*see* **Note 7**).
4. Incubate at 37°C at 250 rpm overnight.
5. Pellet cell by centrifugation the next day at 17,000 × *g* for 5 min or 2100–2700 g for 30 min (*see* **Note 18**).
6. Lyse cells using BugBuster protein extraction reagent. Add DNAse I (20 mg/mL) + $MgCl_2$ (0.5 M) before lysis (*see* **Notes 19 and 20**). See manufacturer's instructions for more details.
7. Mix thoroughly for 30 min and centrifuge at 17,000 × *g* for 5 min or at 2100–2700 × *g* for 15 min at 4°C.
8. Wash inclusion bodies in the pellet by repeating resuspension (*see* **Note 21**) and centrifugation steps in 30 mL triton wash buffer three to four times until the pellet is white and the supernatant clear [29–31].
9. Perform a final wash in resuspension if the supernatant is clear and resuspend the pellet in 10 mL resuspension buffer (*see* **Note 22**).
10. Determine inclusion body protein concentration by solubilizing inclusion bodies as 1/50 or 1/100 dilutions in 6 M guanidine buffer and measure absorbance at 280 nm.
11. Calculate concentration using the molecular weight (MW) and extinction coefficient of the target protein chain. Determine extinction coefficient/MW from the expected protein amino acid sequence using appropriate calculators/software.
12. Determine inclusion body purity by preparing 1 mg/mL solutions in 8 M urea and run 2 μg per lane on SDS-PAGE using 4%–20% Tris-Glycine gels (Fig. 5a) (*see* **Note 23**).
13. Use densitometry on reduced lanes to assess purity as % of target product band of expected MW.
14. Resuspend inclusion bodies in resuspension buffer and keep at 4°C until use, or at −20°C for long-term storage.

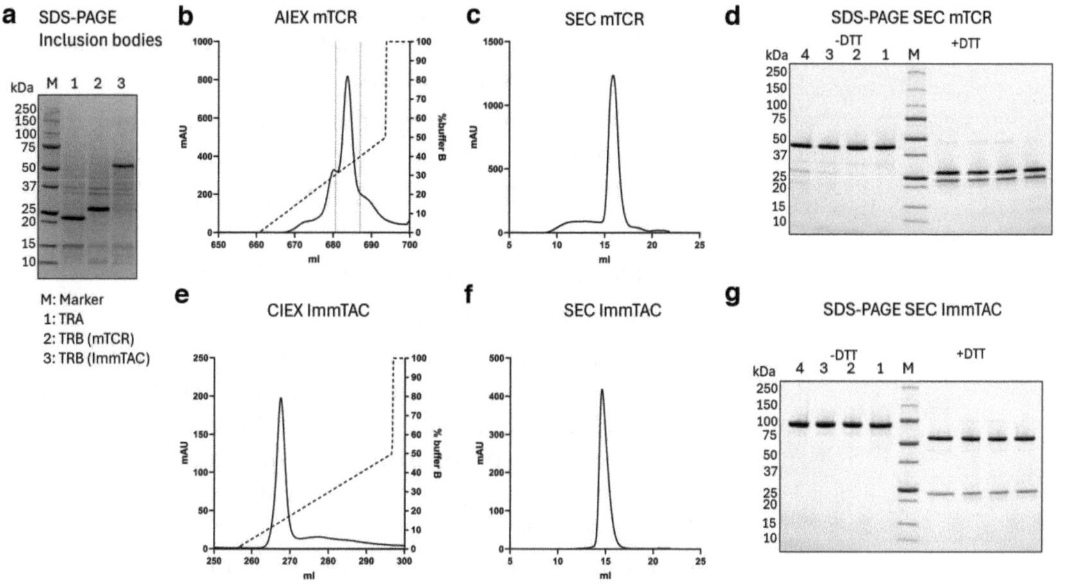

Fig. 5 Purification of mTCRs and ImmTAC molecules. (**a**) SDS-PAGE analysis purified inclusion bodies (IBs). Lane 1 shows alpha chain IBs, lane 2 shows mTCR beta chain IBs and lane 3 shows ImmTAC beta chain IBs. Purity in all lanes is ~70%. (**b**) Anion-exchange chromatography (AIEX) is performed on refolded IBs using a POROS 50HQ™ column (10/100) (Thermo Fisher). Note the dotted lines highlighting the peak fractions analyzed on SDS-PAGE to determine clean fractions (not shown). Clean fractions are pooled for size-exclusion chromatography (SEC). (**c**) SEC of pooled AIEX mTCR fractions using a S200i column (Cytiva). (**d**) The fractions of the main SEC peak are analyzed on SDS-PAGE to determine purity, using approximately 2 μg of SEC fractions in each lane. Purity is around >90% for lanes 1–3 and ~85% for lane 4. (**e**) A cation-exchange chromatography (CIEX) purification of an ImmTAC molecule is shown using a POROS 50HS™ column (10/100) (Thermo Fisher). The preceding AIEX step is not shown, but results are similar to (**b**). (**f**) The main CIEX fractions are analyzed on SDS-PAGE (not shown) and clean fractions are pooled for SEC. (**g**) The purity of the main SEC fractions is determined by using approximately 2 μg of SEC fractions in each lane. It is around ~>90% for lanes 1–4. All gels shown are Mini-PROTEAN® TGX 4%–20% Tris-Glycine gels used with Precision Plus Protein Dual color standard (Bio-Rad). The gel in (**a**) is a TGX precast gel that stained with InstantBlue (Expedeon, Sigma-Aldrich) and (**d**) and (**g**) are TGX stain-free gels. Densitometry on reduced lanes is used to assess purity as % of target product band of expected molecular weight. All gels were imaged and analyzed on a GelDoc Ez Imager and ImageLab 6.0 (Bio-Rad)

3.4 Protein Refolding

After washing and resuspension, inclusion bodies contain the insoluble TCRα, TCRβ, HLA chains or β2M, which need to be dissolved, i.e., denatured and reduced, in a solubilizing agent such as guanidine and refolded in a refold buffer.

1. Use purity corrected inclusion body concentrations for adding protein chains into refold buffer (OD_{280} concentration × % purity = purity corrected concentration).
2. Add guanidine buffer to the inclusion bodies (*see* **Note 24**).
3. Add 20 mM DTT to reduce cysteine residues and incubate at 37°C for 30 min with gentle shaking [9, 30, 32].

4. Use cystamine and cysteamine per 1 L according to mTCR/ImmTAC or HLA refold buffer recipes as redox couple [9] (*see* **Notes 25 and 26**).
5. Add to cold refold buffer just before use and ensure they are fully dissolved.
6. Refold TCRs (mTCR and ImmTAC molecules) by following steps 9–12.
7. Refold for pHLAs by following steps 13–16.
8. Continue for all with step 17.
9. mTCRs and ImmTAC molecules: For a 1 L mTCR refold, mix ~1-2 μmol TCRα chain with ~1 μmol TCRβ chain. For a 1 L ImmTAC refold, mix 1 μmol TCRα chain with ~1 μmol TCRβ chain in approximately 15 mL guanidine buffer to achieve final purity corrected concentration of 4–5 mg/mL [9, 33] (*see* **Note 27**).
10. Aim for final purity corrected concentration of 60–75 mg per 1 L refold buffer. Typically, 75 mg total for mTCR (45 mg alpha + 30 mg beta) and 85 mg total for ImmTAC molecules (25 mg alpha and 60 mg beta) (*see* **Note 28**).
11. Stir cold refold buffer in a beaker using a magnetic stirrer to create a small vortex, but do not allow to foam.
12. Pour the dissolved α- and β-chain inclusion body mixtures (in guanidine buffer) in one gentle go to prevent foaming ("rapid" dilution) into the cold refold buffer.
13. HLAs: For pHLA complexes, add synthetic peptide dissolved in DMSO to the cold refolding buffer (*see* **Note 29**). For a 1 L pHLA refold, mix ~2 μmol HLA heavy chain (in 20 mL guanidine buffer) with 2.5 μmol β2M (in 10 mL guanidine buffer) and ~4 μmol peptide (in 0.5 mL DMSO) [21, 31, 34, 35].
14. Aim for final total concentration of ~90 mg per 1 L refold buffer (*see* **Note 30**).
15. Stir cold refold buffer in a beaker using a magnetic stirrer to create a small vortex, but do not allow to foam.
16. Pour the HLA-heavy chain/β2M/peptide inclusion body mixtures in one gentle go to prevent foaming ("rapid" dilution) into the cold refold buffer.
17. Allow the refold mixture to stir for ~10–15 min.
18. Mix and dialyse against at least 10–20 volumes of ultrapure water (18 MΩ-cm at 25°C) water overnight, followed by 10–20 volumes of 10–20 mM Tris pH 8 at 4°C, stirring gently overnight until the conductivity of the refolds is $</=$2.0 mS/cm [9, 34].

3.5 Protein Purification

The refolded mTCRs, ImmTAC molecules and pHLAs are purified using standard column chromatography techniques, i.e., anion-exchange chromatography (AIEX) and size-exclusion chromatography (SEC). ImmTAC molecules require an additional purification step prior to SEC, typically utilizing cation-exchange chromatography (CIEX) [8, 9, 36]. Biotinylation is performed prior to SEC, if required.

1. Collect the refolds and filter through a 0.45 μm filter [37] (*see* **Note 31**).
2. Purification steps for mTCRs: AIEX and SEC. Follow steps 6–10.
3. Purification steps for ImmTACs: AIEX, CIEX, and SEC. Follow steps 11–19.
4. Purification steps for pHLA: AIEX and SEC. Follow steps 20–24.
5. Continue for all with step 25.
6. mTCRs: Purify the refolded mTCRs by AIEX using a POROS 50HQ column (10/100) using a gradient of 0%–50% AIEX buffer B in four to five column volumes. The mTCR product typically elutes in the middle of this gradient (Fig. 5b) (see Fig. 1a in [9] for example clone NY-ESO-1-1G4).
7. Confirm product-containing fractions by analysing AIEX peak fractions using ~5 μL per fraction for SDS-PAGE (*see* **Note 32**).
8. Pool appropriate peak fractions and concentrate to ~0.5–1 mL using spin concentrators not exceeding ~10 mg/mL.
9. Perform SEC using a S200i 10/300 column, using isocratic elution over ~1 column volume, eluting in 0.5–1 mL fractions using PBS as running buffer [8, 33] (Fig. 5c) (*see* **Note 33**) (see Fig. 1b in [9] for example clone NY-ESO-1-1G4).
10. Determine protein purity by analysing SEC peak fractions using 2 μg per lane for SDS-PAGE (Fig. 5d) (*see* **Note 32)** (see Fig. 2 in [9], for example, clone NY-ESO-1-1G4).
11. ImmTAC molecules: Purify the refolded ImmTAC molecules by AIEX using a POROS 50HQ column (10/100) using a gradient of 0%–50% AIEX buffer B over four to five column volumes. ImmTAC typically elutes in the middle of this gradient.
12. Confirm product-containing fractions by analysing AIEX peak fractions using 2–5 μL per fraction for SDS-PAGE (*see* **Note 32**).
13. Prior to CIEX, dilute the protein sample from AIEX at least 20× in 40 mM CIEX buffer A with pH in the range of 5–6. Additional work may be required to determine the optimal pH for performing the CIEX step.

14. For ImmTAC molecules only, perform CIEX using POROS 50HS 10/100 column with a gradient over 0%–50% CIEX buffer B in four to five column volumes. ImmTAC molecules typically elute in the first half of the gradient (Fig. 5e).
15. Confirm product-containing fractions by analysing CIEX peak fractions ~5 μL per fraction for SDS-PAGE (*see* **Note 32**).
16. Pool relevant CIEX fractions and adjust pH to 8 with 2 M Tris stock (pH 8.1).
17. Concentrate the CIEX pool to ~0.5–1.0 mL using spin concentrators not exceeding ~8 mg/mL.
18. Perform SEC using a S200i 10/300 column, using isocratic elution over ~1 column volume, eluting in 0.5–1 mL fractions using PBS as running buffer [8, 33] (Fig. 5f) (*see* **Note 33**).
19. Confirm product-containing fractions by analysing SEC peak fractions using 2 μg per lane for SDS-PAGE (Fig. 5g) (*see* **Note 32**).
20. pHLAs: Purify the refolded pHLA proteins by AIEX using a POROS 50HQ column (10/100) using a gradient of 0%–50% AIEX buffer B in four to five column volumes. The pHLA complex typically elutes in the first half of the gradient.
21. Confirm product-containing fractions by analysing AIEX peak fractions using ~5 μL per fraction for SDS-PAGE (*see* **Note 32**).
22. Concentrate peak fractions to ~0.5–1 mL using spin concentrators.
23. Perform SEC using a S75i 10/300 column, eluting in 0.5–1 mL fractions using PBS as running buffer [8, 33] (*see* **Note 33**).
24. Confirm product-containing fractions by analysing SEC peak fractions using 2 μg per lane for SDS-PAGE (*see* **Note 32**).
25. Perform—if required—in vitro site-specific biotinylation of the biotin ligase (BirA) tag (AviTag) <u>prior to SEC</u> (for mTCR and pHLA after AIEX, for ImmTAC molecules after CIEX), using a ratio of 1 mg BirA enzyme per 100 mg TCR/pHLA using BirA500 or Bulk BirA biotin-protein ligase biotinylation kit [38, 39].
26. Determine protein concentration and yield, typically described as mg protein per 1 L refold buffer, using 280 nm absorbance measurements.

3.6 Biacore

Determine affinity and kinetics of refolded mTCR or ImmTAC using SPR experiments on Biacore [8, 9, 29, 37, 40], using Biacore T200 or Biacore 8K (or equivalent) with their respective evaluation software. Binding interaction kinetics represent how fast a TCR binds to the pHLA (association rate constant $k_{a/on}$) and how fast the TCR/pHLA complex dissociates (dissociation rate constant $k_{d/off}$). These experiments require that the sensorgram shows sufficient curvature during association and dissociation [41]. Binding affinity defines how strong a complex is, i.e., how much of the TCR/pHLA complex is formed at equilibrium/steady state, where association balances dissociation (dissociation equilibrium constant, K_D). These experiments require that the sensorgram reach steady state during sample injections [41]. Wildtype TCRs usually bind to pHLA at μM affinities and ImmTAC are high-affinity molecules which bind to pHLA in the pM to nM affinity range. Choose Biacore systems, program and fitting method based on expected affinity of molecules and throughput (*see* **Note 34**). Use PBS with 0.005% (v/v) surfactant P20 as running buffer. Add 500 μL P20 (10% stock) to 1000 mL of PBS for a final concentration of 0.005% (v/v) (*see* **Note 35**).

1. Perform all experiments at 25°C (*see* **Note 36**).
2. Dilute high-concentrated stock solution of samples into running buffer (*see* **Note 37**).
3. To mobilize pHLA, dilute biotinylated pHLAs to 1 μg/mL in running buffer and immobilize on a streptavidin-coated CM5 sensor chip using EDC/NHS chemistry.
4. Flow biotinylated pHLAs at a flow rate of 10 μL/min over individual flow cells at a concentration of ~1 μg/mL typically.
5. Inject 20 μL 5 μM biotin over all flow cells after pHLA loading to block unoccupied streptavidin sites.
6. For single injections, steady state experiments and multi-cycle kinetics on Biacore T200, load ~200–1000 response units (RU).
7. For single-cycle and multi-cycle kinetic analysis of high-affinity TCRs on the Biacore 8K, immobilize 50–200 RU of pHLA.
8. Include a flow cell with biotin alone to function as a control, which is used as reference.
9. When using T200, use flow cell one (Fc1) as the biotin control, and load pHLAs on flow cell two (Fc2), flow cell three (Fc3), and flow cell four (Fc4) (*see* **Note 38**).
10. When using the 8K, use flow cell one as the biotin control and flow cell two to load pHLA.
11. Always use reference-subtraction in the analysis.
    a. T200: Fc2-1 (flow cell two minus flow cell one), Fc3-1 and Fc4-1
    b. 8K: Fc2-1 (in each of the eight channels)

Based on expected strength of interaction (e.g., Fig. 6a) choose between (1) single, manual injections on a Biacore T200 if binding and binding strength to a pHLA are unknown, (2) steady state affinity analysis

a

TCR	Affinity generation	HLA-KRAS-G12D K_D [M]	HLA-KRAS-WT K_D [M]
JDI	Parental	6.30±0.30E-05	No binding
JDIa9bwt	First	2.61±0.03E-07	2.58±0.07E-05
JDIa41b1	Second	7.43±0.18E-10	3.00E-06
JDIa96b35-U	Third	6.37±1.03E-11	4.25±0.15E-07

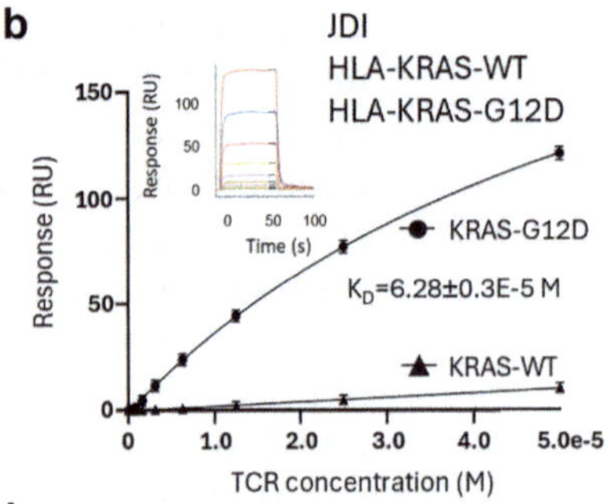

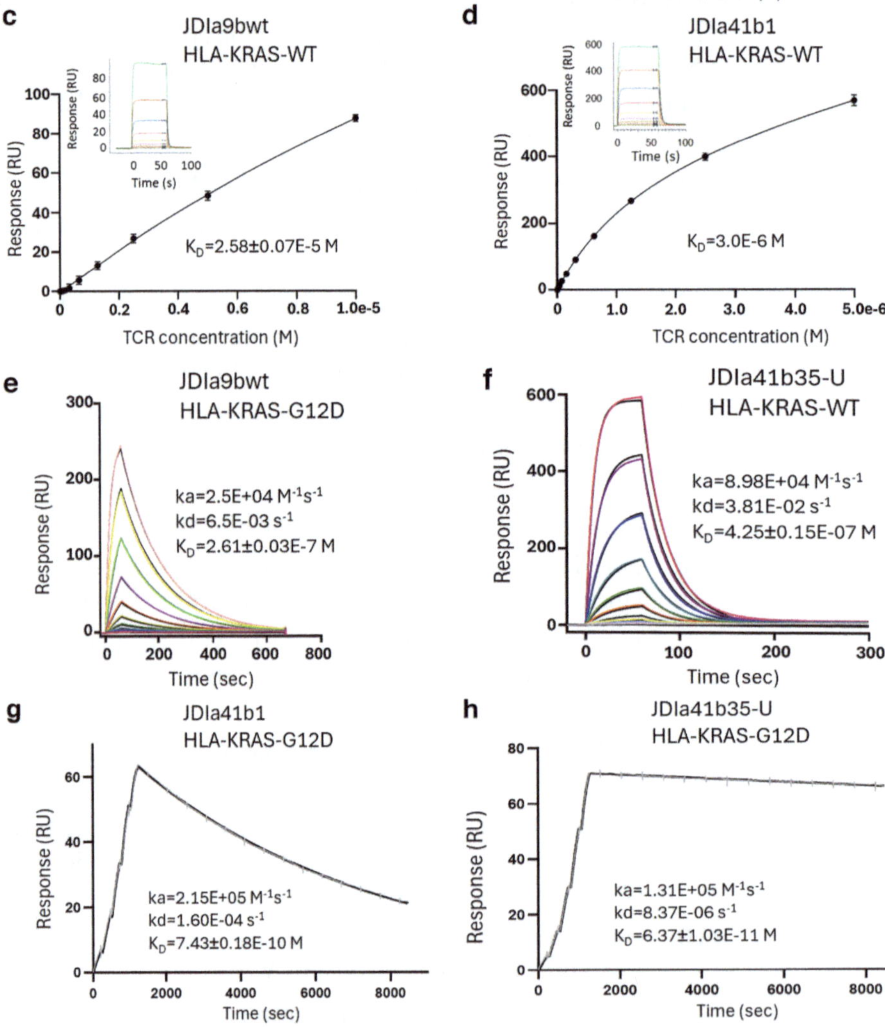

Fig. 6 Affinity (K_D) measurements using Biacore T200 and 8K (modified from former publication [40]), of TCR JDI and its affinity-enhanced variations JDIa9bwt, JDIa41b1, and JDIa41b35-U, to HLA-A*11-KRAS-G12D and HLA-A*11-KRAS-WT. (**a**) Details of the TCRs and their respective affinities are listed. All experiments are performed in duplicates and the K_D values are mean $\pm$ SD. Steady state analyses and multicycle kinetics (MCK) are performed on a Biacore T200, and single-cycle kinetics (SCK) experiments on a Biacore 8K. (**b–d**) Steady state binding analyses show SPR sensorgrams in insets. (**b**) JDI TCR steady state binding analysis using

(Fig. 6b–d) (see Fig. 3 in [9], for example, clone NY-ESO-1-1G4) and multi-cycle kinetics (MCK) (Fig. 6e, f) for low- to intermediate-affinity interactions (see **Notes 39–41 and 43**), or (3) single-cycle kinetics (SCK) (Fig. 6g, h) for high-affinity interactions.

i. Single injection
   1. Set flow rate to 10 μL/min and inject the TCR at its top concentration for 60 s.
   2. For wildtype TCRs, the typical affinity range is ~1–300 μM.
   3. Aim for the top concentration to be approximately 10× of the expected K_D (50–100 μM as a starting point; increase if necessary).

ii. Steady state affinity (Fig. 6b–d)
   1. Prepare serial dilutions in tubes or 96-well plates.
   2. Flow serial dilutions of the mTCR over the immobilized pHLA and determine the response values at equilibrium for each concentration.
   3. Determine binding affinities against the cognate pHLA by performing 5–12-point steady state analysis using TCR concentrations ranging from highest concentrated samples to blank sample (buffer) (*see* **Note 42**).
   4. Perform steady state affinity analysis using the Biacore T200 evaluation software [9] (*see* **Notes 43 and 44**).

iii. SCK (Fig. 6g, h)
   1. For SCK, use a wide range of sample concentrations centered on the anticipated value for the K_D [33].
   2. Flow serial (increasing) dilutions of the high-affinity mTCR or ImmTAC molecules over the immobilized pHLA and determine the $k_{a/on}$ using a 5-point titration with top concentration at least 50× higher than K_D and long dissociation time.

Fig. 6 (continued) ten 2-fold dilutions starting at highest concentration of 50 μM. (**c**) Steady state binding analysis of JDla9bwt binding to HLA-A*11-KRAS-WT measured using ten 2-fold dilutions starting at top TCR concentration of 10 μM. (**d**) Steady state binding analysis of JDla41b1 TCR binding to HLA-A*11-KRAS-WT measured using ten 2-fold dilutions starting at highest TCR concentration of 5 μM. (**e and f**) MCK analyses. (**e**) MCK analysis of JDla9bwt TCR binding to HLA-A*11-KRAS-G12D measured using ten 2-fold dilutions starting at top concentration of 2 μM. (**f**) MCK analysis of JDla41b35-U (also referred to as IMC-KRAS-G12D in [40]) binding to HLA-A*11-KRAS-WT measured using ten 2-fold dilutions starting at highest TCR concentration of 1 μM. (**g and h**) SCK analyses. (**g**) SCK analysis of JDla41b1 TCR binding to HLA-A*11-KRAS-G12D using five injections at 2.5 nM, 5 nM, 10 nM, 20 nM, and 40 nM TCR concentration. (**h**) SCK analysis of JDla41b35-U binding to HLA-A*11-KRAS-G12D using five injections at 2.5 nM, 5 nM, 10 nM, 20 nM, and 40 nM TCR concentration. See [40] for more details

3. Set contact time to 200 s, dissociation time to 7200 s and flow rate: 60 μL/min for both T200 and 8K (*see* **Note 45**).
4. SCK analysis is performed with Biacore Insight Evaluation software (Fig. 6g, h).

4 Notes

1. Sequences of NY-ESO-1-1G4 and its ligand are listed in FASTA format below (PDB: 2BNR) [18]. TRAC (T48C), TRBC (S57C, C75A) and the TCR CDRs as defined by IMTG are underlined.

```
>NY-ESO-1-1G4_TRA (TCR alpha chain)
KQEVTQIPAALSVPEGENLVLNCSFTDSAIYNLQWFRQDPGKGLT-
SLLLIQSSQREQTSGRLNASLDKSSGRSTLYIAASQPGDSATYLCAVRPTS-
GGSYIPTFGRGTSLIVHPYIQNPDPAVYQLRDSKSSDKSVCLFTDFDSQTN-
VSQSKDSDVYITDKCVLDMRSMDFKSNSAVAWSNKSDFACANAFNNSIIP-
EDTFFPS
```

```
>NY-ESO-1-1G4_TRB (TCR beta chain)
NAGVTQTPKFQVLKTGQSMTLQCAQDMNHEYMSWYRQDPGMGLRLIHYSV-
GAGITDQGEVPNGYNVSRSTTEDFPLRLLSAAPSQTSVYFCASSYVGNT-
GELFFGEGSRLTVLEDLKNVFPPEVAVFEPSEAEISHTQKATLVCLAT-
GFYPDHVELSWWVNGKEVHSGVCTDPQPLKEQPALNDSRYALSSRLRV-
SATFWQDPRNHFRCQVQFYGLSENDEWTQDRAKPVTQIVSAEAWGRAD
```

```
>HLA-A*0201
GSHSMRYFFTSVSRPGRGEPRFIAVGYVDDTQFVRFDSDAASQRMEPRAP-
WIEQEGPEYWDGETRKVKAHSQTHRVDLGTLRGYYNQSEAGSHTVQRMYG-
CDVGSDWRFLRGYHQYAYDGKDYIALKEDLRSWTAADMAAQTTKHKWEAAH-
VAEQLRAYLEGTCVEWLRRYLENGKETLQRTDAPKTHMTHHAVSDHEATL-
RCWALSFYPAEITLTWQRDGEDQTQDTELVETRPAGDGTFQKWAAVVVPS-
GQEQRYTCHVQHEGLPKPLTLRWEP
```

```
>Beta-2-microglobulin
MIQRTPKIQVYSRHPAENGKSNFLNCYVSGFHPSDIEVDLLKNGERIEKVE-
HSDLSFSKDWSFYLLYYTEFTPTEKDEYACRVNHVTLSQPKIVKWDRDM
```

```
>NY-ESO-1-SLL peptide
SLLMWITQC
```

2. The pGMT7 nucleotide sequence in the GenBank format is shown below [42]. Other T7 expression plasmids such as pET23d (Novagen) can also be used.

```
LOCUS       pGMT7     3111 bp DNA    circular SYN 13-SEP-2024
DEFINITION  synthetic circular DNA
ACCESSION   .
VERSION     .
KEYWORDS    .
SOURCE      synthetic DNA construct
  ORGANISM  synthetic DNA construct
REFERENCE   1 (bases 1 to 3111)
  AUTHORS   .
```

```
            FEATURES             Location/Qualifiers
                 source          1..3111
                                 /mol_type="other DNA"
                                 /organism="synthetic DNA construct"
                 promoter        20..38
                                 /label=T7 promoter
                                 /note="promoter for bacteriophage
                                 T7 RNA polymerase"
                 RBS             70..92
                 terminator      190..237
                                 /label=T7 terminator
                 rep_origin      complement(466..894)
                                 /direction=LEFT
                                 /label=f1 ori
                 promoter        921..1025
                                 /gene="bla"
                                 /label=AmpR promoter
                 CDS             1026..1886
                                 /codon_start=1
                                 /gene="bla"
                                 /product="beta-lactamase"
                                 /label=AmpR
                 rep_origin      2057..2645
                                 /direction=RIGHT
                                 /label=ori
                 misc_feature    2831..2970
                                 /label=bom
ORIGIN
        1 gatctcgatc ccgcgaaatt aatacgactc actataggga gaccacaacg gtttccctct
       61 agaaataatt ttgtttaact ttaagaagga gatatacata tgggatccat ggtaagcttg
      121 aattccgatc cggctgctaa caaagcccga aaggaagctg agttggctgc tgccaccgct
      181 gagcaataac tagcataacc ccttggggcc tctaaacggg tcttgagggg ttttttgctg
      241 aaaggaggaa ctatatccgg ataattcttg aagacgaaag ggcctcgtga tacgcctatt
      301 tttataggtt aatgtcatga taataatggt ttcttagacg tcaggtggca cttttcgggg
      361 aaatgtgcgc ggaaccccta tttgtttatt tttctaaata cattcaaata tgtatccgct
      421 catgagacaa taaccctgat aaatgcttca ataatatttt gttaaaattc gcgttaaatt
      481 tttgttaaat cagctcattt tttaaccaat aggccgaaat cggcaaaatc ccttataaat
      541 caaaagaata gaccgagata gggttgagtg ttgttccagt ttggaacaag agtccactat
      601 taaagaacgt ggactccaac gtcaaagggc gaaaaaccgt ctatcagggc gatggcccac
      661 tacgtgaacc atcaccctaa tcaagttttt tggggtcgag gtgccgtaaa gcactaaatc
      721 ggaaccctaa agggagcccc cgatttagag cttgacgggg aaagccggcg aacgtggcga
      781 gaaaggaagg gaagaaagcg aaaggagcgg gcgctagggc gctggcaagt gtagcggtca
      841 cgctgcgcgt aaccaccaca cccgccgcgc ttaatgcgcc gctacagggc gcgtcaggtg
      901 gcacttttcg gggaaatgtg cgcggaaccc ctatttgttt atttttctaa atacattcaa
      961 atatgtatcc gctcatgaga caataaccct gataaatgct tcaataatat tgaaaaagga
     1021 agagtatgag tattcaacat ttccgtgtcg cccttattcc cttttttgcg gcattttgcc
     1081 ttcctgtttt tgctcaccca gaaacgctgg tgaaagtaaa agatgctgaa gatcagttgg
     1141 gtgcacgagt gggttacatc gaactggatc tcaacagcgg taagatcctt gagagttttc
     1201 gccccgaaga acgttttcca atgatgagca cttttaaagt tctgctatgt ggcgcggtat
     1261 tatcccgtgt tgacgccggg caagagcaac tcggtcgccg catacactat tctcagaatg
     1321 acttggttga gtactcacca gtcacagaaa agcatcttac ggatggcatg acagtaagag
     1381 aattatgcag tgctgccata accatgagtg ataacactgc ggccaactta cttctgacaa
     1441 cgatcggagg accgaaggag ctaaccgctt ttttgcacaa catgggggat catgtaactc
     1501 gccttgatcg ttgggaaccg gagctgaatg aagccatacc aaacgacgag cgtgacacca
```

```
1561 cgatgcctgc agcaatggca acaacgttgc gcaaactatt aactggcgaa ctacttactc
1621 tagcttcccg gcaacaatta atagactgga tggaggcgga taaagttgca ggaccacttc
1681 tgcgctcggc ccttccggct ggctggttta ttgctgataa atctggagcc ggtgagcgtg
1741 ggtctcgcgg tatcattgca gcactggggc cagatggtaa gccctcccgt atcgtagtta
1801 tctacacgac ggggagtcag gcaactatgg atgaacgaaa tagacagatc gctgagatag
1861 gtgcctcact gattaagcat tggtaactgt cagaccaagt ttactcatat atactttaga
1921 ttgatttaaa acttcatttt taatttaaaa ggatctaggt gaagatcctt tttgataatc
1981 tcatgaccaa aatcccttaa cgtgagtttt cgttccactg agcgtcagac cccgtagaaa
2041 agatcaaagg atcttcttga gatccttttt ttctgcgcgt aatctgctgc ttgcaaacaa
2101 aaaaaccacc gctaccagcg gtggtttgtt tgccggatca agagctacca actctttttc
2161 cgaaggtaac tggcttcagc agagcgcaga taccaaatac tgtccttcta gtgtagccgt
2221 agttaggcca ccacttcaag aactctgtag caccgcctac atacctcgct ctgctaatcc
2281 tgttaccagt ggctgctgcc agtggcgata agtcgtgtct taccgggttg gactcaagac
2341 gatagttacc ggataaggcg cagcggtcgg gctgaacggg gggttcgtgc acacagccca
2401 gcttggagcg aacgacctac accgaactga gatacctaca gcgtgagcta tgagaaagcg
2461 ccacgcttcc cgaagggaga aaggcggaca ggtatccggt aagcggcagg gtcggaacag
2521 gagagcgcac gagggagctt ccagggggaa acgcctggta tctttatagt cctgtcgggt
2581 ttcgccacct ctgacttgag cgtcgatttt tgtgatgctc gtcagggggg cggagcctat
2641 ggaaaaacgc cagcaacgcg gcctttttac ggttcctggc cttttgctgg ccttttgctc
2701 acatgttctt tcctgcgtta tcccctgatt ctgtggataa ccgtattacc gcctttgagt
2761 gagctgatac cgctcgccgc agccgaacga ccgagcgcag cgagtcagtg agcgaggaag
2821 cggaagagcg cctgatgcgg tattttctcc ttacgcatct gtgcggtatt tcacaccgca
2881 atggtgcact ctcagtacaa tctgctctga tgccgcatag ttaagccagt atacactccg
2941 ctatcgctac gtgactgggt catggctgcg ccccgacacc cgccaacacc cgctgacgcg
3001 ccctgacggg cttgtctgct cccggcatcc gcttacagac aagctgtgac cgtctccggg
3061 agctgcatgt gtcagaggtt ttcaccgtca tcaccgaaac gcgcgaggca g
//
```

3. The sequence of the phagemid pEX746 [12] containing the NY-ESO-1-SLL-1G4 clone is shown in the GenBank format below.

```
LOCUS       pEX746-NY-ESO-1-SLL-1G4 5486 bp DNA
            circular SYN 28-OCT-2024
DEFINITION  synthetic circular DNA.
ACCESSION   .
VERSION     .
KEYWORDS    pEX764_NYESO-1G4
SOURCE      synthetic DNA construct
  ORGANISM  synthetic DNA construct
REFERENCE   1 (bases 1 to 5486)
  AUTHORS   .
FEATURES             Location/Qualifiers
    source           1..5486
                     /lab_host="Escherichia coli"
                     /mol_type="other DNA"
                     /organism="synthetic DNA construct"
    promoter         38..142
                     /gene="bla"
                     /label=AmpR promoter
    CDS              143..1003
                     /gene="bla"
                     /product="beta-lactamase"
                     /label=AmpR
```

```
rep_origin      complement(1045..1554)
                /direction=LEFT
                /label=M13 ori
rep_origin      1665..2253
                /direction=RIGHT
                /label=pUC19ori
protein_bind    2314..2335
                /label=CAP binding site
                /bound_moiety="E. coli catabolite
                activator protein"
promoter        2350..2380
                /label=lac promoter
protein_bind    2388..2404
                /label=lac operator
                /bound_moiety="lac repressor
                encoded by lacI"
primer_bind     2407..2426
                /label=YOL13
primer_bind     2412..2428
                /label=M13 rev
misc_feature    2424..2443
                /label=LacZ_alpha_fragment
sig_peptide     2483..2548
                /gene="pelB (fragment)"
                /label=pelB signal sequence
CDS             2549..2869
                /label=TRAV
misc_feature    2624..2641
                /label=CDR1-alpha
misc_feature    2693..2710
                /label=CDR2-alpha
misc_feature    2816..2854
                /label=CDR3-alpha
CDS             2888..3169
                /label=TRAC
variation       3029..3031
                /label=T48C
                /note="Stabilising cysteine"
sig_peptide     3216..3272
                /label=gIII Leader
                /note="gIII Leader"
CDS             3273..3604
                /label=TRBV
misc_feature    3345..3359
                /label=CDR1 beta
misc_feature    3411..3428
                /label=CDR2 beta
misc_feature    3540..3575
                /label=CDR3 beta
CDS             3606..3995
                /label=TRBC
```

```
                     variation       3774..3776
                                     /label=S57C
                                     /note="Stabilising cysteine"
                     variation       3828..3830
                                     /label=C75A
                     CDS             4029..5240
                                     /label=M13 geneIII
                     primer_bind     complement(4205..4224)
                                     /label=YOL22
                     promoter        complement(5265..5283)
                                     /label=T7 promoter
                     primer_bind     complement(5287..5303)
                                     /label=M13 fwd
                     misc_featur     5288..5446
                                     /label=lacZ_alpha_fragment

ORIGIN
    1 gttaactacg tcaggtggca cttttcgggg aaatgtgcgc ggaaccccta tttgtttatt
   61 tttctaaata cattcaaata tgtatccgct catgagacaa taaccctgat aaatgcttca
  121 ataatattga aaaaggaaga gtatgagtat tcaacatttc cgtgtcgccc ttattccctt
  181 ttttgcggca ttttgccttc ctgtttttgc tcacccagaa acgctggtga aagtaaaaga
  241 tgctgaagat cagttgggtg cacgagtggg ttacatcgaa ctggatctca acagcggtaa
  301 gatccttgag agttttcgcc ccgaagaacg ttttccaatg atgagcactt ttaaagttct
  361 gctatgtggc gcggtattat cccgtgttga cgccgggcaa gagcaactcg gtcgccgcat
  421 acactattct cagaatgact tggttgagta ctcaccagtc acagaaaagc atcttacgga
  481 tggcatgaca gtaagagaat tatgcagtgc tgccataacc atgagtgata acactgcggc
  541 caacttactt ctgacaacga tcggaggacc gaaggagcta accgcttttt tgcacaacat
  601 gggggatcat gtaactcgcc ttgatcgttg ggaaccggag ctgaatgaag ccataccaaa
  661 cgacgagcgt gacaccacga tgcctgtagc aatggcaaca acgttgcgca aactattaac
  721 tggcgaacta cttactctag cttcccggca acaattaata gactggatgg aggcggataa
  781 agttgcagga ccacttctgc gctcggccct tccggctggc tggtttattg ctgataaatc
  841 tggagccggt gagcgtgggt ctcgcggtat cattgcagca ctggggccag atggtaagcc
  901 ctcccgtatc gtagttatct acacgacggg gagtcaggca actatggatg aacgaaatag
  961 acagatcgct gagataggtg cctcactgat taagcattgg taactgtcag accaagttta
 1021 ctcatatata ctttagattg atttaccccg gttgataatc agaaaagccc caaaaacagg
 1081 aagattgtat aagcaaatat ttaaattgta aacgttaata ttttgttaaa attcgcgtta
 1141 aatttttgtt aaatcagctc attttttaac caataggccg aaatcggcaa aatcccttat
 1201 aaatcaaaag aatagcccga gatagggttg agtgttgttc cagtttggaa caagagtcca
 1261 ctattaaaga acgtggactc caacgtcaaa gggcgaaaaa ccgtctatca gggcgatggc
 1321 ccactacgtg aaccatcacc caaatcaagt tttttggggt cgaggtgccg taaagcacta
 1381 aatcggaacc ctaaagggag cccccgattt agagcttgac ggggaaagcg aacgtggcga
 1441 gaaaggaagg gaagaaagcg aaaggagcgg gcgctagggc gctggcaagt gtagcggtca
 1501 cgctgcgcgt aaccaccaca cccgccgcgc ttaatgcgcc gctacagggc gcgtaaaagg
 1561 atctaggtga agatcctttt tgataatctc atgaccaaaa tcccttaacg tgagttttcg
 1621 ttccactgag cgtcagaccc cgtagaaaag atcaaaggat cttcttgaga tccttttttt
 1681 ctgcgcgtaa tctgctgctt gcaaacaaaa aaaccaccgc taccagcggt ggtttgtttg
 1741 ccggatcaag agctaccaac tctttttccg aaggtaactg gcttcagcag agcgcagata
 1801 ccaaatactg ttcttctagt gtagccgtag ttaggccacc acttcaagaa ctctgtagca
 1861 ccgcctacat acctcgctct gctaatcctg ttaccagtgg ctgctgccag tggcgataag
 1921 tcgtgtctta ccgggttgga ctcaagacga tagttaccgg ataaggcgca gcggtcgggc
 1981 tgaacggggg gttcgtgcac acagcccagc ttggagcgaa cgacctacac cgaactgaga
 2041 tacctacagc gtgagctatg agaaagcgcc acgcttcccg aagggagaaa ggcggacagg
 2101 tatccggtaa gcggcagggt cggaacagga gagcgcacga gggagcttcc agggggaaac
 2161 gcctggtatc tttatagtcc tgtcgggttt cgccacctct gacttgagcg tcgatttttg
```

2221 tgatgctcgt caggggggcg gagcctatgg aaaaacgcca gcaacgcggc ctttttacgg
2281 ttcctggcct tttgctggcc ttttgctcac atgtaatgtg agttagctca ctcattaggc
2341 accccaggct ttacacttta tgcttccggc tcgtatgttg tgtggaattg tgagcggata
2401 acaatttcac acaggaaaca gctatgacca tgattacgcc aagctacgta taataatatt
2461 ctatttcaag gagacagtca taatgaaata cctattgcct acggcagccg ctggattgtt
2521 attactcgcg gcccagccgg ccatggccca ggaggtgaca cagattcctg cagctctgag
2581 tgtcccagaa ggagaaaact tggttctcaa ctgcagtttc actgatagcg ctatttacaa
2641 cctccagtgg tttaggcagg accctgggaa aggtctcaca tctctgttgc ttattcagtc
2701 aagtcagaga gagcaaacaa gtggaagact taatgcctcg ctggataaat catcaggacg
2761 tagtacttta tacattgcag cttctcagcc tggtgactca gccacctacc tctgtgctgt
2821 gaggcccaca tcaggaggaa gctacatacc tacatttgga agaggaacca gccttattgt
2881 tcatccgtat atccagaacc ctgaccctgc cgtgtaccag ctgagagact ctaaatccag
2941 tgacaagtct gtctgcctat tcaccgattt tgattctcaa acaaatgtgt cacaaagtaa
3001 ggattctgat gtgtatatca cagacaaatg tgtgctagac atgaggtcta tggacttcaa
3061 gagcaacagt gctgtggcct ggagcaacaa atctgacttt gcatgtgcaa acgccttcaa
3121 caacagcatt attccagaag acaccttctt ccccagccca gaaagttcct aataacctag
3181 gttaattaag aattctttaa gaaggggata tacatatgaa aaaattatta ttcgcaattc
3241 ctttagttgt tcctttctat tctcacagcg cgggtgtcac tcagacccca aaattccagg
3301 tcctgaagac aggacagagc atgacactgc agtgtgccca ggatatgaac catgaataca
3361 tgtcctggta tcgacaagac ccaggcatgg ggctgaggct gattcattac tcagttggtg
3421 ctggtatcac tgaccaagga gaagtcccca atggctacaa tgtctccaga tcaaccacag
3481 aggatttccc gctcaggctg ctgtcggctg ctccctccca gacatctgtg tacttctgtg
3541 ccagcagtta cgtcgggaac accggggagc tgttttttgg agaaggctct aggctgaccg
3601 tactggagga cctgaaaaac gtgttcccac ccgaggtcgc tgtgtttgag ccatcagaag
3661 cagagatctc ccacacccaa aaggccacac tggtgtgcct ggccacaggc ttctaccccg
3721 accacgtgga gctgagctgg tgggtgaatg ggaaggaggt gcacagtggg gtctgcacag
3781 acccgcagcc cctcaaggag cagcccgccc tcaatgactc cagatacgct ctgagcagcc
3841 gcctgagggt ctcggccacc ttctggcagg acccccgcaa ccacttccgc tgtcaagtcc
3901 agttctacgg gctctcggag aatgacgagt ggacccagga tagggccaaa cccgtcaccc
3961 agatcgtcag cgccgaggcc tggggtagag cagacgcggc cgcatctaga attcaccatc
4021 atcactagac tgttgaaagt tgtttagcaa aaccccatac agaaaattca tttactaacg
4081 tctggaaaga cgacaaaact ttagatcgtt acgctaacta tgagggctgt ctgtggaatg
4141 ctacaggcgt tgtagtttgt actggtgacg aaactcagtg ttacggtaca tgggttccta
4201 ttgggcttgc tatccctgaa aatgagggtg gtggctctga gggtggcggt tctgagggtg
4261 gcggttctga gggtggcggt actaaacctc ctgagtacgg tgatacacct attccgggct
4321 atacttatat caaccctctc gacggcactt atccgcctgg tactgagcaa aaccccgcta
4381 atcctaatcc ttctcttgag gagtctcagc ctcttaatac tttcatgttt cagaataata
4441 ggttccgaaa taggcagggg gcattaactg tttatacggg cactgttact caaggcactg
4501 accccgttaa aacttattac cagtacactc ctgtatcatc aaaagccatg tatgacgctt
4561 actggaacgg taaattcaga gactgcgctt tccattctgg ctttaatgag gatccattcg
4621 tttgtgaata tcaaggccaa tcgtctgacc tgcctcaacc tcctgtcaat gctggcggcg
4681 gctctggtgg tggttctggt ggcggctctg agggtggtgg ctctgagggt ggcggttctg
4741 agggtggcgg ctctgaggga ggcggttccg gtggtggctc tggttccggt gattttgatt
4801 atgaaaagat ggcaaacgct aataaggggg ctatgaccga aaatgccgat gaaaacgcgc
4861 tacagtctga cgctaaaggc aaacttgatt ctgtcgctac tgattacggt gctgctatcg
4921 atggtttcat tggtgacgtt tccggccttg ctaatggtaa tggtgctact ggtgattttg
4981 ctggctctaa ttcccaaatg gctcaagtcg gtgacggtga taattcacct ttaatgaata
5041 atttccgtca atatttacct tccctccctc aatcggttga atgtcgccct tttgtcttta
5101 gcgctggtaa accatatgaa ttttctattg attgtgacaa aataaactta ttccgtggtg
5161 tctttgcgtt tcttttatat gttgccacct ttatgtatgt attttctacg tttgctaaca
5221 tactgcgtaa taaggagtct taataaggta ccctctagtc aaggcctata gtgagtcgta
5281 ttacggactg gccgtcgttt tacaacgtcg tgactgggaa aaccctggcg ttacccaact
5341 taatcgcctt gcagcacatc ccccttttcgc cagctggcgt aatagcgaag aggcccgcac
5401 cgatcgccct tcccaacagt tgcgcagcct gaatggcgaa tggcgcttcg cttggtaata
5461 aagcccgctt cggcgggctt tttttt
//

4. Humanized UCHT1 sequences in the GenBank format are shown below [14].

```
LOCUS       UCHT1_heavy_chain   366 bp DNA    linear
            MAM 18-OCT-2024
DEFINITION  UCHT1 heavy chain humanised.
ACCESSION   .
VERSION     .
KEYWORDS    .
SOURCE      synthetic DNA construct
 ORGANISM   synthetic DNA construct
REFERENCE   1 (bases 1 to 366)
  AUTHORS   Shalaby MR, Shepard HM, Presta L, Rodrigues
            ML, Beverley PC, Feldmann M, Carter P.
  TITLE     Development of humanized bispecific
            antibodies reactive with cytotoxic
            lymphocytes and tumor cells overexpressing
            the HER2 protooncogene.
  JOURNAL   J Exp Med 1992 Jan 1;175:217-25
  PUBMED    1346155
FEATURES             Location/Qualifiers
     source          1..366
                     /mol_type="genomic DNA"
                     /organism="synthetic DNA construct"
     misc_feature    75..99
                     /label=CDR1
     misc_feature    151..171
                     /label=CDR2
     misc_feature    289..333
                     /label=CDR3
ORIGIN
    1 gaggtgcagc tcgtggagtc tggtggaggt ctggtgcagc ctggaggctc actccgtttg
   61 tcctgcgcag cttctggtta ctcattcact ggctacacca tgaactgggt gcgtcaggcc
  121 ccaggaaagg gccttgagtg ggttgcactt attaatcctt acaaaggtgt tactacctac
  181 gccgatagcg tcaagggccg tttcactata agcgtagata aatccaaaaa cacagcctac
  241 ctgcaaatga acagcctgcg tgctgaggac actgccgtct attattgtgc tagaagcgga
  301 tactacggcg atgacgactg gtattttgac gtctggggtc aaggaaccct ggtcaccgtc
  361 tcctcg
//
```

```
LOCUS       UCHT1_light_chain   327 bp DNA    linear
            MAM 18-OCT-2024
DEFINITION  UCHT1 light chain humanised.
ACCESSION   .
VERSION     .
KEYWORDS    .
SOURCE      synthetic DNA construct
 ORGANISM   synthetic DNA construct
REFERENCE   1 (bases 1 to 327)
  AUTHORS   Shalaby MR, Shepard HM, Presta L, Rodrigues
            ML, Beverley PC, Feldmann M, Carter P.
```

```
      TITLE     Development of humanized bispecific
                antibodies reactive with cytotoxic lym-
                phocytes and tumor cells overexpressing
                the HER2 protooncogene.
      JOURNAL   J Exp Med 1992 Jan 1;175:217-25
      PUBMED    1346155
    FEATURES             Location/Qualifiers
        source           1..327
                         /mol_type="other DNA"
                         /organism="synthetic DNA construct"
        misc_feature     79..96
                         /label=CDR1
        misc_feature     148..168
                         /label=CDR2
        misc_feature     265..291
                         /label=CDR3

ORIGIN
        1 gacatccaga tgacccagtc cccgtcctcc ctgtctgcct ccgtgggcga cagagtcacc
       61 atcacctgca gggcaagtca ggacattaga aattatttaa actggtatca acagaaacca
      121 ggaaaagcac cgaaactcct gatctactac acatcaagat tagagtcagg agtcccatca
      181 cgcttcagtg gcagtgggtc tggaacagat tatactctca ccattagcag tctgcaaccg
      241 gaggatttcg ccacttacta ctgccaacag ggtaatacgc ttccgtggac gttcggacag
      301 ggcaccaagg tggaaatcaa acggact
//
```

5. Helper phage M13KO7 [43] (e.g., Antibody design labs, Invitrogen, NEB, etc.) can also be used. Phages are pH eluted using, for example, 0.5 M acetic acid (pH 2.8) or 0.01 M HCl (pH 2) or 0.1 M glycine-HCl (pH 2.2) or 0.1 M triethylamine. Readjust the pH to 7.4–8 after elution by addition of 1 M Tris-HCl [44–46].
6. If biotin is a concern in the blocking buffer, MPBS/non-fat dried milk (NFDM) can be dialysed [47] or debiotinylated using streptavidin magnetic beads.
7. Autoinduction media (AIM) can be used instead of 2xYT, which can boost inclusion body yields. For example, AIM Super Broth Base including Trace elements (FORMEDIUM): 35 g/L tryptone, 20 g/L yeast extract, 3.3 g/L NH_4SO_4, 6.8 g/L KH_2PO_4, 7.1 g/L Na_2HPO_4, 0.5 g/L glucose, 2 g/L α-lactose, 0.15 g/L $MgSO_4$, 0.03 g/L trace elements with suitable protocol.
8. The α and β chains of the wildtype TCRs can be isolated from cDNA of T-cell clones or phage display libraries by PCR [48].
9. Methods to clone include direct gene/fragment synthesis combined with either HiFi and Gibson assembly or restriction endonucleases digest and T4-ligation. Alternatively, PCR-based amplification methods with primers and restriction endonucleases and T4-ligation can be used. If restriction digests are used, ensure that the enzyme sites are not present

in the genes of interest. Constructs can be cloned using PCRs with a 5′-primer specific to the TRAV/TRBV chain containing an NdeI site incorporating an ATG start codon, and a 3′ -primer specific to TRAC and TRBC containing a HindIII site and a TAA stop codon [9], as shown below using NY-ESO-1-SLL-1G4 and its HLA as an example. Other sites such as XbaI and EcoRI can also be used.

a) TRAV21

```
           pGMT7          NdeI          TRAV21
CTTTAAGAAGGAGATATACATATGCAGGAGGTGACACAGATT
 L  *  E  G  D  I  H  M  Q  E  V  T  Q  I
```

b) TRBV6-5

```
           pGMT7          NdeI          TRBV6-5
CTTTAAGAAGGAGATATACATATGGGTGTCACTCAGACCCCA
 L  *  E  G  D  I  H  M  G  V  T  Q  T  P
```

c) TRAC

```
               TRAC          HindIII      pGMT7
TTCCCCAGCCCAGAAAGTTCCTAAGCTTGAAATCCGATCCGG
 F  P  S  P  E  S  S  *  A  *  I  P  I  R
```

d) TRBC

```
               TRBC          HindIII      pGMT7
GAGGCCTGGGGTAGAGCAGACTAAGCTTGAATTCCGATCCGG
 E  A  W  G  R  A  D  *  A  *  I  P  I  R
```

e) HLA-A*02

```
          pGMT7          NdeI          HLA-A*02
CTTTAAGAAGGAGATATACATATGGGTTCTCATTCTATGAGG
 L  *  E  G  D  I  H  M  G  S  H  S  M  R

          HLA-A*02          HindIII      pGMT7
CTCACCTTAAGATGGGAGCCGTAAGCTTGAATTCCGATCCGG
 L  T  L  R  W  E  P  *  A  *  I  P  I  R
```

f) β2M

```
           pGMT7          NdeI          β2M
CTTTAAGAAGGAGATATACATATGATTCAAAGAACTCCAAAA
 L  *  E  G  D  I  H  M  I  Q  R  T  P  K

              β2M          HindIII      pGMT7
GTTAAGTGGGATCGAGACATGTAAGCTTGAATTCCGATCCGG
 V  K  W  D  R  D  M  *  A  *  I  P  I  R
```

Mutations TRAC T48C, TRBC S57C, and TRBC C75A can be added via site-directed mutagenesis methods [9, 21, 22]. A possibility is also to add the mutations in TRAC and TRBC to dedicated TRA-chain and TRB-chain expression vectors before subcloning of the TRAV and TRBV with suitable primers [25]. For ImmTAC molecules, the final construct is in the following format: UCHT1-(GGGGS)$_1$-V$_L$-(GGGGS)$_5$-V$_H$-TRBV-TRBC.

The V_L and H_L are linked via a $(GGGGS)_5$ linker. The single-chain UCHT1 is linked via a $(GGGGS)_1$ linker to the N-terminus of TRBV (*see* **Note 4**).

g) 5′-end

```
           pGMT7          NdeI         VL
CTTTAAGAAGGAGATATACATATGGCAATTCAGATGACTCAA
 L  *  E  G  D  I  H  M  A  I  Q  M  T  Q

     VH              (GGGGS)1              TRBV6-5
GTAACAGTGTCCTCCGGTGGGGGGGGATCCGGTGTCACTCAGACCCCA
 V  T  V  S  S  G  G  G  G  S  G  V  T  Q  T  P
```

h) 3′-end

```
            TRBC         HindIII     pGMT7
GAGGCCTGGGGTAGAGCAGACTAAGCTTGAATTCCGATCCGG
 E  A  W  G  R  A  D  *  A  *  I  P  I  R
```

10. For transformations, plasmid DNA is prepared using standard miniprep kits, such as QIAprep Spin Miniprep Kit (Qiagen). It is recommended that 1 μL of 1–10 ng/μL purified plasmid DNA is directly added to 50 μL of *E. coli* cells, but a lower volume of cells can be used per transformation as well.
11. In case of TCRs, better results might be achieved using Rosetta cells with chloramphenicol in the selection media. See Sect. 2.1, point 4.
12. Other strategies can also be applied, e.g., for CDR2α, the core three residues can be randomized and for CDR2β, the core four residues, and CDR3α and CDR3β can be targeted by five to six NNK each, respectively [23, 25].
13. Typically, ten electroporations generate libraries of up to 10^9 clones.
14. Antigen concentrations in each pan need to be optimized for each target. Different selection strategies should be tried.
15. The initial inoculant volume should contain 10× more cells than the library size. An OD_{600} of 0.1 ≈ ~4×10^8 cells/mL, i.e., 100 mL of cells will contain ~4×10^9 cells.
16. The ratio of PEG precipitation buffer to supernatant is reported in varying ratios. One part PEG precipitation buffer and three parts supernatant will work too.
17. A starter culture can be prepared the night before to help grow up speeds on the day. Inoculate 5–10 mL of TYP media with a single colony and grow it overnight at 37°C and 250 rpm. Use 10–100 μL of overnight culture on the day to inoculate the larger volume.
18. Cell pellets can be stored at −20°C or −80°C.
19. The recommended volume of BugBuster for efficient lysis of *E. coli* is 5 mL of reagent per gram of wet cell pellet (see user

manual TB245, Novagen, Merck Millipore). Other bacterial cell lysis methods are sonication, cell disruptors or freeze/thaw in a lysis buffer (e.g., 10 mM Tris, 10 mM $MgCl_2$, 150 mM NaCl, 10% glycerol).

20. DNAse I (20 mg/mL) and $MgCl_2$ (0.5 M) must be added to the BugBuster solution. RNase at 20 μg/mL or lysozyme at 100 μg/mL are optional additions to lysis buffers [31, 35].

21. Resuspension can be performed using pipetting up and down using sterile pipettes. Depending on the volume, larger volume pipettes are easier to use. The pellet can also be dislodged by a single-use spatula first and the solution can also be vortexed.

22. The supernatant should be clear. If not, repeat resuspension and centrifugation steps. A slimy inclusion body pellet or a yellow-brown pellet may suggest poor lysis and/or expression.

23. Other gels can be used, but do not use guanidine buffer to denature inclusion bodies for SDS-PAGE analysis as this causes smearing.

24. Alternatively, a buffer containing 8 M urea can be used [20].

25. Redox concentrations/ratios are different depending on the protein to be refolded and variations might improve refolds. Varying the urea and arginine can also be evaluated [49].

26. Alternative redox couples may be used during refolding such as reduced 5 mM glutathione (GSH)/0.5 mM oxidized glutathione, 0.74 g/L mercaptoethylamine with 0.83 g/L cysteine, as well as DTT or β-mercaptoethanol/atmospheric oxygen, and cysteine in reduced and oxidized forms [22, 35].

27. Chains can also be added using deferred incubation steps. The α chain inclusion bodies are added first, incubated at 37°C for 30 min with 10 mM dithiothreitol present, then added to cold refold buffer (50 mM Tris, pH 8.1, 2 mM EDTA, 2.5 M urea, 6 mM cysteamine hydrochloride, and 4 mM cystamine) and incubated for a further 15 min. The β chain inclusion bodies are incubated separately at 37°C for 15 min in the presence of 10 mM dithiothreitol, and then combined with the α chain inclusion bodies [34].

28. Varying the α:β ratios and/or amount added might improve refolding.

29. The peptides (>95% purity) are chemically synthesized and dissolved in DMSO at 4 mg/mL prior use [50].

30. Varying HLA:β2M ratios and/or amount added might improve refolding, e.g., 1 μmol HLA heavy hain with 0.7 μmol β2M and 2 μmol peptide [20].

31. Any low protein binding filter membrane such as PES can be used. A pre-filtration step and other pre-filters of 0.8 μm or 1.2 μm can be applied if direct filtration through 0.45 μm pre-filter is challenging. Refolds often have visible, white precipitate.

32. Run reduced and non-reduced samples. DTT or β-mercaptoethanol can be used to reduce samples. 10% Bis-Tris gels [35] or other gels can also be used (Fig. 5a).

33. At this stage, the sample can be buffer exchanged directly into other buffers for downstream applications such as different Biacore buffers other than PBS, such as, 10 mM HEPES pH 7.4, 150 mM NaCl, 3 mM EDTA, 0.005% (v/v) surfactant P20 [34, 37].

34. For low- to intermediate-affinity molecules, such as most wildtype TCRs, use an instrument like a Biacore T200. The Biacore T200 is also capable of analyzing high-affinity interactions. For high-throughput screening across the full affinity range, including high-affinity ImmTAC molecules, use instruments like a Biacore 8K.

35. Always use matched sample and running buffers.

36. Experiments can also be performed at 37°C.

37. Use buffer exchange methods such as dialysis or spin columns if dilution is not possible.

38. A non-cognate pHLA should also be used as a control. Specific TCRs should not bind to non-cognate pHLAs, and "background" or very low-level binding could suggest aggregation or non-specific protein-protein interactions.

39. The K_D value is often not known. In such cases, run pilot experiments with widely spaced concentrations. For example, five concentrations in the range 2 to 1500 nM [41] or lower/higher. Wildtype TCRs can have low affinities in the 5–300 μM range.

40. For an interaction with $K_D = 1$ nM, use concentrations from 0.1 to 10 nM, and for an interaction with $K_D = 1$ μM, use concentrations from 0.1 μM to 10 μM, etc., always including a blank (zero-concentration) [41].

41. Slow ImmTAC/HLA off-rates ($k_{d/off}$) can be calculated using two-hour dissociation phase data from a final injection.

42. Perform a minimum of two measurements to obtain k_D values.

43. In some instances, low affinity TCRs may provide measurable on-rate ($k_{a/on}$) and off-rate ($k_{d/off}$), which means steady state is not attained. In those cases, measuring multi-cycle kinetics (MCK) (Fig. 6e, f) will be a more appropriate way to determine the K_D. This can generally be done with the same dataset intended for steady-state measurements.

44. Cross-reactivity can be assessed at this stage. Load non-cognate HLAs or appropriate HLA mixes. Load up to 5000 RU [40]. Inject TCR/ImmTAC molecules at a high concentration to identify any potential cross-reactivity.
45. Contact time, dissociation time and flow rate can be optimized for each experiment.

Acknowledgments

We would like to thank Tris Vaughan and Robert Wilkinson for useful discussions. The authors would also like to gratefully acknowledge the feedback and helpful comments made by Chandramouli Chillakuri, Peter Molloy, Jakub Jaworski, Miriam Hock, Nathaniel Liddy, and Emma Baston.

References

1. Shiravand Y, Khodadadi F, Kashani SMA et al (2022) Immune checkpoint inhibitors in cancer therapy. Curr Oncol 29:3044–3060. https://doi.org/10.3390/curroncol29050247
2. Cappell KM, Kochenderfer JN (2023) Long-term outcomes following CAR T cell therapy: what we know so far. Nat Rev Clin Oncol 20:359–371. https://doi.org/10.1038/s41571-023-00754-1
3. Dumontet C, Reichert JM, Senter PD et al (2023) Antibody–drug conjugates come of age in oncology. Nat Rev Drug Discov 22:641–661. https://doi.org/10.1038/s41573-023-00709-2
4. Tsuchikama K, Anami Y, Ha SYY, Yamazaki CM (2024) Exploring the next generation of antibody–drug conjugates. Nat Rev Clin Oncol **21**:203–223. https://doi.org/10.1038/s41571-023-00850-2
5. Uhlén M, Fagerberg L, Hallström BM et al (2015) Tissue-based map of the human proteome. Science 347:1260419–1260419. https://doi.org/10.1126/science.1260419
6. The human cell in plasma membrane - The Human Protein Atlas. https://www.proteinatlas.org/humanproteome/subcellular/plasma+membrane. Accessed 16 Jul 2024
7. Hu Z, Yuan J, Long M et al (2021) The Cancer Surfaceome Atlas integrates genomic, functional and drug response data to identify actionable targets. Nat Cancer 2:1406–1422. https://doi.org/10.1038/s43018-021-00282-w
8. Liddy N, Bossi G, Adams KJ et al (2012) Monoclonal TCR-redirected tumor cell killing. Nat Med 18:980–987. https://doi.org/10.1038/nm.2764
9. Boulter JM, Glick M, Todorov PT et al (2003) Stable, soluble T-cell receptor molecules for crystallization and therapeutics. Protein Eng Des Sel 16:707–711. https://doi.org/10.1093/protein/gzg087
10. Hughes SJ, Johnston LH, Carlos A, de Smith GL (1991) Vaccinia virus encodes an active thymidylate kinase that complements a cdc8 mutant of Saccharomyces cerevisiae. J Biol Chem 266:20103–20109. https://doi.org/10.1016/s0021-9258(18)54896-1
11. Rosenberg AH, Lade BN, Dao-shan C et al (1987) Vectors for selective expression of cloned DNAs by T7 RNA polymerase. Gene 56:125–135. https://doi.org/10.1016/0378-1119(87)90165-x
12. Karsten JB, Bent AT, Eamon MP, YI L (2014) T cell receptor display, US20140371085A1
13. Beverley PCL, Callard RE (1981) Distinctive functional characteristics of human "T" lymphocytes defined by E rosetting or a monoclonal anti-T cell antibody. Eur J Immunol 11:329–334. https://doi.org/10.1002/eji.1830110412
14. Shalaby MR, Shepard HM, Presta L et al (1992) Development of humanized bispecific antibodies reactive with cytotoxic lymphocytes and tumor cells overexpressing the HER2 protooncogene. J Exp Medicine 175:217–225. https://doi.org/10.1084/jem.175.1.217
15. Kristensen P, Winter G (1998) Proteolytic selection for protein folding using filamentous

bacteriophages. Fold Des 3:321–328. https://doi.org/10.1016/s1359-0278(98)00044-3
16. Lefranc M-P (2011) IMGT unique numbering for the variable (V), constant (C), and groove (G) domains of IG, TR, MH, IgSF, and MhSF. Cold Spring Harb Protoc 2011:pdb.ip85. https://doi.org/10.1101/pdb.ip85
17. Garboczi DN, Ghosh P, Utz U et al (1996) Structure of the complex between human T-cell receptor, viral peptide and HLA-A2. Nature 384:134–141. https://doi.org/10.1038/384134a0
18. Chen J-L, Stewart-Jones G, Bossi G et al (2005) Structural and kinetic basis for heightened immunogenicity of T cell vaccines. J Exp Med 201:1243–1255. https://doi.org/10.1084/jem.20042323
19. Zhao Y, Bennett AD, Zheng Z et al (2007) High-affinity TCRs generated by phage display provide CD4+ T cells with the ability to recognize and kill tumor cell lines. J Immunol 179:5845–5854. https://doi.org/10.4049/jimmunol.179.9.5845
20. Garboczi DN, Hung DT, Wiley DC (1992) HLA-A2-peptide complexes: refolding and crystallization of molecules expressed in Escherichia coli and complexed with single antigenic peptides. Proc National Acad Sci 89:3429–3433. https://doi.org/10.1073/pnas.89.8.3429
21. Cole DK, Miles KM, Madura F et al (2014) T-cell receptor (TCR)-peptide specificity overrides affinity-enhancing TCR-major histocompatibility complex interactions. J Biol Chem 289:628–638. https://doi.org/10.1074/jbc.m113.522110
22. Karsten JB, Meir G (2003) Soluble T cell receptor, WO2003020763
23. Li Y, Moysey R, Molloy PE et al (2005) Directed evolution of human T-cell receptors with picomolar affinities by phage display. Nat Biotechnol 23:349–354. https://doi.org/10.1038/nbt1070
24. Gao GF, Willcox BE, Wyer JR et al (2000) Classical and nonclassical class I major histocompatibility complex molecules exhibit subtle conformational differences that affect binding to CD8αα. J Biol Chem 275:15232–15238. https://doi.org/10.1074/jbc.275.20.15232
25. Dunn SM, Rizkallah PJ, Baston E et al (2006) Directed evolution of human T cell receptor CDR2 residues by phage display dramatically enhances affinity for cognate peptide-MHC without increasing apparent cross-reactivity. Protein Sci 15:710–721. https://doi.org/10.1110/ps.051936406
26. Oates J, Hassan NJ, Jakobsen BK (2015) ImmTACs for targeted cancer therapy: why, what, how, and which. Mol Immunol 67:67–74. https://doi.org/10.1016/j.molimm.2015.01.024
27. Lee CMY, Iorno N, Sierro F, Christ D (2007) Selection of human antibody fragments by phage display. Nat Protoc 2:3001–3008. https://doi.org/10.1038/nprot.2007.448
28. Robertson IB, Mulvaney R, Dieckmann N et al (2023) Tuning the potency and selectivity of ImmTAC molecules by affinity modulation. Clin Exp Immunol 215:105–119. https://doi.org/10.1093/cei/uxad120
29. Willcox BE, Gao GF, Wyer JR et al (1999) TCR binding to peptide-MHC stabilizes a flexible recognition interface. Immunity 10:357–365. https://doi.org/10.1016/s1074-7613(00)80035-7
30. Willcox BE, Gao GF, Wyer JR et al (1999) Production of soluble αβ T-cell receptor heterodimers suitable for biophysical analysis of ligand binding. Protein Sci 8:2418–2423. https://doi.org/10.1110/ps.8.11.2418
31. Cole DK, Rizkallah PJ, Boulter JM et al (2007) Computational design and crystal structure of an enhanced affinity mutant human CD8 αα coreceptor. Proteins Struct Funct Bioinform 67: 65–74. https://doi.org/10.1002/prot.21176
32. Chancellor A, Simmons RA, Khanolkar RC et al (2023) Promiscuous recognition of MR1 drives self-reactive mucosal-associated invariant T cell responses. J Exp Med 220:e20221939. https://doi.org/10.1084/jem.20221939
33. Raman MCC, Rizkallah PJ, Simmons R et al (2016) Direct molecular mimicry enables off-target cardiovascular toxicity by an enhanced affinity TCR designed for cancer immunotherapy. Sci Rep-UK 6:18851. https://doi.org/10.1038/srep18851
34. Cole DK, Yuan F, Rizkallah PJ et al (2009) Germ line-governed recognition of a cancer epitope by an immunodominant human T-cell receptor. J Biol Chem 284:27281–27289. https://doi.org/10.1074/jbc.m109.022509
35. Cole DK, Rizkallah PJ, Gao F et al (2006) Crystal structure of HLA-A*2402 complexed with a telomerase peptide. Eur J Immunol 36:170–179. https://doi.org/10.1002/eji.200535424
36. Purbhoo MA, Boulter JM, Price DA et al (2001) The human CD8 coreceptor effects cytotoxic T cell activation and antigen sensitivity primarily by mediating complete phosphorylation of the t cell receptor ζ chain. J Biol Chem 276:32786–32792. https://doi.org/10.1074/jbc.m102498200
37. Cole DK, Dunn SM, Sami M et al (2008) T cell receptor engagement of peptide-major histocompatibility complex class I does not modify CD8 binding. Mol Immunol 45:2700–2709. https://doi.org/10.1016/j.molimm.2007.12.009
38. O'Callaghan CA, Byford MF, Wyer JR et al (1999) BirA enzyme: production and

application in the study of membrane receptor–ligand interactions by site-specific biotinylation. Anal Biochem 266:9–15. https://doi.org/10.1006/abio.1998.2930
39. Coles CH, Mulvaney RM, Malla S et al (2020) TCRs with distinct specificity profiles use different binding modes to engage an identical peptide–HLA complex. J Immunol 204:ji1900915. https://doi.org/10.4049/jimmunol.1900915
40. Poole A, Karuppiah V, Hartt A et al (2022) Therapeutic high affinity T cell receptor targeting a KRASG12D cancer neoantigen. Nat Commun 13: 5333. https://doi.org/10.1038/s41467-022-32811-1
41. Cytivia kinetics and affinity measurements with Biacore systems. https://cdn.cytivalifesciences.com/api/public/content/digi-33041-pdf. Accessed 19 Jul 2024
42. Michael BJ (2009) Gamma-delta T cell receptors, US20090105133
43. Vieira J, Messing J (1987) Production of single-stranded plasmid DNA. Methods Enzym 153:3–11. https://doi.org/10.1016/0076-6879(87)53044-0
44. Alfaleh MA, Alsaab HO, Mahmoud AB et al (2020) Phage display derived monoclonal antibodies: from bench to bedside. Front Immunol **11**:1986. https://doi.org/10.3389/fimmu.2020.01986
45. França RKA, Studart IC, Bezerra MRL et al (2023) Progress on phage display technology: tailoring antibodies for cancer immunotherapy. Viruses 15:1903. https://doi.org/10.3390/v15091903
46. Kim H, Ho M (2018) Isolation of antibodies to heparan sulfate on glypicans by phage display. Curr Protoc Protein Sci 94:e66. https://doi.org/10.1002/cpps.66
47. Hoffman WL, Jump AA (1989) Inhibition of the streptavidin-biotin interaction by milk. Anal Biochem 181:318–320. https://doi.org/10.1016/0003-2697(89)90250-9
48. Moysey R, Vuidepot A-L, Boulter JM (2004) Amplification and one-step expression cloning of human T cell receptor genes. Anal Biochem 326:284–286. https://doi.org/10.1016/j.ab.2003.12.011
49. Madura F, Rizkallah PJ, Legut M et al (2019) TCR-induced alteration of primary MHC peptide anchor residue. Eur J Immunol 49:1052–1066. https://doi.org/10.1002/eji.201948085
50. Aleksic M, Liddy N, Molloy PE et al (2012) Different affinity windows for virus and cancer-specific T-cell receptors: Implications for therapeutic strategies. Eur J Immunol 42:3174–3179. https://doi.org/10.1002/eji.201242606

Check for updates

Chapter 5

Design and Evaluation of Domain-Rearranged Bispecific Antibody Fragments for Functional Optimization

Ryutaro Asano

Abstract

Bispecific antibody (BsAb) fragments, such as bispecific tandem scFv and bispecific diabody (Db), have a convenient size that allows for rapid tissue penetration and high target retention. Among BsAb fragments, the bispecific Db and bispecific tandem single-chain Fv (taFv) formats are the most widely used. To prevent homodimer formation, a single-chain Db (scDb) was developed by connecting two hetero-scFv fragments with an additional middle linker. The domain order consisting of BsAb fragments can be varied, and several reports have suggested that this order affects both expression levels and function. However, the order of the most functional domains varies depending on the antibody clones used to construct BsAb fragments and their targets, making it difficult to predict. Both bispecific scDb and taFv formats have eight possible domain orders, which also affect expression levels and function. Here, we demonstrated a method for rearranging the domain order and functionally evaluating bispecific scDb and taFv formats.

Keywords Bispecific antibody, Bispecific diabody, Single-chain diabody, Bispecific tandem scFv, Functional domain order

1 Introduction

Advances in antibody engineering techniques have enabled the generation of relatively small bispecific antibodies (BsAbs), also referred to as BsAb fragments, consisting of only two fragment variable (Fv). Compared with other BsAbs, such as IgG-like BsAbs produced via quadroma or recombinant technologies, these BsAb fragments have a convenient size for rapid tissue penetration and high target retention [1, 2]. Among BsAb fragments, bispecific diabody (Db) [3] and bispecific tandem single-chain Fv (taFv) [4] formats are the most widely used, and their features and structural differences have been reviewed [5–7]. Although highly dependent on the antibody clone used, Dbs are more likely to be produced as soluble proteins by bacteria than taFvs. However, because Dbs are heterodimers consisting of two hetero-scFv fragments, non-active homodimers may also form simultaneously with the active heterodimers. To prevent

Masahiro Yasunaga (ed.), *Bispecific Antibodies*, Methods in Molecular Biology, vol. 2997,
https://doi.org/10.1007/978-1-0716-5037-0_5,
© The Author(s), under exclusive license to Springer Science+Business Media, LLC, part of Springer Nature 2026

homodimer formation, single-chain Db (scDb) was developed by connecting the two hetero-scFv fragments with an additional middle linker [8]. In contrast, taFvs can be produced intrinsically from a single open reading frame as a single species. Furthermore, the two binding sites in a taFv molecule can rotate and bend freely, which may enhance the likelihood of simultaneous binding of two antigen epitopes on two different cell surfaces. Blinatumomab (Blincyto®) is an example of a bispecific taFv and the first T-cell engaging BsAb to be approved by the US Food and Drug Administration [9].

The domain order consisting of BsAb fragments can be varied, and several studies suggest that this order affects both expression levels and function [10, 11]. The bispecific Db format has four possible domain orders, and cytotoxic enhancement has been achieved in a bispecific Db targeting human epidermal growth factor receptor on cancer cells and CD3 on T cells by rearranging the domain order [12]. However, the order of the most functional domains varies depending on the antibody clones used to construct BsAb fragments and targets [13], making it difficult to predict. Both bispecific scDb and taFv formats have eight possible domain orders, each of which can affect expression levels and function [14]. Here, we demonstrate a method for rearranging the domain order and functionally evaluating bispecific scDb and taFv formats.

2 Materials

2.1 Preparation of Domain-Rearranged BsAb Fragments

1. Gene cassettes for BsAb fragments.
2. T7 promoter-based expression vector.
3. *Escherichia coli* strain.
4. 2 × YT broth.
5. Ampicillin.
6. Isopropyl-β-D-thiogalactopyranoside (IPTG).
7. Metal-immobilized affinity chromatography resin.
8. Gel filtration column.
9. Phosphate-buffered saline (PBS).

2.2 Evaluation of BsAb Fragments

1. Culture flasks (25 and 75 cm^2).
2. Culture medium.
3. Anti-CD3 monoclonal antibody.
4. Peripheral blood mononuclear cells (PBMCs).
5. Fetal bovine serum (FBS).
6. Penicillin.
7. Streptomycin.

8. Interleukin-2.
9. Target cancer cell line. Herein, human bile duct carcinoma (TFK-1) is used as an example [15].
10. 96-well half-area flat-bottomed plate.
11. Purified BsAb fragments.
12. PBS.
13. MTS reagent.

3 Methods

Although standard protocols for microbial and mammalian expression systems can be applicable to the expression of bispecific scDb and taFv, the preparation using the *E. coli* expression system is demonstrated here as an example. To determine the most appropriate domain order for each bispecific Db and taFv, therapeutic efficacy should be evaluated preferentially. Affinity maturation of selected BsAb fragments with a suitable domain order is effective [16, 17]; however, binding affinity to individual targets often shows little correlation with therapeutic efficacy when screening for optimal domain orders [12, 13, 18]. Evaluation of cross-linking ability between soluble target antigens using surface plasmon resonance spectroscopy has also shown limited correlation with therapeutic efficacy [12]. Conversely, evaluating cross-linking ability using whole target cells and atomic force microscopy may reflect therapeutic efficacy, though it requires some specialized techniques and equipment [19]. The following sections describe the selection of the functional domain order of T-cell-engaging BsAb fragments as a representative example, using an in vitro cancer growth inhibition assay.

3.1 Design and Preparation of Domain-Rearranged BsAb Fragments

1. Select two kinds of Fv domains to construct bispecific Db and taFv of interest (Fig. 1a).
2. Connect the genes for VH and VL in the domain orders, as described in Fig. 1b.
3. Add a pelB signal peptide at the N-terminus for secretion and a c-Myc tag and His tag at the C-terminus for detection and purification (Fig. 1b, *see* **Note 1**).
4. Prepare gene cassettes based on 3 (*see* **Note 2**).
5. Insert the gene cassette into a T7 promoter-based expression vector and transform *E. coli* (*see* **Note 3**).
6. Culture transformed *E. coli* at 28°C in 2 × YT broth supplemented with 100 μg/mL ampicillin.
7. Induce protein expression by adding IPTG to a final concentration of 1 mM when the culture reaches $O.D._{600\ nm} = 0.8$.

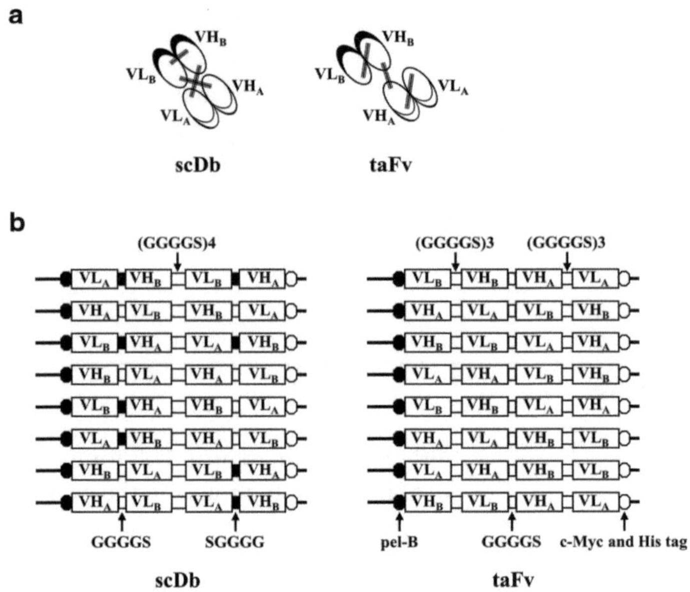

Fig. 1 Design of domain-rearranged bispecific antibody fragments. (**a**) Schematic representation of single-chain diabody (scDb) and tandem single-chain Fv (taFv) structures. (**b**) Construction of gene cassettes for scDb and taFv, illustrating the eight possible domain orders for expression in *Escherichia coli*. Each glycine-rich peptide linker is shown as a representative example

8. Incubate the culture at 20°C for 48 h.
9. Isolate crude purified BsAb fragments from the bacterial supernatant using immobilized metal affinity chromatography (*see* **Note 4**).
10. Equilibrate a gel filtration column with PBS.
11. Load 5 mL of crude purified BsAb fragments and fractionate BsAb fragment monomers at a flow rate of 2.5 mL/min (*see* **Note 5**).

3.2 Evaluation of BsAb Fragments to Determine Functional Domain Order

1. Precoat a 25 cm^2 culture flask with anti-CD3 monoclonal antibody (10 mg/mL) for 1 h.
2. Add PBMCs to the flask at a density of 1 × 10^6 cells/mL to induce lymphokine-activated killer cells with the T-cell phenotype (T-LAK) cells (*see* **Note 6**).
3. Culture the cells in a medium supplemented with 10% FBS, 100 U/mL penicillin, 100 μg/mL streptomycin, and 100 IU/mL recombinant human interleukin-2 at 37°C for 2 days.
4. Transfer the culture to a larger culture flask (75 cm^2) and continue incubation for eight days under the same conditions, replacing the medium every 2 days.

5. Seed TFK-1 cells (5×10^3 cells in 100 μL) into each well of a 96-well half-area flat-bottomed plate.
6. Culture the TFK-1 cells in a medium supplemented with 10% FBS, 100 U/mL penicillin, and 100 μg/mL streptomycin at 37°C for 1 day.
7. Remove the culture medium by aspiration.
8. Add 100 μL of T-LAK cells plus varying concentrations of BsAb fragments to each well (*see* **Note 7**).
9. Culture at 37°C for 2 days in a medium supplemented with 10% FBS, 100 U/mL penicillin, and 100 μg/mL streptomycin.
10. Remove the culture medium by aspiration.
11. Wash three times with PBS to remove effector cells and dead target cells.
12. Add 90 μL of culture medium and 10 μL of MTS reagent to each well.
13. Incubate at 37°C until the A_{490} of the control reaches approximately 1.
14. Measure absorbance at 490 nm using a microplate reader.
15. Calculate the percentage of growth inhibition of target cancer cells as follows (*see* **Note 8**):

Percentage growth inhibition of target cells = [1 − (A_{490} of experiment − A_{490} of background)/(A_{490} of control − A_{490} of background)] × 100

4 Notes

1. Signal peptides and tags are modified according to the expression hosts and purification methods, respectively. Besides *E. coli*, *Pichia pastoris*, and *Brevibacillus choshinensis* have been reported as microbial expression hosts for BsAb fragments [14].
2. Gene cassettes for the BsAb fragment with signal peptides and tags can be prepared using a standard overlap PCR method or can be obtained from a gene synthesis service. In both cases, codon optimization tailored to the expression host can increase the yield.
3. Other promoter-based expression vectors can also be used for this purpose. The preparation of an Fc fusion of BsAb fragments with rearranged domain orders is another option [20]. Fc-fused BsAb fragments reflect the functional domain order, and BsAb fragments can also be prepared from the Fc fusion format by specific protease digestion.

4. Fusion of tags for purification may negatively affect molecular folding and function [21]. This may hamper the identification of functional domain order. One option is functional integration of *Staphylococcus aureus* protein A (protein A) binding ability into BsAb fragments via mutagenesis, allowing purification via protein A in the same way as conventional intact IgG antibodies [22].
5. The final yield and multimerization states vary depending on the expression system and domain order [14]. These are important criteria for determining the most functional domain order for BsAb fragments.
6. Unstimulated PBMCs can also be used to determine the functional domain order, especially when evaluating antibody-dependent cellular cytotoxicity involving NK cells. One advantage of using T-LAK cells is the large expansion of effector cells, which has been used in many studies.
7. The standard ratio of T-LAK cells to cancer cells is 5:1. This ratio can be adjusted based on T-LAK cells or BsAb fragment efficacy.
8. Evaluation of cytokine secretion from T-LAK cells induced by BsAb fragments using ELISA correlates with therapeutic efficacy [12, 23]. However, determining the time course and selecting the appropriate cytokines for evaluation remain challenging.

References

1. Robinson MK, Doss M, Shaller C, Narayanan D, Marks JD, Adler LP, Gonzalez Trotter DE, Adams GP (2005) Quantitative immuno-positron emission tomography imaging of HER2-positive tumor xenografts with an iodine-124 labeled anti-HER2 diabody. Cancer Res 65(4):1471–1478. https://doi.org/10.1158/0008-5472.CAN-04-2008
2. Sundaresan G, Yazaki PJ, Shively JE, Finn RD, Larson SM, Raubitschek AA, Williams LE, Chatziioannou AF, Gambhir SS, Wu AM (2003) 124I-labeled engineered anti-CEA minibodies and diabodies allow high-contrast, antigen-specific small-animal PET imaging of xenografts in athymic mice. J Nucl Med 44(12):1962–1969
3. Holliger P, Prospero T, Winter G (1993) "Diabodies": small bivalent and bispecific antibody fragments. Proc Natl Acad Sci U S A 90(14):6444–6448
4. Schlereth B, Fichtner I, Lorenczewski G, Kleindienst P, Brischwein K, da Silva A, Kufer P, Lutterbuese R, Junghahn I, Kasimir-Bauer S, Wimberger P, Kimmig R, Baeuerle PA (2005) Eradication of tumors from a human colon cancer cell line and from ovarian cancer metastases in immunodeficient mice by a single-chain Ep-CAM-/CD3-bispecific antibody construct. Cancer Res 65(7):2882–2889
5. Kriangkum J, Xu B, Nagata LP, Fulton RE, Suresh MR (2001) Bispecific and bifunctional single chain recombinant antibodies. Biomol Eng 18(2):31–40
6. Kufer P, Lutterbuse R, Baeuerle PA (2004) A revival of bispecific antibodies. Trends Biotechnol 22(5):238–244
7. Kontermann RE (2005) Recombinant bispecific antibodies for cancer therapy. Acta Pharmacol Sin 26(1):1–9
8. Kipriyanov SM, Moldenhauer G, Schuhmacher J, Cochlovius B, Von der Lieth CW, Matys ER, Little M (1999) Bispecific tandem diabody for tumor therapy with improved antigen binding and pharmacokinetics. *J Mol Biol* 293(1):41–56
9. Nunez-Prado N, Compte M, Harwood S, Alvarez-Mendez A, Lykkemark S, Sanz L,

Alvarez-Vallina L (2015) The coming of age of engineered multivalent antibodies. Drug Discov Today 20(5):588–594. https://doi.org/10.1016/j.drudis.2015.02.013
10. Lu D, Jimenez X, Witte L, Zhu Z (2004) The effect of variable domain orientation and arrangement on the antigen-binding activity of a recombinant human bispecific diabody. Biochem Biophys Res Commun 318(2):507–513
11. Brischwein K, Schlereth B, Guller B, Steiger C, Wolf A, Lutterbuese R, Offner S, Locher M, Urbig T, Raum T, Kleindienst P, Wimberger P, Kimmig R, Fichtner I, Kufer P, Hofmeister R, da Silva AJ, Baeuerle PA (2006) MT110: a novel bispecific single-chain antibody construct with high efficacy in eradicating established tumors. Mol Immunol 43(8):1129–1143. https://doi.org/10.1016/j.molimm.2005.07.034
12. Asano R, Kumagai T, Nagai K, Taki S, Shimomura I, Arai K, Ogata H, Okada M, Hayasaka F, Sanada H, Nakanishi T, Karvonen T, Hayashi H, Katayose Y, Unno M, Kudo T, Umetsu M, Kumagai I (2013) Domain order of a bispecific diabody dramatically enhances its antitumor activity beyond structural format conversion: the case of the hEx3 diabody. Protein Eng Des Sel 26(5):359–367. https://doi.org/10.1093/protein/gzt009
13. Kuwahara A, Nagai K, Nakanishi T, Kumagai I, Asano R (2020) Functional domain order of an anti-EGFR x anti-CD16 bispecific diabody involving NK cell activation. Int J Mol Sci 21(23):8914. https://doi.org/10.3390/ijms21238914
14. Asano R, Kuroki Y, Honma S, Akabane M, Watanabe S, Mayuzumi S, Hiyamuta S, Kumagai I, Sode K (2018) Comprehensive study of domain rearrangements of single-chain bispecific antibodies to determine the best combination of configurations and microbial host cells. Mabs 10(6):854–863. https://doi.org/10.1080/19420862.2018.1476815
15. Saijyo S, Kudo T, Suzuki M, Katayose Y, Shinoda M, Muto T, Fukuhara K, Suzuki T, Matsuno S (1995) Establishment of a new extrahepatic bile duct carcinoma cell line, TFK-1. Tohoku J Exp Med 177(1):61–71
16. Asano R, Hosokawa K, Taki S, Konno S, Shimomura I, Ogata H, Okada M, Arai K, Onitsuka M, Omasa T, Nakanishi T, Umetsu M, Kumagai I (2020) Build-up functionalization of anti-EGFR x anti-CD3 bispecific diabodies by integrating high-affinity mutants and functional molecular formats. Sci Rep 10(1):4913. https://doi.org/10.1038/s41598-020-61840-3
17. Nakanishi T, Maru T, Tahara K, Sanada H, Umetsu M, Asano R, Kumagai I (2013) Development of an affinity-matured humanized anti-epidermal growth factor receptor antibody for cancer immunotherapy. Protein Eng Des Sel 26(2):113–122. https://doi.org/10.1093/protein/gzs088
18. Asano R, Nagai K, Makabe K, Takahashi K, Kumagai T, Kawaguchi H, Ogata H, Arai K, Umetsu M, Kumagai I (2018) Structural considerations for functional anti-EGFR x anti-CD3 bispecific diabodies in light of domain order and binding affinity. Oncotarget 9(17):13884–13893. https://doi.org/10.18632/oncotarget.24490
19. Maejima A, Ishibashi K, Kim H, Kumagai I, Asano R (2021) Evaluation of intercellular cross-linking abilities correlated with cytotoxicities of bispecific antibodies with domain rearrangements using AFM force-sensing. Biosens Bioelectron 178:113037. https://doi.org/10.1016/j.bios.2021.113037
20. Asano R, Ikoma K, Kawaguchi H, Ishiyama Y, Nakanishi T, Umetsu M, Hayashi H, Katayose Y, Unno M, Kudo T, Kumagai I (2010) Application of the Fc fusion format to generate tag-free bi-specific diabodies. FEBS J 277(2):477–487. https://doi.org/10.1111/j.1742-4658.2009.07499.x
21. Kuwahara A, Ikebukuro K, Asano R (2023) Protein engineering of antibody fragments for pharmaceutical production. Appl Phys Rev 10(3):031311. https://doi.org/10.1063/5.0158032
22. Kuwahara A, Nazuka M, Kuroki Y, Ito K, Watanabe S, Kumagai I, Asano R (2023) Functional integration of protein A binding ability to antibody fragments for convenient and tag-free purification. Bioengineered 14(1):2259093. https://doi.org/10.1080/21655979.2023.2259093
23. Asano R, Shimomura I, Konno S, Ito A, Masakari Y, Orimo R, Taki S, Arai K, Ogata H, Okada M, Furumoto S, Onitsuka M, Omasa T, Hayashi H, Katayose Y, Unno M, Kudo T, Umetsu M, Kumagai I (2014) Rearranging the domain order of a diabody-based IgG-like bispecific antibody enhances its antitumor activity and improves its degradation resistance and pharmacokinetics. MAbs 6(5):1243–1254. https://doi.org/10.4161/mabs.29445

Chapter 6

Production of Bispecific Antibody Fragments from Fc Fusion

Ryutaro Asano

Abstract

Bispecific antibody (BsAb) fragments, such as bispecific tandem single-chain Fv and bispecific diabodies, have a convenient size for rapid tissue penetration and high target retention. Their smaller size also allows them to be expressed in cost-effective microorganisms. However, the production of BsAb fragments using *Escherichia coli* is often hampered by their expression as insoluble aggregates. In addition, BsAb fragments generally require the fusion of affinity peptide tags, which might negatively impact molecular folding and function and increase immunogenicity, as most of these tags are derived from non-human sequences. The production of BsAb fragments from their Fc fusion with a protease recognition site inserted around the hinge region is an ideal strategy. Protein A purification can be applied to Fc fusion proteins, and BsAb fragments can be prepared by protease digestion. Papain is also applicable for preparing BsAb fragments from Fc fusion; however, undesired digestion due to low sequence specificity is sometimes a problem. Herein, a method using a human rhinovirus 3C protease recognition site is presented as an example of the production of BsAb fragments from Fc fusion.

Keywords Bispecific antibody fragment, Bispecific diabody, Bispecific tandem scFv, Fc fusion, Protease digestion

1 Introduction

Advances in antibody engineering techniques have enabled the generation of various engineered bispecific antibody (BsAb) formats with improved properties. Some engineered BsAbs have been developed from only two antibody fragments, such as Fv, single-chain Fv (scFv) [1, 2], and the variable domain of the heavy chain of the heavy-chain antibody [3]. The most common BsAbs, also referred to as BsAb fragments, are bispecific tandem scFv (taFv) [4], bispecific diabody (Db) [5], and single-chain Db (scDb) [6]. Compared with immunoglobulin G (IgG)-like BsAbs produced by quadroma or recombinant technologies, these BsAb fragments have a convenient size for rapid tissue penetration and high target retention [7, 8]. Their smaller size also allows them to be expressed in microorganisms such as *Escherichia coli*, which is expected to lead to cost-effective production.

Masahiro Yasunaga (ed.), *Bispecific Antibodies*, Methods in Molecular Biology, vol. 2997,
https://doi.org/10.1007/978-1-0716-5037-0_6,
© The Author(s), under exclusive license to Springer Science+Business Media, LLC, part of Springer Nature 2026

However, the production of BsAb fragments using *E. coli* is often hampered by their expression as insoluble aggregates, such as inclusion bodies, which require additional refolding procedures [9–12]. Although efficient refolding methods have been proposed for scFv [12, 13], they require strict control of introducing additives such as oxidizing reagents, and it is uncertain whether they can be efficiently applied to all scFvs and BsAb fragments. Additionally, BsAb fragments generally require the fusion of artificial affinity peptide tags, such as polyhistidine, hemagglutinin, and FLAG tags, for initial purification, unlike IgG-like BsAbs with an Fc region, which have affinity for *Staphylococcus aureus* protein A (protein A) and thus are applicable to conventional protein A purification. However, tag fusion may negatively affect molecular folding, function, and immunogenicity because most of these tags are derived from non-human sequences [14].

Producing BsAb fragments from their Fc fusion format with a protease recognition site inserted around the hinge region is an ideal strategy. Application to pharmaceutical production may not be realistic, but this strategy is expected to contribute to the detailed characterization of the BsAb fragments of interest. Protein A purification can generally be applied to Fc fusion proteins, and BsAb fragments can be prepared by protease digestion. Papain, an endopeptidase that digests the hinge region, has traditionally been used to prepare Fab fragments. Papain is also applicable for preparing BsAb fragments from the Fc fusion format; however, undesired digestion due to low sequence specificity is sometimes a problem [15]. Herein, a method using the human rhinovirus 3C (HRV3C) protease recognition site is presented as an example of the production of BsAb fragments from the Fc fusion format. This has already been applied to the preparation and characterization of bispecific taFv [16], Db and scDb [15], tetrabodies [17], and trispecific antibodies [18].

2 Materials

2.1 Preparation of Fc-Fused BsAb Fragments

1. Mammalian expression vector
2. Gene cassette for Fc-fused BsAb fragments
3. HEK293 cells
4. Transfection reagent
5. Serum-free medium
6. 1M Tris-HCl (pH 9.0)
7. Empty column (0–2 ml bed volume)
8. Protein A-immobilized resin
9. Wash buffer (50 mM Tris-HCl and 200 mM NaCl, pH 8.0)

10. Elution buffer (100 mM glycine-HCl, pH 2.7)
11. Phosphate-buffered saline (PBS)
12. Gel for sodium dodecyl sulfate-polyacrylamide gel electrophoresis (SDS-PAGE)

2.2 Preparation of BsAb Fragments by Specific Protease Digestion

1. Glutathione *S*-transferase (GST)-fused HRV3C protease
2. 50 mM dithiothreitol
3. 50 mM ethylenediamine tetraacetic acid
4. Gel for SDS-PAGE
5. Empty column (0–2 ml bed volume)
6. Glutathione-immobilized resin
7. Protein A-immobilized resin
8. PBS
9. Gel filtration column

3 Methods

The standard protocols apply to expression using mammalian expression systems and purification through protein A of BsAbs with the Fc region, which can be referred to in other sections in this special issue. Briefly, after constructing mammalian expression vectors harboring the gene cassette for Fc-fused BsAb fragments, mammalian cells such as Chinese hamster ovary and HEK293 cells are transfected with these vectors. Culture supernatants from stably expressed clones or transiently expressed cells are applied to the protein A-immobilized column, and crude purified BsAb with the Fc region can be obtained. The following sections describe the detailed design of the hinge to insert a protease recognition site (Fig. 1), the preparation of Fc-fused BsAb fragments using transiently expressed HEK293 cells as an example, and the preparation of BsAb fragments from the Fc fusion format by protease digestion (Fig. 2).

3.1 Design and Preparation of Fc-Fused BsAb Fragments

1. Genes encoding the BsAb fragment of interest were selected, designed, and obtained (Fig. 1a, *see* **Note 1**).
2. Connect the gene for BsAb fragments to genes for the human IgG1 Fc region via a recognition site (LEVLFQGP) for the HRV3C protease (Fig. 1b, *see* **Note 2**).
3. Insert the resulting gene cassette into the mammalian expression vector for the Fc-fused BsAb fragment.
4. Transfect it into HEK293 cells using the transfection reagent.
5. Cultivate in serum-free medium for 5 to 7 days.

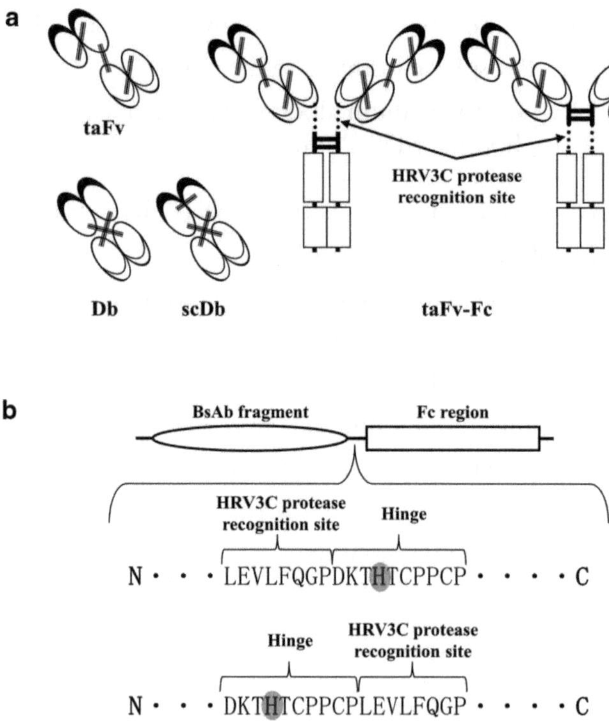

Fig. 1 Design of Fc-fused BsAb fragments. (**a**) Schematic diagrams of BsAb fragments and its Fc fusion format. taFv, tandem single-chain Fv (taFv); Db, diabody; single-chain Db (scDb). (**b**) The two designs for inserting a protease recognition site into the hinge region for constructing the Fc fusion format. To reduce the unexpected cleavage caused by dissolved active oxygen, a H237Y mutation into the hinge region is effective [19, 20]. The numbering of residues was based on the study of Kabat et al. [21]

6. Collect the supernatant and add 5% volume of 1 M Tris-HCl to adjust the pH.
7. Fill the empty column with protein A-immobilized resin.
8. Apply 6 column volumes of wash buffer.
9. Apply the supernatant to the column.
10. Apply 6 column volumes of wash buffer.
11. Elute the crude purified Fc-fused BsAb with 2 column volumes of elution buffer.
12. Add 5% volume of 1 M Tris-HCl to the eluate fraction.
13. Dialyze the eluate fraction with PBS.
14. Confirm the preparation of BsAb fragments using SDS-PAGE under reduced and non-reduced conditions.

3.2 Preparation of BsAb Fragments by Specific Protease Digestion

1. Add 2 μl of GST-fused HRV3C protease, 40 μl of 50 mM dithiothreitol, and 40 μl of 50 mM EDTA to 2 ml of protein A-purified Fc-fused BsAb.

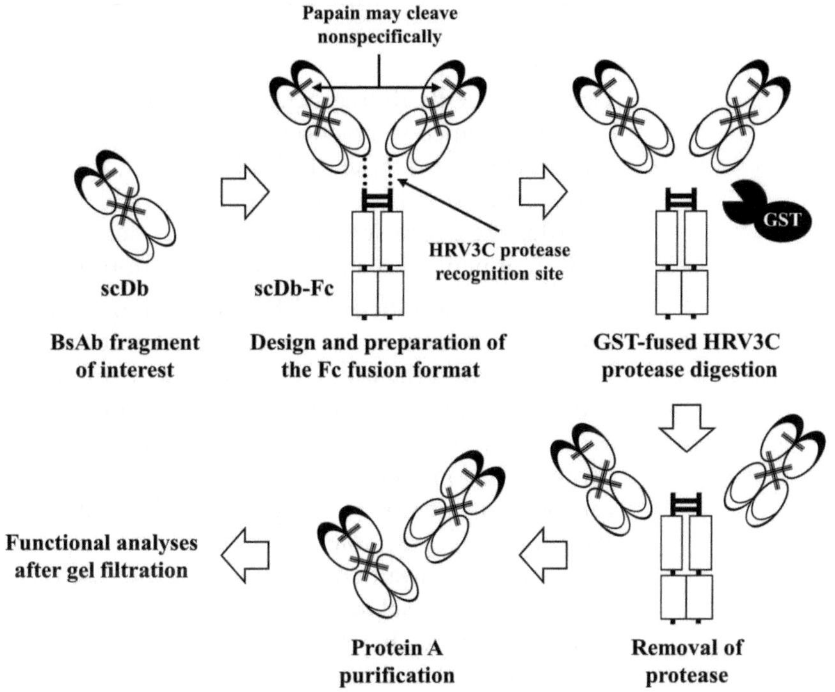

Fig. 2 Scheme for preparation of BsAb fragments from the Fc fusion format employing specific protease digestion

2. Incubate at 4°C for 16 h.
3. Confirm the cleavage efficiency using SDS-PAGE under reduced and non-reduced conditions (*see* **Note 3**).
4. Add to an empty column packed with 0.5 ml of glutathione-immobilized resin to remove GST-fused HRV3C protease.
5. Collect the flow-through fractions.
6. Apply it to an empty column packed with protein A-immobilized resin to remove the digested Fcs and undigested Fc-fused BsAb fragments.
7. Wash the column with the addition of PBS (*see* **Note 4**).
8. Confirm the preparation of BsAb fragments using SDS-PAGE under reduced and non-reduced conditions.
9. Apply the flow-through fraction to the gel filtration column (*see* **Note 5**).
10. Collect each eluted fraction and confirm the function according to the purpose (*see* **Note 6**).

4 Notes

1. Genes for the BsAb fragment can be prepared using a standard overlap polymerase reaction method or can be obtained from a gene synthesis service. In both cases, codon optimization to fit the expression host can increase the yield. Another option is to order the entire synthetic gene after designing the Fc-fused BsAb fragment with a protease recognition site.
2. The origins and subclasses of the Fc region and the recognition sites for proteases can be freely altered according to their purpose. However, the use of Fc regions with affinity for protein A is desirable for convenient initial purification. Although many protease recognition sites are applicable, GST-fused HRV3C protease is commercially available (PreScission protease) and can be easily removed using glutathione-immobilized resin. Papain is also applicable; however, its low sequence specificity may result in unwanted digestion. In the case of scDb-Fc, the middle linker of scDb was unexpectedly digested with papain [15]. To reduce the cleavage caused by dissolved active oxygen, an H237Y mutation in the hinge region is effective [19, 20]. If the recognition site for the protease is inserted after the hinge region, dimers of BsAb fragments linked by the hinge region can be prepared like the $F(ab')_2$ fragment (Fig. 1a) [16]. In the case of a BsAb fragment consisting of two individual chains, such as Db, fuse the Fc region to only one of the chains. After constructing individual expression vectors for each chain, co-transfect them into mammalian expression host cells.
3. If cleavage efficiency is insufficient, increase incubation time.
4. The BsAb fragments of interest may be present in the washing fraction. In this case, use a mixture of the flow-through and wash fractions for the next step. Digested Fc fragments and undigested Fc-fused BsAb fragments can be eluted using the IgG elution buffer.
5. The column is equilibrated with PBS, and then 0.25 ml of purified BsAb is loaded onto the column at a flow rate of 0.5 ml/min. Gel filtration is optional; however, Fc-fused BsAb fragments tend to form multimeric structures, and a gel filtration step is required to prepare the monomers of interest. Furthermore, multimeric fractions are predicted to form unexpectedly uniform structures, such as dimers and tetramers, and often show high activity [16, 17].
6. The preparation of BsAb fragments from the Fc fusion format enables the preparation of BsAb fragments that do not contain any artificial affinity peptide tags and sometimes results in increased activity. Although the reasons for this increased activity

are unclear, local structural perturbations due to the absence of tags may affect the activity of the BsAb fragments [15]. This method can also be applied to prepare more complicated engineered antibody fragments, such as trispecific antibodies [18].

References

1. Bird RE, Hardman KD, Jacobson JW et al (1988) Single-chain antigen-binding proteins. Science 242(4877):423–426
2. Huston JS, Levinson D, Mudgett-Hunter M et al (1988) Protein engineering of antibody binding sites: recovery of specific activity in an anti-digoxin single-chain Fv analogue produced in *Escherichia coli*. Proc Natl Acad Sci U S A 85(16):5879–5883
3. Harmsen MM, Fijten HP, Dekker A et al (2008) Passive immunization of pigs with bispecific llama single-domain antibody fragments against foot-and-mouth disease and porcine immunoglobulin. Vet Microbiol 132(1–2):56–64
4. Schlereth B, Fichtner I, Lorenczewski G et al (2005) Eradication of tumors from a human colon cancer cell line and ovarian cancer metastases in immunodeficient mice by a single-chain Ep-CAM-/CD3-bispecific antibody construct. Cancer Res 65(7):2882–2889
5. Holliger P, Prospero T, Winter G (1993) "Diabodies": small bivalent and bispecific antibody fragments. Proc Natl Acad Sci U S A 90(14):6444–6448
6. Brusselbach S, Korn T, Volkel T et al (1999) Enzyme recruitment and tumor cell killing in vitro by a secreted bispecific single-chain diabody. Tumor Target 4(2):115–123
7. Robinson MK, Doss M, Shaller C et al (2005) Quantitative immuno-positron emission tomography imaging of HER2-positive tumor xenografts with an iodine-124 labeled anti-HER2 diabody. Cancer Res 65(4):1471–1478
8. Sundaresan G, Yazaki PJ, Shively JE et al (2003) 124I-labeled engineered anti-CEA minibodies and diabodies allow high-contrast, antigen-specific small-animal PET imaging of xenografts in athymic mice. J Nucl Med 44(12):1962–1969
9. Peipp M, Valerius T (2002) Bispecific antibodies targeting cancer cells. Biochem Soc Trans 30(4):507–511
10. Helfrich W, Kroesen BJ, Roovers RC et al (1998) Construction and characterization of a bispecific diabody for retargeting T cells to human carcinomas. Int J Cancer 76(2):232–239
11. Arndt MA, Krauss J, Kipriyanov SM et al (1999) A bispecific diabody that mediates natural killer cell cytotoxicity against xenotransplantated human Hodgkin's tumors. Blood 94(8):2562–2568
12. Tsumoto K, Shinoki K, Kondo H et al (1998) Highly efficient recovery of functional single-chain Fv fragments from inclusion bodies overexpressed in Escherichia coli by controlled introduction of oxidizing reagent–application to a human single-chain Fv fragment. J Immunol Methods 219(1–2):119–129
13. Umetsu M, Tsumoto K, Hara M et al (2003) How additives influence the refolding of immunoglobulin-folded proteins in a stepwise dialysis system. Spectroscopic evidence for highly efficient refolding of a single-chain Fv fragment. J Biol Chem 278(11):8979–8987
14. Kuwahara A, Ikebukuro K, Asano R (2023) Protein engineering of antibody fragments for pharmaceutical production. Appl Phys Rev 10(3):031311
15. Asano R, Ikoma K, Kawaguchi H et al (2010) Application of the Fc fusion format to generate tag-free bi-specific diabodies. FEBS J 277(2):477–487
16. Asano R, Ikoma K, Shimomura I et al (2011) Cytotoxic enhancement of a bispecific diabody by format conversion to tandem single-chain variable fragment (taFv) the case of the hEx3 diabody. J Biol Chem 286(3):1812–1818
17. Asano R, Ikoma K, Sone Y et al (2010) Highly enhanced cytotoxicity of a dimeric bispecific diabody, the hEx3 tetrabody. J Biol Chem 285(27):20844–20849
18. Kimura K, Kuwahara A, Suzuki S et al (2023) Cancer therapeutic trispecific antibodies recruiting both T and natural killer cells to cancer cells. Oncol Rep 50(6):212
19. Yan BX, Boyd D, Kaschak T et al (2012) Engineering upper hinge improves stability and effector function of a human IgG1. J Biol Chem 287(8):5891–5897
20. Suzuki S, Annaka H, Konno S et al (2018) Engineering the hinge region of human IgG1 Fc-fused bispecific antibodies to improve fragmentation resistance. Sci Rep 8(1):17253
21. Kabat EA, Wu TT, Perry HM et al (1991) Sequences of proteins of immunological interest, 5th edn. National Institutes of Health, Bethesda, MD

Check for updates

Chapter 7

Enhanced Chemical Method for Bispecific Antibody Production: Leveraging Site-Specific Conjugation and Stable Crosslinker Technology

Tomohiro Fujii and Yutaka Matsuda

Abstract

Bispecific antibodies (BisAbs) represent a major breakthrough in biotherapeutics by combining the functionalities of two different antibodies to unlock their potential for a wide range of applications. Conventional BisAb production has relied on labor-intensive and complex genetic engineering methods. This chapter introduces a cutting-edge protocol that circumvents the need for genetic modifications by using chemical conjugation strategies. We present a streamlined, fully chemical methodology for generating BisAbs that leverages the precision of the AJICAP second-generation technology for accurate site-specific conjugation. In addition, this chapter explores the application of haloketone chemistry as an improved and more effective replacement for traditional maleimide-based crosslinking techniques for the fusion of two antibody-related proteins. The analytical performance and results of the newly developed BisAb were thoroughly investigated, highlighting its high efficiency and effectiveness. This innovative approach represents a significant step forward in the production of BisAbs and offers a simpler, more accessible, and potentially scalable alternative to traditional genetic engineering techniques.

Keywords Bispecific antibodies, Antibody-drug conjugates, AJICAP, Site-specific conjugation technology, Maleimide alternatives, Haloketone

1 Introduction

Bispecific antibodies (BisAbs) are genetically engineered antibodies designed to simultaneously bind to two different targets or epitopes [1, 2]. This dual specificity makes BisAbs highly versatile, with applications ranging from cancer therapy, where they can direct immune cells such as T and NK cells to tumors, to diagnostic tools. Despite the diversity of BisAb formats, more than 100 have been identified, their development is often hampered by the complexity of engineering processes and optimization of cell-culture conditions [2]. Factors such as the affinity balance between the two binding regions, their spatial arrangement, and the properties of the linkers

Masahiro Yasunaga (ed.), *Bispecific Antibodies*, Methods in Molecular Biology, vol. 2997,
https://doi.org/10.1007/978-1-0716-5037-0_7,
© The Author(s), under exclusive license to Springer Science+Business Media, LLC, part of Springer Nature 2026

connecting them play critical roles in the functionality of BisAbs. Traditional antibody-engineering techniques are not only slow but also involve cumbersome steps for introducing specific tags or modifications [3].

Recent advancements have led to the emergence of chemical methods for generating BisAbs, and they offer promising alternatives to biological engineering [4, 5]. However, these methods are primarily suited for research purposes and lack scalability and robustness for broader applications.

Our chapter introduces a novel approach using Fc-affinity peptide reagents under the banner of AJICAP second-generation technology (Fig. 1) [6]. This technology facilitates the non-genetic site-specific production of antibody-drug conjugates (ADCs), demonstrating the creation of ADCs with precise conjugation at two different sites. Comparative studies of these site-specific ADCs have shown that they possess a broader therapeutic window than those produced using conventional random conjugation methods [7]. To develop a chemically conjugated BisAb, we systematically addressed the challenges associated with traditional conjugation techniques [8]. Our focus was on identifying a reliable crosslinker that would allow for the efficient

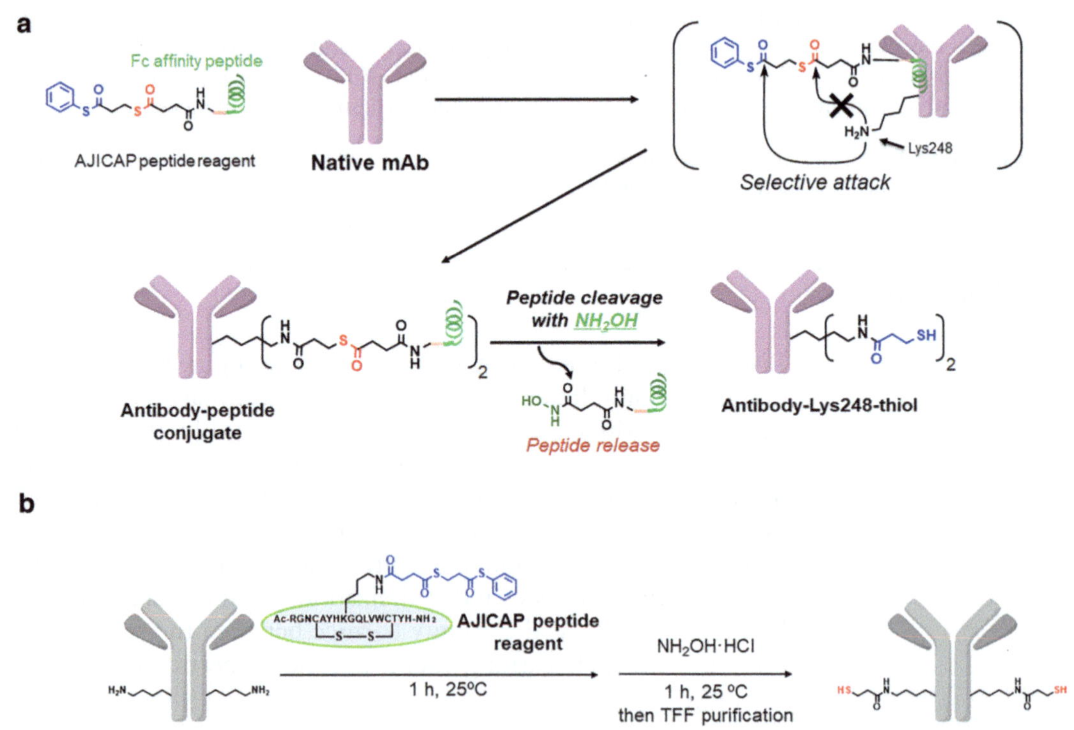

Fig. 1 Overview of AJICAP second-generation technology: (**a**) Reaction mechanism of AJICAP second-generation technology and (**b**) site-specific modification of teplizumab

coupling of an antibody with a second targeting molecule without the drawbacks associated with maleimide-based linkers, which are prone to hydrolysis and can lead to the detachment of the linked payload [9]. We analyzed click chemistry as an alternative because of its robustness and efficiency in creating stable linkages. This necessitates the use of a crosslinker capable of converting a specific lysine residue on the antibody into a functional group suitable for click reactions. Despite the widespread use of maleimide-dibenzocyclooctyne (DBCO) for these purposes, the instability of the thiol-maleimide linkages poses a significant risk of premature payload release [10]. To overcome these limitations, we selected α-haloketones, a group known for forming stable, irreversible linkages with thiols [8, 11]. This choice was driven by the need for a crosslinker that combines reactivity, stability, and ease of synthesis. α-Haloketones, when paired with bioorthogonal groups such as cycloalkynes, offer a promising route to creating reagents that are both reactive and stable, a combination rarely achieved with traditional crosslinkers (Figs. 2 and 3) [8, 11]. In advancing the

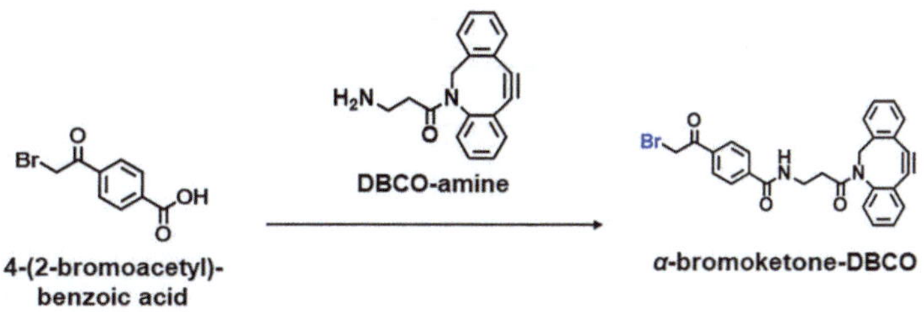

Fig. 2 α-Bromoketone-DBCO synthesis

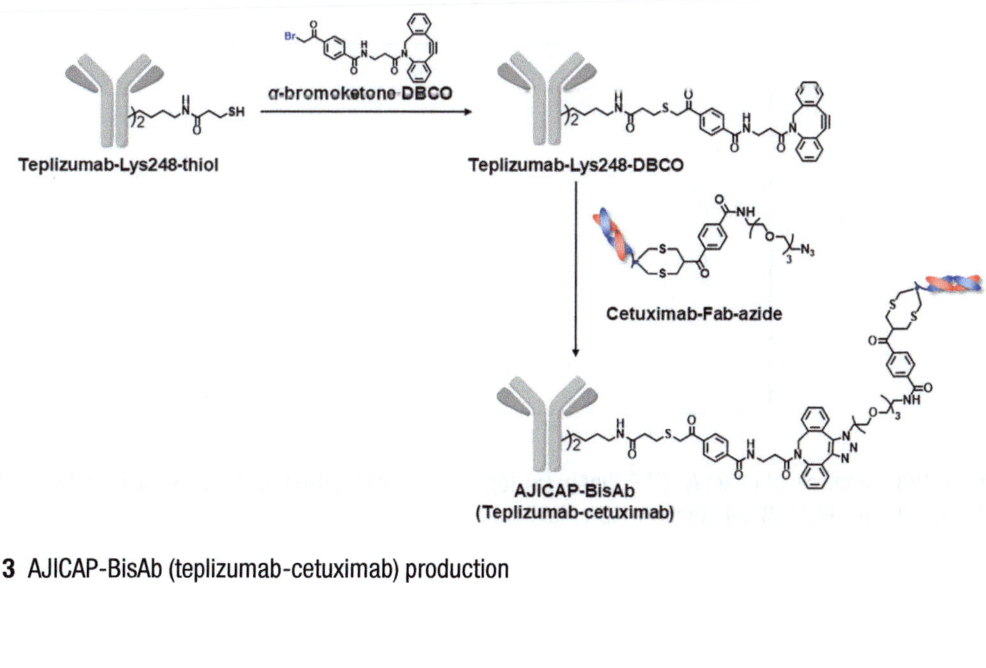

Fig. 3 AJICAP-BisAb (teplizumab-cetuximab) production

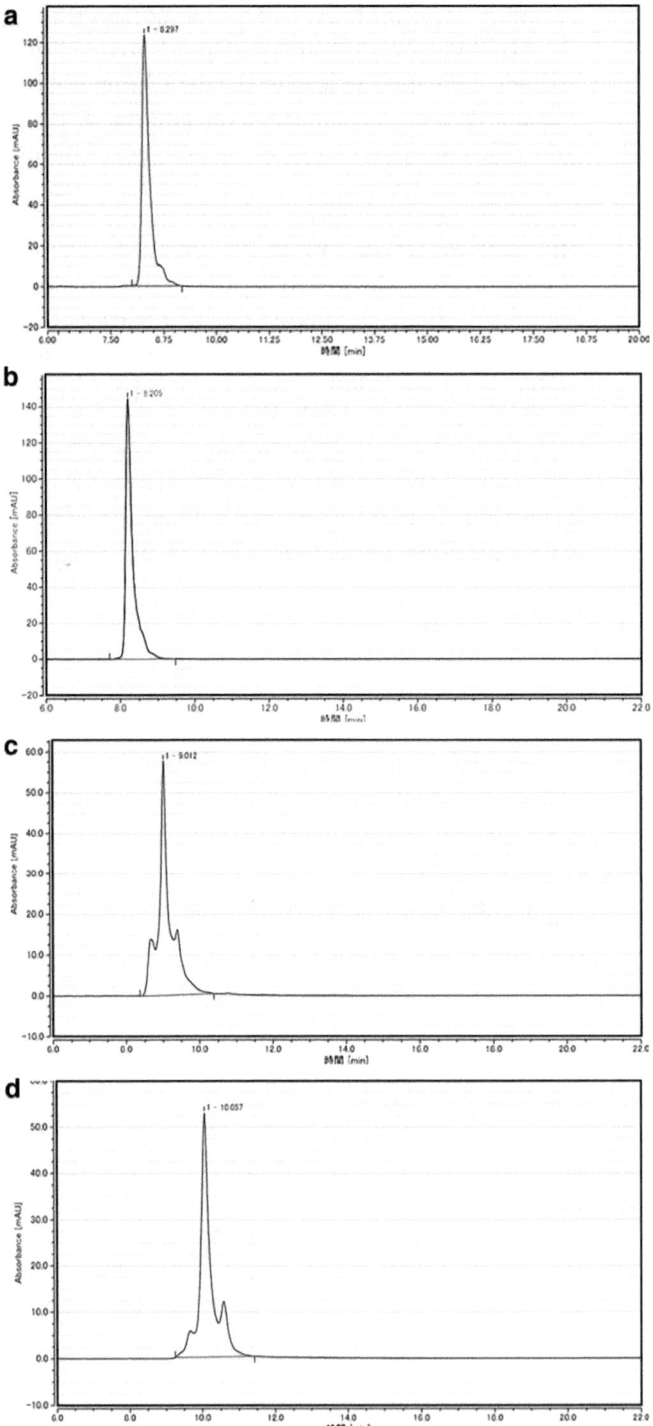

Fig. 4 RP-HPLC analysis of intact AJICAP-BisAb. (**a**) Teplizumab, (**b**) teplizumab-Lys248-thiol, (**c**) teplizumab-Lys248-DBCO, (**d**) AJICAP-BisAb (teplizumab-cetuximab)

production of BisAbs through fully chemical means, our research led to the development and validation of an α-bromoketone-DBCO crosslinker (Fig. 2). This approach uses a readily available commercial compound to establish a stable and efficient crosslinking mechanism. The method outlined below leverages AJICAP second-generation technology alongside the α-bromoketone-DBCO crosslinker (Figs. 3 and 4), providing a streamlined, non-genetic pathway for the site-specific conjugation of antibodies.

2 Materials

2.1 Site-Specific Antibody Modification

1. 10 mg/mL Anri-CD3 antibody. We use human IgG1 teplizumab (Tzield®) as an example
2. AcONa buffer: 20 mM acetate, pH 5.5 by using double-deionized water. Filter-sterilize the solution by using a 0.22 μm membrane filter (*see* **Note 1**).
3. AJICAP peptide reagent: Solid-phase peptide synthesis methodology (*see* **Note 2**).
4. *N,N*-dimethylformamide (DMF).
5. $NH_2OH{\cdot}HCl$ (Hydroxylamine hydrochloride) (*see* **Note 3**).
6. PBSE buffer: 50 mM PBS, pH7.4, 10 mM EDTA.
7. Tangential flow filtration (TFF) system with a Sartocon Slice 200 Eco Hydrosart membrane (molecular weight cutoff of 30 kDa) (*see* **Note 4**).
8. Thermomixer for temperature- and agitation-controlled experiments.
9. UV spectrophotometer (e.g., Nanodrop 2000C or Solo-VPE system).

2.2 Prcparation of α-Bromoketone-DBCO

1. DBCO-amine.
2. 4-(2-bromoacetyl)benzoic acid.
3. Dicyclohexylcarbodiimide.
4. 1-Hydroxybenzotriazole.
5. CH_2Cl_2.
6. Hexane:AcOEt (1:1) for column chromatography.
7. Biotage Isolera one flash chromatography system
8. Thin-layer chromatography (TLC).
9. Rotavapor® R-100 rotary evaporators with I-100 controller and V-100 pump

2.3 Modification of Antibody Thiol by α-Bromoketone-DBCO

1. α-Haloketone-DBCO.
2. Teplizumab-Lys248-thiol.
3. Centripure P50 desalting column (purchased from emp Biotech GmbH).
4. Thermomixer for temperature- and agitation-controlled experiments.
5. UV spectrophotometer (e.g., Nanodrop 2000C or Solo-VPE system).

2.4 Click Reaction to Produce

AJICAP

AJICAP-BisAb

1. DBCO-modified teplizumab.
2. Cetuximab-Fab-azide (*see* **Note 5**).
3. Thermomixer for temperature- and agitation-controlled experiments.
4. Centripure P50 desalting column (purchased from emp Biotech GmbH).
5. UV spectrophotometer (e.g., Nanodrop 2000C or Solo-VPE system).
6. Size-exclusion chromatography (SEC) purification column: Superdex 200 Increase 10/300GL
7. SEC purification Buffer: PBS, pH 7.4
8. SEC purification system: ÄKTA purifie
9. Detector Settings: Wavelengths 215 nm and 280 nm for detection

2.5 RP-HPLC Analysis Under Non-Reducing Conditions

1. Column: Sepax Proteomix RP-1000 5 μm 2.1 × 50 mm (Sepax) (*see* **Note 6**).
2. HPLC System with binary gradient pump, temperature-controlled column compartment, autosampler and a diode array detector was used to monitor absorbance at 280 nm, with 450 nm as the reference wavelength (e.g., Agilent 1260).
3. Mobile Phase A (MPA): 0.1% trifluoroacetic acid (TFA) and 2% acetonitrile in water; mobile phase B (MPB): 0.1% TFA in acetonitrile (*see* **Note 7**).
4. Prepare samples of antibody, antibody-thiol, antibody-DBCO, and BisAb at 1 mg/mL concentration, with each sample volume set for injection at 20 μL.

3 Methods

The protocol was refined for teplizumab, an IgG1 antibody. For alternative antibodies, proportional quantities of the reagent can be adjusted to ascertain the necessary volume for successful conjugation. For information on the reaction's compatibility, please consult additional literature [6, 12].

3.1 Site-Specific Antibody Modification Using AJICAP Peptide Reagent with Subsequent Linker Cleavage for Peptide Release

1. Sterilize the AcONa buffer and dissolve teplizumab to prepare a 10 mg/mL solution (*see* **Note 8**).
2. Add a 20 mM solution of AJICAP peptide reagent in DMF to the teplizumab solution, using 4.0 equivalents relative to the amount of teplizumab and ensuring that the final concentration of DMF is less than 8% (v/v).
3. Incubate the reaction mixture at 20°C for 1 h (*see* **Note 9**).
4. After 1 h of incubation, confirm successful conjugation using LC-Q-TOF MS (*see* **Note 9**).
5. Add an excess of $NH_2OH{\cdot}HCl$ to the reaction mixture. This step introduces NH_2OH to cleave the linkage between the AJICAP peptide reagent and antibody.
6. Stir the mixture for an additional 1 h to complete the cleavage process.
7. Set up the TFF system with a Sartocon Slice 200 Eco Hydrosart membrane (30 kDa) (*see* **Note 10**).
8. Purify the reaction mixture using TFF with 20 mM AcONa buffer (pH 5.5) as the diafiltration buffer, adjusting the antibody concentration to 20 mg/mL.
9. Perform a buffer exchange using the TFF system with PBSE buffer as the new diafiltration buffer, adjusting the antibody concentration to 7.5 mg/mL.
10. Measure the antibody concentration using a UV spectrophotometer with the A280 method. Calculate the recovery amount of antibody-thiol by multiplying the recovery volume by the determined antibody concentration.
11. Confirm successful conjugation using RP-HPLC (*see* **Note 9**).

3.2 Preparation of α-Bromoketone-DBCO

1. Precisely weigh 217.4 mg of DBCO-Amine, along with 184.3 mg of DCC and 23.3 mg of HOBt.
2. Under a fume hood for safety, dissolve 193.0 mg of 4-(2-bromoacetyl)benzoic acid in 9 mL of CH_2Cl_2 in a reaction flask.
3. Add the pre-weighed DBCO-Amine, DCC, and HOBt to the solution of 4-(2-bromoacetyl)benzoic acid.

4. Stir the resulting mixture at room temperature for 1 h, ensuring the reaction flask is tightly sealed and maintained in an inert atmosphere.
5. Filter the reaction mixture to remove any solid byproducts, collecting the filtrate (mother liquor) for further purification.
6. Transfer the mother liquor into a round-bottom flask and concentrate the solution using a rotary evaporator under reduced pressure until most of the CH_2Cl_2 is removed.
7. Use a rotary evaporator to concentrate the filtrate under reduced pressure until most of the solvent is removed.
8. Set up a column chromatography apparatus, using a Hexane:AcOEt (1:1) mixture as the eluting solvent.
9. Identify and collect fractions containing the desired product through TLC analysis.
10. Further concentrate the collected fractions using a rotary evaporator to remove any remaining solvent.
11. Dry the concentrated product in a vacuum pump to remove all traces of the solvent.
12. Weigh the final dried product to determine the total yield, which should be around 341.9 mg of α-bromoketone-DBCO.

3.3 Modification of Antibody Thiol by α-Bromoketone-DBCO

1. Adjust the pH of the PBSE buffer to 7.4 for use in both the reaction and purification.
2. Dissolve α-haloketone-DBCO in DMA to a concentration of 20 mM.
3. Prepare a 5 mg/mL solution of teplizumab-Lys248-thiol in PBSE buffer.
4. Add six equivalents of the 20 mM α-haloketone-DBCO solution to the teplizumab solution, ensuring thorough mixing.
5. Incubate the mixture at 25°C for 1 h to allow conjugation.
6. Equilibrate a Centripure P50 desalting column with PBSE buffer, according to the manufacturer's guidelines.
7. Load the reaction mixture onto the equilibrated column for purification.
8. Elute the column with PBSE buffer and collect the conjugated product.
9. Confirm successful conjugation and assess purity using RP-HPLC

3.4 Click Reaction to Produce AJICAP-BisAb

1. Dissolve DBCO-modified teplizumab to a final concentration of 2 mg/mL in PBSE buffer with pH adjusted to 7.4.
2. Prepare a separate solution of cetuximab-Fab-azide at 0.1 mM concentration in PBSE buffer.

3. Add four equivalents of the cetuximab-Fab-azide solution to the DBCO-modified teplizumab solution, ensuring accurate molar ratios for effective conjugation.
4. Incubate the combined solution at 25°C, with gentle shaking for 5 h to allow the click chemistry reaction to proceed. Confirm successful conjugation and assess purity using RP-HPLC or LC-Q-TOF MS.

3.5 SEC Purification of AJICAP-BisAb

1. System Setup: Prepare the ÄKTA purifier system for a Superdex 200 Increase 10/300GL column.
2. Buffer Preparation: PBS buffer adjusted to pH 7.4.
3. Column Equilibration: Equilibrate the SEC column with the prepared PBS buffer. Set the flow rate to 0.4 mL/min, allowing sufficient time for thorough column equilibration before sample application.
4. Sample Injection: Inject the click reaction mixture into the ÄKTA purifier system to avoid introducing air bubbles.
5. Purification Run: Conduct SEC purification at a constant flow rate of 0.4 mL/min. Monitor the process to ensure stable operation throughout purification.
6. Detection and Analysis: Detect eluted fractions using a dual-wavelength detector set to 215 and 280 nm.
7. Fraction Collection: Collect fractions containing AJICAP-BisAb as they elute from the column on the basis of the detector signals.
8. Collect the purified AJICAP-BisAb from the desalting process and confirm its formation using RP-HPLC.
9. Determine the concentration of the purified AJICAP-BisAb using a UV spectrophotometer with the A280 method. Calculate the recovery amount of antibody-thiol by multiplying the recovery volume by the determined antibody concentration.

3.6 RP-HPLC Analysis under Non-Reducing Conditions

1. Configure the HPLC system with a Sepax Proteomix RP-1000 5 μm 2.1 × 50 mm column (*see* **Note 6**).
2. Set the column temperature to 80°C and the flow rate to 0.5 mL/min. Prepare MPA and MPB.
3. Prepare samples of antibody, antibody-thiol, antibody-DBCO, and BisAb at 1 mg/mL concentration, with each sample volume set for injection at 20 μL.
4. Initiate the chromatographic run with a total duration of 22 min consisting of a 1 min isocratic hold at 30% MPB, a 15 min linear gradient from 30% to 45% MPB, a 3 min wash using 95% MPB, and a 3 min re-equilibration at 30% MPB.

RP-HPLC analysis under non-reducing conditions proved to be efficient for UV-based analysis, whereas hydrophobic interaction chromatography was unable to effectively separate DAR. Generally, UV-based analysis provides more accurate DARs than Q-TOF MS, which can be influenced by ionization efficiency [11]. RP-HPLC is a potential alternative method for analyzing complex antibody conjugates, as exemplified in this study using AJICAP-BisAb (Fig. 4).

4 Notes

1. The pH is a crucial factor for the reaction's effectiveness. It is advised to maintain the pH within the range of 4.7 to 6.0 for optimal results [12].
2. For comprehensive details, including the methodology for solid-phase peptide synthesis followed by linker installation, refer to the cited literature [6].
3. To regulate the pH, use of hydroxylamine hydrochloride salt is necessary. The application of aqueous hydroxylamine is not recommended because of its tendency to increase the pH.
4. Alternative ultrafiltration/diafiltration systems such as PendoTECH®, Pellicon®, and Vivaflow® are also suitable for this procedure.
5. In the site-specific production of cetuximab Fab-azide, the Thiobridge modification technology was used [15]. This sophisticated chemical conjugation technique facilitates the crosslinking of two thiol groups, each bearing two leaving groups, by using a specifically designed reagent. The process begins with the enzymatic digestion of the anti-EGFR antibody cetuximab [14], yielding a Fab fragment endowed with two accessible sulfhydryl groups. After this preparation, Thiobridge modification was applied sequentially, thereby incorporating an azide group into the cetuximab Fab. For detailed information, please see the references [8].
6. For insight into the specific characteristics of this analysis, please refer to [13].
7. For this analysis, mobile phase containing only FA without TFA is also compatible. However, to improve HPLC separation, TFA was included as a component of the mobile phase. It is important to note that to optimize ionization efficiency, the concentration of TFA should be maintained at less than 0.5%.
8. The exact concentration is not critical. Peptide conjugation to Lys248 showed consistent results across the tested range of concentrations [12]. The reaction was complete even at a low

antibody concentration (1 mg/mL). Even with a prolonged reaction time (>16 h), the target conversion was achieved. This antibody modification reaction was highly compatible over a wide range of reaction temperatures (4–37°C) (entries 10 and 11). This wide tolerability may allow the application of this reaction to the production of unusual mAbs with unfavorable solubility.

9. Hydrophobic interaction chromatography is also useful for this monitoring technique. For detailed information, please see the references [16].
10. For small-scale preparation, gel filtration is a viable option. For detailed information, please see the references [17].

References

1. Esfandiari A, Cassidy S, Webster RM (2022) Bispecific antibodies in oncology. Nat Rev Drug Discov 21(6):411–412. https://doi.org/10.1038/d41573-022-00040-2
2. Underwood DJ, Bettencourt J, Jawad Z (2022) The manufacturing considerations of bispecific antibodies. Expert Opin Biol Ther 22(8):1043–1065. https://doi.org/10.1080/14712598.2022.2095900
3. Yamazaki S, Matsuda Y (2022) Tag-free enzymatic modification for antibody–drug conjugate production. Chem Select 7(48). https://doi.org/10.1002/slct.202203753
4. Szijj P, Chudasama V (2021) The renaissance of chemically generated bispecific antibodies. Nat Rev Chem 5(2):78–92. https://doi.org/10.1038/s41570-020-00241-6
5. Fujii T, Matsuda Y (2023) Novel formats of antibody conjugates: recent advances in payload diversity, conjugation, and linker chemistry. Expert Opin Biol Ther 23(11):1053–1065. https://doi.org/10.1080/14712598.2023.2276873
6. Fujii T, Matsuda Y, Seki T, Shikida N, Iwai Y, Ooba Y, Takahashi K, Isokawa M, Kawaguchi S, Hatada N, Watanabe T, Takasugi R, Nakayama A, Shimbo K, Mendelsohn BA, Okuzumi T, Yamada K (2023) AJICAP second generation: improved chemical site-specific conjugation technology for antibody-drug conjugate production. Bioconjug Chem 34(4):728–738. https://doi.org/10.1021/acs.bioconjchem.3c00040
7. Watanabe T, Arashida N, Fujii T, Shikida N, Ito K, Shimbo K, Seki T, Iwai Y, Hirama R, Hatada N, Nakayama A, Okuzumi T, Matsuda Y (2024) Exo-cleavable linkers: enhanced stability and therapeutic efficacy in antibody-drug conjugates. J Med Chem 67(20):18124–18138. https://doi.org/10.1021/acs.jmedchem.4c01251
8. Fujii T, Ito K, Takahashi K, Aoki T, Takasugi R, Seki T, Iwai Y, Watanabe T, Hirama R, Tsumura R, Fuchigami H, Yasunaga M, Matsuda Y (2023) Bispecific antibodies produced via chemical site-specific conjugation technology: AJICAP second-generation. ACS Med Chem Lett 14(12):1767–1773. https://doi.org/10.1021/acsmedchemlett.3c00414
9. Szijj PA, Bahou C, Chudasama V (2018) Minireview: addressing the retro-michael instability of maleimide bioconjugates. Drug Discov Today Technol 30:27–34. https://doi.org/10.1016/j.ddtec.2018.07.002
10. Matsuda Y, Shikida N, Hatada N, Yamada K, Seki T, Nakahara Y, Endo Y, Shimbo K, Takahashi K, Nakayama A, Mendelsohn BA, Fujii T, Okuzumi T, Hirasawa S (2024) AJICAP-M: traceless affinity peptide mediated conjugation technology for site-selective antibody-drug conjugate synthesis. Org Lett 26(27):5597–5601. https://doi.org/10.1021/acs.orglett.4c00878
11. Erian A, Sherif S, Gaber H (2003) The chemistry of α-haloketones and their utility in heterocyclic synthesis. Molecules 8(11):793–865. https://doi.org/10.3390/81100793
12. Watanabe T, Fujii T, Stofleth JT, Takasugi R, Takahashi K, Matsuda Y (2023) Scale-up synthesis of site-specific antibody–drug conjugates using AJICAP second-generation technology. Org Process Res Dev 27(6):1136–1143. https://doi.org/10.1021/acs.oprd.3c00111
13. Matsuda Y, Leung M, Tawfiq Z, Fujii T, Mendelsohn BA (2021) In-situ reverse phased HPLC analysis of intact antibody-drug

conjugates. Anal Sci 37(8):1171–1176. https://doi.org/10.2116/analsci.20P424

14. Badescu G, Bryant P, Bird M, Henseleit K, Swierkosz J, Parekh V, Tommasi R, Pawlisz E, Jurlewicz K, Farys M, Camper N, Sheng X, Fisher M, Grygorash R, Kyle A, Abhilash A, Frigerio M, Edwards J, Godwin A (2014) Bridging disulfides for stable and defined antibody drug conjugates. Bioconjug Chem 25(6):1124–1136. https://doi.org/10.1021/bc500148x
15. Kuo WT, Lin WC, Chang KC, Huang JY, Yen KC, Young IC, Sun YJ, Lin FH (2015) Quantitative analysis of ligand-EGFR interactions: a platform for screening targeting molecules. PLoS One 10(2):e0116610. https://doi.org/10.1371/journal.pone.0116610
16. Fujii T, Reiling C, Quinn C, Kliman M, Mendelsohn BA, Matsuda Y (2021) Physical characteristics comparison between maytansinoid-based and auristatin-based antibody-drug conjugates. Explor Target Antitumor Ther 2(6):576–585. https://doi.org/10.37349/etat.2021.00064
17. Yamazaki S, Shikida N, Takahashi K, Matsuda Y, Inoue K, Shimbo K, Mihara Y (2021) Lipoate-acid ligase a modification of native antibody: synthesis and conjugation site analysis. Bioorg Med Chem Lett 51:128360. https://doi.org/10.1016/j.bmcl.2021.128360

Check for updates

Chapter 8

Laboratory-Scale Production of Recombinant IgG Antibodies

Takahiro Anzai and Masahiro Yasunaga

Abstract

The development of hybridoma technology by Köhler and Milstein revolutionized the field of antibody research, paving the way for the creation of monoclonal antibodies. Due to their ability to specifically bind target molecules via antigen–antibody reactions, antibodies have become a major focus in therapeutic development. However, antibodies obtained from immunized animals such as mice may cause immunogenicity or side effects when administered to humans. While humanized mice offer a solution, their high-cost limits widespread adoption. Advances in genetic engineering technology have made it possible to create chimeric antibodies or humanized antibodies and have greatly reduced immunogenicity. In this chapter, we explain in detail a laboratory-scale protocol to produce recombinant IgG antibodies at the milligram scale, which is essential for in vivo research, using transient expression in CHO cells.

Keywords Monoclonal antibody, Recombinant antibody, Antibody engineering, Animal study, Transient expression, CHO cells, Antibody production, Milligram-scale, IgG antibody

1 Introduction

The establishment of hybridoma technology by Köhler and Milstein enabled the production of monoclonal antibodies [1]. Due to their ability to specifically bind to target molecules through antigen–antibody reactions, the development of therapeutic antibodies has progressed rapidly.

Antibodies are typically produced by immunizing animals such as mice. When such host-derived antibodies are administered to humans, they are recognized as foreign substances, leading to immunogenicity and potentially causing adverse effects [2]. Immunizing humans to produce human-derived antibodies is ethically impossible. Although technologies such as the development of humanized mice have been advanced [3, 4], these mice are expensive and not yet a common technology.

To overcome this challenge, genetic engineering techniques have been employed to modify antibodies. These modifications include the creation of chimeric antibodies, where the variable regions of

Masahiro Yasunaga (ed.), *Bispecific Antibodies*, Methods in Molecular Biology, vol. 2997,
https://doi.org/10.1007/978-1-0716-5037-0_8,

© The Author(s), under exclusive license to Springer Science+Business Media, LLC, part of Springer Nature 2026

host-derived antibodies are combined with the constant regions of human antibodies, and humanized antibodies, where only the complementarity-determining regions (CDRs) of host-derived antibodies are grafted onto human antibody frameworks [5, 6]. These developments have significantly reduced the immunogenicity of antibodies in humans.

Numerous therapeutic antibodies developed through these methods are already on the market, saving the lives of many patients. The global market for antibody therapeutics in 2023 was approximately 218.6 billion dollars [7]. Further market expansion is anticipated, driven not only by the development of single monoclonal antibodies but also by next-generation antibodies such as antibody–drug conjugates and bispecific antibodies [8, 9].

In laboratory-level antibody development, in vivo experiments to evaluate the pharmacokinetics and efficacy of candidate antibodies that could be developed into seeds are essential. These experiments require purified recombinant antibodies in the range of a few milligrams to several hundred milligrams to evaluate antibodies at appropriate doses and with appropriate dosing schedules [10, 11], and the authors have experienced considerable difficulties in preparing such quantities. Recently, highly productive systems like the ExpiCHO system have been developed, lowering the hurdles for recombinant antibody preparation [12]. In this chapter, we introduce a protocol for producing several milligrams to several hundred milligrams of antibodies, which are necessary for in vitro and in vivo animal studies at research facilities, based on our experience [13–16].

2 Materials

2.1 Antibody Expression Plasmid Construction

2.1.1 Reagents and Kits

1. Artificial gene synthesis fragment
2. pcDNA3.3 or pcDNA3.4 vector
3. High Fidelity DNA Polymerase
4. PCR Primers
5. PCR product purification/Gel extraction kit
6. In-Fusion HD Enzyme Premix
7. Ampicillin Sodium Salt (100 mg/mL stock solution in 70% ethanol)
8. LB Broth
9. Agar
10. Chemically Competent *E. coli* cells
11. Plasmid Miniprep Kit
12. Agarose for electrophoresis grade
13. DNA gel loading dye

14. DNA ladder/molecular weight marker
15. PCR tubes
16. 1.5 mL micro tubes
17. 15 mL conical tubes
18. Disposable dishes
19. Bacterial cell spreaders
20. Sterile, nuclease-free water

2.1.2 Equipment

1. NanoDrop Spectrophotometer
2. Centrifuge
3. Bacterial Incubator/Shaker
4. Heat block
5. Thermal Cycler
6. Microcentrifuge
7. Agarose Gel Electrophoresis System
8. UV Transilluminator
9. Vortex mixer
10. Pipettes and sterile tips

2.2 Preparation of Large-Scale Plasmid DNA for Transfection

2.2.1 Reagents and Kits

1. LB Broth
2. Ampicillin Sodium Salt (100 mg/mL stock solution in 70% ethanol)
3. Flasks
4. 15 and 50 mL conical tubes
5. 1.5 mL micro tubes
6. Plasmid Miniprep Kit
7. Plasmid Midiprep or Megaprep Kit

2.2.2 Equipment

1. NanoDrop Spectrophotometer
2. Centrifuge
3. Bacterial Incubator/Shaker
4. Vortex mixer
5. Pipettes and sterile tips

2.3 Transient Transfection of CHO Cells Using the ExpiCHO Expression System

2.3.1 Reagents and Kits

1. ExpiCHO Expression System Kit (ExpiCHO-S cells, ExpiCHO Expression Medium, ExpiFectamine CHO Reagent, ExpiFectamine CHO Enhancer, ExpiCHO Feed)
2. OptiPRO SFM
3. Trypan Blue Solution, 0.4%
4. Erlenmeyer cell culture flasks with vent cap

5. 15 and 50 mL conical tubes
6. 1.5 mL micro tubes
7. Syringes
8. 0.22 μm pore size syringe filters

2.3.2 Equipment

1. CO_2 incubator (37°C, 8% CO_2, humidity of around 95%, and 32°C, 5% CO_2, humidity of around 95%)
2. Orbital shaker (Fig. 1) (110–125 rpm)
3. Cell counter
4. Vortex mixer
5. Pipettes and sterile tips
6. Pipette controller and Serological Pipettes

2.4 Purification of Transiently Expressed Antibody from Culture Supernantant

1. Protein A or Protein G agarose
2. Econo-Pac column
3. Ammonium sulfate

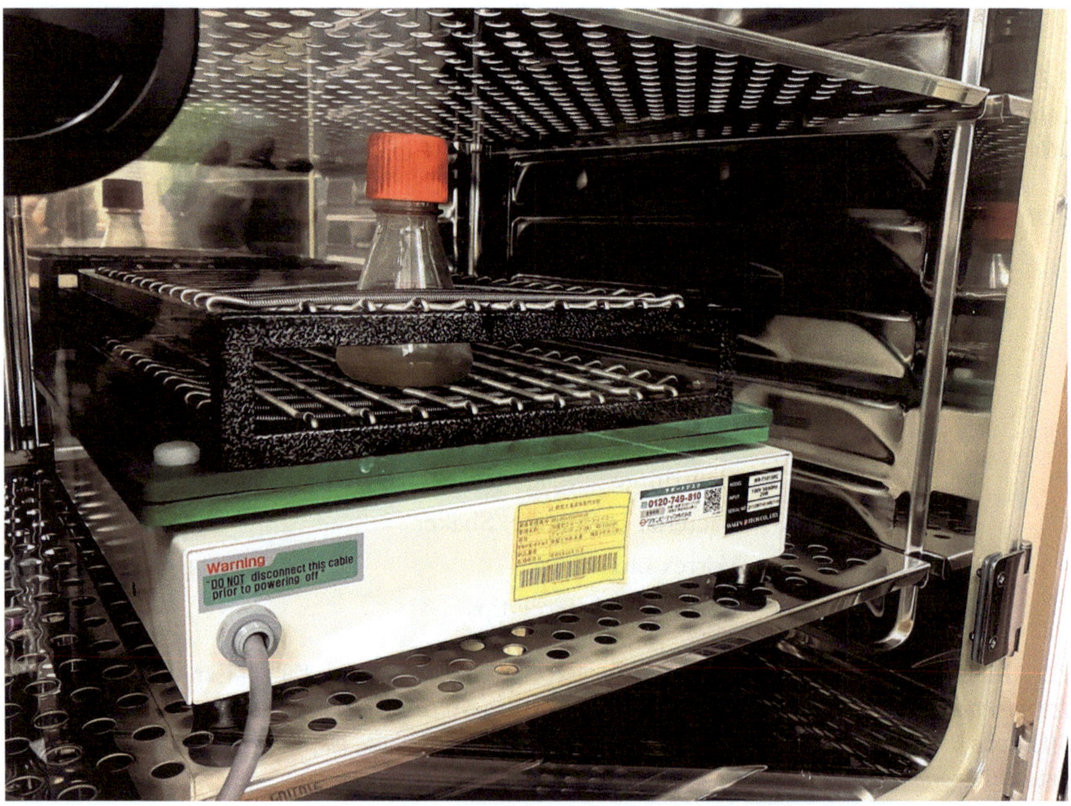

Fig. 1 An orbital shaker installed in a CO_2 incubator. The one used is from WAKEN BIETEC (WB-T101SRC). The figure shows a 125 mL flask being cultured

2.4.1 Reagents and Kits

4. 0.22 μm pore size bottle-top filter
5. Equilibration Buffer: 0.05 M Tris-HCl (pH 7.6), 0.15 M NaCl, 0.75 M ammonium sulfate
6. Elution Buffer: 0.1 M Glycine-HCl (pH 3.0)
7. Neutralization buffer: 1 M Tris-HCl (pH 8.0)
8. PBS
9. 1.5 mL micro tubes
10. 15 and 50 mL conical tubes
11. Syringes
12. 0.22 μm pore size syringe filters
13. Magnetic stir bar
14. Beaker (100–1000 mL size, depends on culture volume)
15. Sterile, nuclease-free water
16. Ultrafiltration device

2.4.2 Equipment

1. Column stand
2. NanoDrop Spectrophotometer
3. Centrifuge
4. Magnetic stir
5. Pipettes and sterile tips
6. Pipette controller and Serological Pipettes

3 Methods

The following protocol explains the production of a biosimilar of trastuzumab as an example of a recombinant antibody.

3.1 Construction of Antibody Expression Plasmids

This section details the design and construction of heavy and light chain expression vectors.

1. Obtain the amino acid sequences of the heavy and light chains of trastuzumab from an appropriate database (e.g., IMGT, DrugBank, or the original patent) (*see* **Note 1**).
2. Design artificial DNA fragments for mammalian expression based on the amino acid sequences obtained in Step 1 (*see* **Note 2**).
3. Construct an antibody expression plasmid using synthesized artificial genes (heavy and light chains) (*see* **Note 3**). If performing PCR, use a high-fidelity DNA polymerase to minimize errors. An example PCR reaction condition is: 30 cycles of 98°C for 10 s; 60°C for 5 s, 68°C for 30 s; and a final

extension of 68°C for 1 min. Adjust the annealing temperature and extension time according to the primer design and the length of the target sequence. The final reaction volume is typically 10 μL.

4. Use a PCR cleanup kit according to the manufacturer's protocol to purify the amplified DNA fragments.
5. Prepare the vector side. Amplify using inverse PCR. The vector side can also be prepared by performing restriction enzyme treatment.
6. Perform agarose gel electrophoresis, cut out the amplified bands, and purify using a gel extraction kit according to the manufacturer's protocol (*see* **Note 4**).
7. Perform In-Fusion reaction. For example, mix 100 ng of each insert (heavy or light chain) and 100 ng of linearized vector. Add 1 μL of 5× In-Fusion HD Enzyme Premix to each reaction for a 5 μL total reaction volume. Incubate at 50°C for 15 minutes using a thermal cycler (*see* **Note 5**).
8. Transform the ligation product (or In-Fusion reaction mixture) into chemically competent *E. coli* cells (e.g., DH5α or a similar strain suitable for cloning).
9. Plate the transformed cells on 1.5% LB agar plates containing 100 μg/mL ampicillin for selection of transformants and incubate at 37°C overnight.
10. Select several colonies and plant them in a separate tube containing 3 mL of LB medium with 100 μg/mL ampicillin as an antibiotic. Incubate with shaking (e.g., 200–250 rpm) these tubes at 37°C for approximately 16 h (*see* **Note 6**).
11. Extract plasmid DNA from each culture using a commercially available miniprep kit, following the manufacturer's instructions.
12. Verify the sequence of each plasmid (heavy and light chain) by DNA sequencing. Ensure that the sequence from the start of the signal peptide to the stop codon is correct and free of mutations.

3.2 Preparation of Large-Scale Plasmid DNA for Transfection

This protocol requires a total of 1 μg of plasmid DNA (combined heavy and light chain plasmids) per 1 mL of CHO cell culture. For example, a 200 mL transfection will require 200 μg of total plasmid DNA.

1. Amplify the transformed *E. coli* that contains the plasmid (sequence-verified in Sect. 3.1) by culturing (*see* **Note 7**).
2. Prepare large quantities of each plasmid (heavy and light chain) using a commercially available midiprep or maxiprep kit, following the manufacturer's instructions (*see* **Note 8**).

3. Determine the concentration and purity of the plasmid DNA using a spectrophotometer.

3.3 Transient Transfection of CHO Cells Using the ExpiCHO Expression System

This protocol utilizes the Thermo Fisher Scientific ExpiCHO Expression System (*see* **Note 9**). Here, we describe a modified protocol for a 200 mL transfection (*see* **Note 10**).

1. The day before transfection, seed ExpiCHO-S cells at a density of 3–4 × 10^6 viable cells/mL in ExpiCHO Expression Medium (*see* **Note 11**). Perform trypan blue staining and use a cell counter to check the number of cells and cell viability.
2. On the day of transfection, dilute the cells to a density of 6 × 10^6 viable cells/mL in 200 mL of ExpiCHO Expression Medium pre-warmed to 37°C.
3. Prepare the plasmid DNA mix. For a 200 mL transfection, a total of 200 μg of plasmid DNA will be needed (*see* **Note 12**). This typically consists of an equimolar ratio of heavy and light chain plasmids, but optimization may be required (*see* **Note 13**).
4. Prepare the DNA-OptiPRO SFM complex. In a sterile 50 mL conical tube, add the 200 μg of plasmid DNA mixture. Then, add OptiPRO SFM to a final volume of 8 mL. Gently invert the tube several times to mix.
5. Prepare the ExpiFectamine CHO Reagent-OptiPRO SFM mixture. Gently invert the ExpiFectamine CHO Reagent vial 4–5 times to mix. In a separate sterile 15 mL conical tube, add 640 μL of ExpiFectamine CHO Reagent to 7.4 mL of OptiPRO SFM. Gently invert the tube several times to mix.
6. Combine the complexes. Add the ExpiFectamine CHO Reagent-OptiPRO SFM mixture (from step 5) to the DNA-OptiPRO SFM mixture (from step 4) in the 50 mL tube. Invert the tube immediately to mix and create the ExpiFectamine CHO reagent/plasmid DNA complex solution.
7. Incubate the ExpiFectamine CHO reagent/plasmid DNA complex solution at room temperature for 5 min.
8. Add the ExpiFectamine CHO reagent/plasmid DNA complex solution to the ExpiCHO-S cells in the flask (*see* **Note 14**).
9. Incubate the flask in a humidified 37°C incubator with 8% CO_2 on an orbital shaker platform rotating at 110–125 rpm for 18 to 22 h (*see* **Note 15**).
10. After 18 to 22 h culture, add 1.2 mL of ExpiCHO Enhancer and 32 mL of ExpiCHO Feed per 200 mL of culture medium.
11. Incubate the flask in a humidified 32°C incubator with 5% CO_2 on an orbital shaker platform rotating at 110–125 rpm for further 4 days.

12. On day 5 post-transfection, add another 32 mL of ExpiCHO Feed per 200 mL of culture medium.
13. Incubate the flask in a humidified 32°C incubator with 5% CO_2 on an orbital shaker platform rotating at 110–125 rpm for further 6–9 days.
14. Harvest the culture supernatant containing the secreted antibody typically 14 days post-transfection, depending on the specific antibody and culture performance (*see* **Note 16**).
15. Monitor antibody titer and cell viability to determine the optimal harvest time (*see* **Note 17**).

3.4 Purification of Transiently Expressed Antibody from Culture Supernatant

This section describes the purification of the antibody from the harvested culture supernatant using Protein A affinity chromatography.

1. Transfer the harvested culture supernatant to centrifuge tubes and centrifuge at 6000 rpm for 10 min to remove cells and debris.
2. Transfer the clarified supernatant to a beaker and stir with a magnetic stirrer. Add ammonium sulfate to the supernatant at a ratio of 1/10 of the supernatant volume (w/v).
3. Filter the solution using a bottle-top filter for clarification (*see* **Note 18**).
4. Pack the Protein A resin into an Econo-Pac column (*see* **Note 19**).
5. After washing the resin with 50 mL water, equilibrate the column by flowing through 20 mL of equilibration buffer twice.
6. Load the filtered culture supernatant onto the equilibrated column. Collect the flow-through in a separate container in case the antibody does not bind to the column.
7. Wash the column twice with 20 mL of equilibration buffer to remove non-specifically bound proteins.
8. After washing, add 7.2 mL of elution buffer to the column and collect the eluate in a 15 mL tube containing 800 μL of neutralization buffer. Once all the elution buffer has passed through the column, immediately cap the 15 mL tube and invert it several times to mix and neutralize the solution.
9. Measure the absorbance at 280 nm using a NanoDrop spectrophotometer. If protein is detected, repeat the elution step by adding more elution buffer to the column. Stop collecting the eluate when the absorbance reaches a value with two decimal places.
10. Pool all collected fractions containing the antibody, concentrate, and buffer-exchange to PBS using ultrafiltration device (*see* **Note 20**).

11. Antibodies that can be cryopreserved should be frozen, and those that cannot should be stored at 4°C.
12. The prepared antibodies are used for the experiments requiring the antibodies, respectively (*see* **Note 21**)

4 Notes

1. The amino acid sequence of the antibody can be obtained from public databases or published literature. For example, sequence information for the anti-HER2 antibody trastuzumab can be found in the DrugBank database (https://go.drugbank.com/drugs/DB00072). It is recommended to cross-reference the sequence with multiple sources when possible, to ensure accuracy.
2. Based on the amino acid sequence obtained in FASTA format, design an artificial DNA fragment. For gene synthesis, a service such as gBlocks Gene Fragments from Integrated DNA Technologies (IDT) or a vendor that provides a similar service can be used. To ensure proper secretion and expression, add an appropriate signal peptide sequence to the N-terminus of both the heavy and light chains. Effective signal peptide sequences can be derived from published literature. For example, the following signal peptides optimized for heavy and light chains in CHO cell expression, as reported [17] can be used:

   Heavy Chain (H5): MDWTWRFLFVVAAATGVQS (ATGGACTGGACCTGGAGGTTCCTCTTTGTGGTGGCAGCAGCTACAGGTGTCCAGTCC)

   Light Chain (L1): MDMRVPAQLLGLLLLWLSGARC (ATGGACATGAGGGTCCCTGCTCAGCTCCTGGGGCTCCTGCTGCTCTGGCTCTCAGGTGCCAGATGT)

   These sequences are added to the 5′ end of the respective heavy and light chain coding sequences. When converting amino acid sequences to nucleotide sequences, it is recommended to perform codon optimization for the host organism (in this case, Cricetulus griseus (CHO) cells) to enhance expression levels. Several companies that offer gene synthesis services have released web tools for codon optimization, and it is a good idea to use these tools at the same time when ordering gene synthesis. For example, codon optimization is possible at the following site. https://en.vectorbuilder.com/tool/codon-optimization.html
3. If 15 bp of DNA sequence is added to the vector side to be inserted when designing the artificial gene, it can be constructed without performing PCR in this step.

4. When amplifying a vector using PCR, if the vector to be used as a template and the amplified product are not separated by gel extraction, it will be difficult to isolate the plasmid with the inserted insert when transforming.
5. In-Fusion cloning is a convenient method, but other cloning strategies, such as restriction enzyme-based cloning are also acceptable. If the bands can be confirmed by agarose gel electrophoresis, the concentration of the sample after purification has not been measured and adjusted. In this case, the reaction has been carried out using 0.3 μL of 5× In-Fusion HD Enzyme Premix, 0.6 μL of insert, and 0.6 μL of vector.
6. By performing **Note 4**, most colonies will be derived from plasmids with the insert correctly inserted. Therefore, it is usually sufficient to pick up a maximum of three colonies.
7. If a replica of the colony picked up in step 3.1.8 is kept, the trouble of performing a new transformation can be avoided.
8. Since Midi prep and Maxi prep take a long time, they can also be prepared by doing several Mini preps for convenience. For example, 20 μg of plasmid can be obtained per column from a 5 mL culture, so 20 μg of plasmid can be obtained by using 10 cultures and 10 Mini prep columns.
9. We have considered using several different kits, but we use this system because it has the most stable antibody production.
10. For larger-scale transfection, we recommend using a 1 L flask containing 200 mL of cell culture medium or two 500 mL flasks containing 100 mL of cell culture medium to ensure optimal mixing and gas exchange.
11. Working stocks of ExpiCHO-S cells are prepared at passage 3 and stored at 1×10^7 cells/mL. Since the doubling time is 16 h, the required number of cells should be calculated in advance, and an experimental schedule for transfection should be established. For example, start the culture with one vial of stock in 30 mL of ExpiCHO Medium in a 125 mL Erlenmeyer flask. After 96 h, the cell density will reach 9.6×10^6 cells/mL in 30 mL. Transfer the cells to a 500 mL flask and add 60 mL of ExpiCHO Medium, resulting in 90 mL of cell culture at 3.2×10^6 cells/mL. After an additional 16 h of culture, the cell density will reach 6.4×10^6 cells/mL. Then, add 90 mL of ExpiCHO Medium to this, resulting in 180 mL of cell culture at 3.2×10^6 cells/mL. Use this cell culture for Step 3.3.1. This means that it takes approximately 6 days from thawing a stock cell until the cells are ready for transfection.
12. The manufacturer's instructions state that the concentration of plasmid DNA should be in the range of 0.5–0.8 μg/mL,

but since antibodies have been obtained at a concentration of around 0.2 μg/mL, we do not usually perform a concentration operation for the plasmid.

13. The manufacturer's instructions state that the ExpiCHO system recommends a 1:1 ratio of heavy and light chain plasmids.
14. When dripping, shake the flask while dripping a few drops at a time to avoid locally high concentrations.
15. The shaking speed is 125 rpm for flasks up to 250 mL and around 110 rpm for those 500 mL or larger.
16. Fig. 2 shows culture volume and antibody production of multiple antibody clones. Even with the expression of the same antibody clone, significant variability in antibody yield can be observed. As the signal sequences are identical, we speculate that variations in cell condition or plasmid quality may contribute to the observed differences in antibody production levels.
17. Fig. 3 shows examples of cell viability, cell count, and antibody production from two independent trials using the same IgG. There does not appear to be a perfect correlation between cell viability and antibody production.
18. When using a diatomaceous earth-derived filter aid such as Celpure (Sigma-Aldrich), it is possible to filter efficiently without clogging.

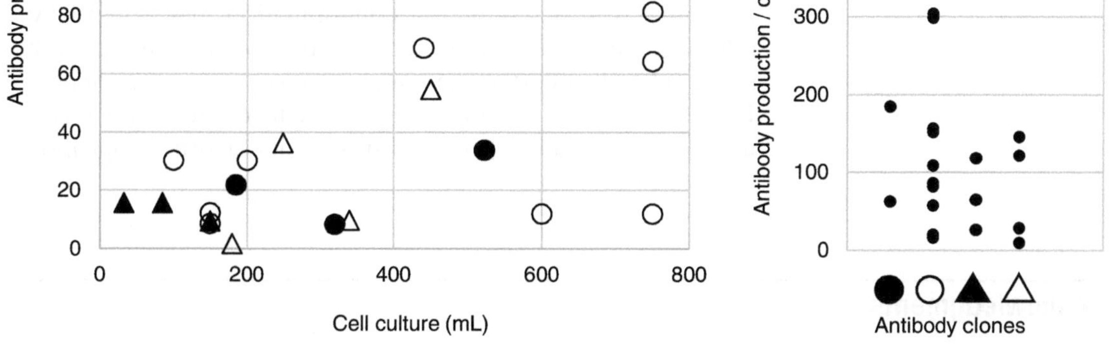

Fig. 2 Culture volume and antibody production of multiple antibody clones. (**a**) Antibody yield after Protein A purification is plotted against the culture volume for four different antibody clones. Black circles, white circles, black triangles, and white triangles represent different antibody clones. (**b**) Antibody production per culture volume for the four different antibody clones is plotted. The data shown in (**b**) were calculated from the results shown in (**a**)

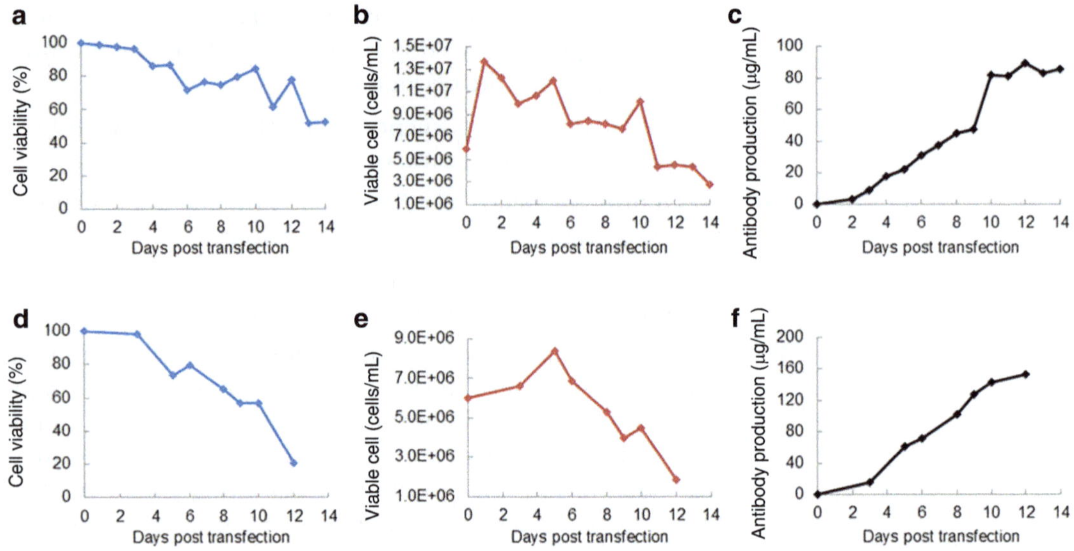

Fig. 3 Monitoring of antibody production in two independent trials. Panels (**a**–**c**) depict the results of the first trial, while panels (**d**–**f**) represent the second trial. (**a**, **d**) Cell viability over the course of the culture. (**b**, **e**) Viable cell density over time. (**c**, **f**) Quantification of antibody concentration in the culture supernatant. Antibody levels were determined using a sandwich ELISA with anti-human IgG antibodies as the capture and detection antibodies

19. The amount of resin follows the amount of antibody produced. For example, in the case of rProtein A Sepharose Fast Flow (Cytiva), 50 mg of human IgG antibody binds per mL of resin, so to purify 50 mg of human IgG, 2 mL of resin is used with a margin.
20. If the amount of antibody produced is high, it is possible to prepare a highly pure antibody by Protein A purification alone. If a higher purity antibody is required, further purification is carried out using gel filtration chromatography.
21. In animal studies, endotoxin contamination may potentially influence the results. However, we have not encountered any issues related to endotoxin contamination in our experience. If necessary, it is recommended to reduce endotoxin levels using commercially available removal kits or other appropriate methods.

Acknowledgments

The authors thank members of the Yasunaga laboratory for helpful discussion, especially Dr. S. Saijou and Dr. S. Hanaoka for valuable comments in the early phase of this study, and M. Shimada for secretarial support. This work was supported in part by research grants from Mukai Science and Technology Foundation, and the Precise Measurement Technology Promotion Foundation to T.A. and the National Cancer Center Research and Development Fund from the National Cancer Center, Japan to T.A. and M.Y.

References

1. Kohler G, Milstein C (1975) Continuous cultures of fused cells secreting antibody of predefined specificity. Nature 256:495–497
2. Harding FA, Stickler MM, Razo J, DuBridge RB (2010) The immunogenicity of humanized and fully human antibodies: residual immunogenicity resides in the CDR regions. Mabs 2(3):256–265
3. Brüggemann M, Neuberger MS (1996) Strategies for expressing human antibody repertoires in transgenic mice. Immunol Today 17(8):391–397
4. Lee EC, Liang Q, Ali H, Bayliss L, Beasley A, Bloomfield-Gerdes T, Bonoli L, Brown R, Campbell J, Carpenter A, Chalk S, Davis A, England N, Fane-Dremucheva A, Franz B, Germaschewski V, Holmes H, Holmes S, Kirby I, Kosmac M, Legent A, Lui H, Manin A, O'Leary S, Paterson J, Sciarrillo R, Speak A, Spensberger D, Tuffery L, Waddell N, Wang W, Wells S, Wong V, Wood A, Owen MJ, Friedrich GA, Bradley A (2014) Complete humanization of the mouse immunoglobulin loci enables efficient therapeutic antibody discovery. Nat Biotechnol 32:356–363
5. Riechmann L, Clark M, Waldmann H, Winter G (1988) Reshaping human antibodies for therapy. Nature 332:323–327
6. Jones PT, Dear PH, Foote J, Neuberger MS, Winter G (1986) Replacing the complementarity-determining regions in a human antibody with those from a mouse. Nature 321:522–525
7. BCC Research (2024). Antibody Drugs: Technologies and Global Markets, Research and Markets. [https://www.researchandmarkets.com/reports/5464887/]
8. Beck A, Goetsch L, Dumontet C, Corvaïa N (2017) Strategies and challenges for the next generation of antibody-drug conjugates. Nat Rev Drug Discov 16:315–337
9. Spiess C, Zhai Q, Carter PJ, Zha J (2015) Alternative molecular formats and therapeutic applications for bispecific antibodies. Mol Immunol 67(2 Pt A):95–106
10. Chapman K, Pullen N, Coney L, Dempster M, Andrews L, Bajramovic J, Baldrick P, Buckley L, Jacobs A, Hale G, Green C, Ragan I, Robinson V (2009) Preclinical development of monoclonal antibodies: considerations for the use of non-human primates. Mabs 1(5):505–516
11. Vazquez-Lombardi R, Nevoltris D, Luthra A, Schofield P, Zimmermann C, Christ D (2018) Transient expression of human antibodies in mammalian cells. Nat Protoc 13(1):99–117
12. Jain NK, Barkowski-Clark S, Altman R, Johnson K, Sun F, Zmuda J, Liu CY, Kita A, Schulz R, Neill A, Ballinger R, Patel R, Liu J, Mpanda A, Huta B, Chiou H, Voegtli W, Panavas T (2017) A high density CHO-S transient transfection system: Comparison of ExpiCHO and Expi293. Protein Expr Purif 134:40–46
13. Yasunaga M, Saijou S, Hanaoka S, Anzai T, Tsumura R, Matsumura Y (2019) Significant antitumor effect of an antibody against TMEM180, a new colorectal cancer-specific molecule. Cancer Sci 110(2):761–770
14. Tsumura R, Anzai T, Manabe S, Takashima H, Koga Y, Yasunaga M, Matsumura Y (2021) Antitumor effect of humanized anti-tissue factor antibody-drug conjugate in a model of peritoneal disseminated pancreatic cancer. Oncol Rep 45(1):329–336
15. Anzai T, Saijou S, Takashima H, Hara M, Hanaoka S, Matsumura Y, Yasunaga M (2022) Identification of CD73 as the antigen of an antigen unknown monoclonal antibody established by exosome immunization, and its antibody-drug conjugate exerts an antitumor effect on glioblastoma cell lines. Pharmaceuticals 15(7):837
16. Tsumura R, Anzai T, Koga Y, Takashima H, Matsumura Y, Yasunaga M (2024) Anti-tissue factor antibody conjugated with monomethylauristatin E or deruxtecan in pancreatic cancer models. Cancer Sci 115(4):1234–1244
17. Haryadi R, Ho S, Kok YJ, Pu HX, Zheng L, Pereira NA, Li B, Bi X, Goh LT, Yang Y, Song Z (2015) Optimization of Heavy Chain and Light Chain Signal Peptides for High Level Expression of Therapeutic Antibodies in CHO Cells. PLoS One 10(2):e0116878

secretarial support. This work was supported in part by research grants from Mukai Science and Technology Foundation, and the Precise Measurement Technology Promotion Foundation to T.A. and the National Cancer Center Research and Development Fund from the National Cancer Center Japan to T.A. and M.O.

References

[illegible]

Part II

Bioassays

Check for updates

Chapter 9

Visualization of Immunological Synapse Formation by Intercellular Cross-Linking of T Cell-Dependent Bispecific Antibodies Using Fluorescence Microscopy

Rikuto Nakamura, Ryo Tsumura, Takahiro Anzai, Ryutaro Asano and Masahiro Yasunaga

Abstract

Immunocytochemistry is a common method to visually detect the localization of target proteins in cells or tissues using fluorescent dyes conjugated to antibodies. Especially in the field of cancer therapeutics, it is used to visualize and analyze the localization of target proteins in cancer cells and their interactions with other immune cells. In this chapter, we describe a protocol to detect how T cell-dependent bispecific antibodies (TDBs) cross-link cancer cells and T cells. We focus on the adhesion molecules involved in the cross-linking between cells by TDBs and provide a quick and easy method for visualization.

Keywords Immunological synapse, Bispecific antibody, Immunocytochemistry, Microscopy

1 Introduction

Immunocytochemistry is an established and widely utilized technique in biomedical research for visualizing and quantifying the localization of proteins within cells and tissues [1–3]. This method leverages reagents conjugated with fluorescent dyes that specifically bind to target proteins, enabling the detection and analysis of protein expression [4]. Since its introduction, fluorescent immunostaining has evolved significantly, incorporating advances such as multiplex labeling, which allows the simultaneous detection of multiple target proteins within a single sample [5, 6]. The technique serves as a foundational tool across diverse research domains, ranging from fundamental cell biology to clinical diagnostics [6–8].

In oncology research, understanding the localization of target proteins within cancer cells and the tumor microenvironment, as well as their interactions with immune cells, is crucial [9–11]. Fluorescent immunostaining is frequently employed in these contexts to elucidate

Masahiro Yasunaga (ed.), *Bispecific Antibodies*, Methods in Molecular Biology, vol. 2997,
https://doi.org/10.1007/978-1-0716-5037-0_9,
© The Author(s), under exclusive license to Springer Science+Business Media, LLC, part of Springer Nature 2026

the complex interactions within tumors, providing insights that are essential for the development of targeted therapies [12, 13].

T cell-dependent bispecific antibodies (TDBs) represent an innovative class of therapeutic antibodies designed to simultaneously bind antigens on T cells and cancer cells [14–16]. By facilitating the formation of immunological synapses (IS), which are specialized adhesion structures between T cells and cancer cells, TDBs enhance the cytotoxic activity of T cells against tumor cells [17–20]. Despite the therapeutic potential of TDBs, the precise molecular mechanisms governing their function remain incompletely understood [21, 22].

Fluorescent immunostaining provides a powerful tool for visualizing TDB-mediated interactions at the cellular level, offering insights into the dynamic processes underlying their cytotoxicity [13]. This chapter details a fluorescent immunostaining protocol designed to detect IS formation mediated by TDBs. The initial steps involve the culture of target cancer cells expressing specific antigens, such as EGFR, and the differentiation of peripheral blood mononuclear cells (PBMCs). Subsequent co-culture of T cells, cancer cells, and TDBs within culture slides is conducted to approximate the tumor microenvironment in vitro. After fixation and permeabilization of the cells, which serve as preparatory steps for fluorescent immunostaining, nonspecific binding sites are blocked using a blocking buffer containing bovine serum albumin (BSA). Sequential application of primary antibodies specific to the target proteins, followed by secondary antibodies conjugated with fluorescent dyes, allows for the accurate evaluation of protein localization.

2 Materials

2.1 Cells

1. Cancer cell expressing antigen for interest (in this case EGFR)
2. Peripheral Blood Mononuclear Cells (PBMCs)

2.2 Cell Culture

1. Thermostatic bath (37°C)
2. CO_2 humidified incubator (37°C, 5% CO_2)
3. Culture plate (6 wells, 100 mm dish)
4. RPMI-1640: 10% fetal bovine serum (FBS), 1% penicillin-streptomycin (*see* **Note 1**)
5. Human anti-CD3 antibody (OKT-3)
6. Human interleukin-2 (IL-2)
7. 0.25% Trypsin-EDTA
8. 0.5% Trypan Blue solution
9. Phosphate buffered saline (PBS)

2.3 Labware for Fluorescent Immunostaining

1. Culture plate (8-well culture slide)
2. Cover slide
3. 4% paraformaldehyde (PFA)
4. Blocking Buffer: 3% BSA (*see* **Note 2**)
5. 0.5% Triton-X reagent
6. Aluminum foil for light shielding
7. Fluoromount-G
8. Fluorescence microscope and appropriate filters

2.4 Reagents for Fluorescent Immunostaining

All reagents for fluorescent immunostaining should be diluted in PBS.

1. Rabbit anti-human EGFR antibody (Abcam, #E235)
2. Rat anti-human CD3 antibody (Abcam, #ab11089)
3. Goat anti-rabbit Alexa Fluor488 antibody (Thermo Fisher Scientific, #A11008)
4. Goat anti-rat Alexa Flour647 antibody (Thermo Fisher Scientific, #A21247)
5. Acti-stainTM^{555} phalloidin (Cytoskeleton, #PHDH1-A)
6. DAPI

3 Methods

3.1 Preparation for Cancer Cell

1. Cultivate the cancer cells of interest in 10 cm cell culture plates to approximately 80% confluence (*see* **Note 3**).
2. Aspirate the medium from the 100 mm cell culture plate and gently add 10 mL of PBS.
3. Aspirate PBS from the plate and wash by adding 10 mL of PBS again.
4. After aspirating PBS from the plate, add 3 mL of trypsin and shake gently so that the trypsin spreads over the entire plate.
5. Place the plate in an incubator (37°C, 5% CO_2) and wait 5–10 min for the cancer cells to detach from the plate (*see* **Note 4**).
6. Add 2 mL of RPMI-1640 to the plate and completely detach the cells from the plate by pipetting to make a cell suspension (*see* **Note 5**).
7. Collect the cell suspension in 15 mL conical centrifuge tubes.
8. Add 5 mL of medium to the plate again, carefully resuspend the cells again, and place them in a 15 mL conical centrifuge tube (*see* **Note 6**).

9. Centrifuge the 15 mL conical centrifuge tube containing the cell suspension (1200 rpm, 4°C, 3 min).
10. Check that the cells have formed a pellet and remove the solution except for the cell pellet by aspiration.
11. Add 5 mL of medium to the cell pellet and suspend well by pipetting to bring the solution to a uniform cell concentration (*see* **Note 7**).
12. Place 50 μL of the cell suspension in a well of a 96-well plate, add 50 μL of trypan blue, and mix by pipetting.
13. The solution is placed in a hemocytometer and the number of live and dead cells is determined (*see* **Note 8**).
14. Seed the cancer cells in 10 cm dishes at approximately 1.0×10^6 cells/well and incubate for 3–4 days.
15. After that, pass the cells every 3–4 days until the desired number of cancer cells is obtained.

3.2 Preparation for Cytotoxic T Cell

1. OKT-3 diluted to 1 μg/mL in PBS is coated at 1 mL/well in 3 wells of a 6-well culture plate and incubated for 2 h.
2. Suspend PBMCs (1.0×10^7 cells/tube) in RPMI and collect them in 15 mL conical centrifuge tubes.
3. Centrifuge (2000 rpm, 4°C, 5 min) to remove the supernatant.
4. Remove the 6-well culture plate coated with OKT-3 from the incubator and remove all the coating liquid.
5. Add 12 mL of medium to a pellet of PBMCs in a 15 mL conical centrifuge tube and suspend well by pipetting to make the solution a uniform cell concentration.
6. Add 4 mL/well of PBMC suspension to the 6-well culture plates.
7. Add IL-2 to 100 IU/mL and incubate for 3–4 days at 37°C 5% CO_2 (*see* **Note 9**).
8. T cells are collected in 15 mL conical centrifuge tubes and centrifuged (2000 rpm, 4°C, 5 min) (*see* **Note 10**).
9. Check that the T cells have formed a pellet and remove the solution except for the cell pellet by aspiration.
10. After this, count the number of cells and viable cells in the same way as in 3.1.
11. Add Medium and IL-2 and pass on (*see* **Note 11**).
12. Maintain a cell concentration of 1.0×10^5–1.0×10^7 cells/mL.

3.3 Co-Culture of Cancer Cell, T Cell, and TDB

1. T cells are collected in 15 mL conical centrifuge tubes and centrifuged (2000 rpm, 4°C, 5 min).
2. Check that the T cells have formed a pellet and remove the solution except for the cell pellet by aspiration.

3. Add 5 mL of Medium to the cell pellet and suspend well by pipetting to bring the solution to a uniform cell concentration.
4. Place 50 μL of the cell suspension in a well of a 96-well plate, add 50 μL of Trypan blue, and mix by pipetting.
5. The solution is placed in a hemocytometer to determine the number of live and dead cells; T cells are resuspended with medium again to a concentration of 3.0×10^6 cells/mL.
6. TDB is prepared by adding medium to twice the target concentration (*see* **Note 12**).
7. Completely remove the medium from the 8-well culture slide using an aspirator.
8. Add 50 μL (= 1.5×10^5 cells/well, E:T=10:1) of T cell suspension to the 8-well culture slide.
9. Incubate for 1–2.5 h at 37°C in 5% CO_2.

3.4 Fluorescent Immunostaining

1. Remove the 8-well culture slide from the incubator, add 700 μL of 4% paraformaldehyde, and leave at room temperature for 15 min.
2. Gently remove the solution from the inside of the well and wash the well with PBS three times.
3. Add an appropriate amount of blocking buffer to the well and rinse it, then add 200 μL of blocking buffer and leave it at room temperature for 45 min.
4. Wash the well three times with PBS, then add 200 μL of 0.5% Triton-X solution and leave at room temperature for 5 min.
5. After one wash with PBS, add rat anti-CD3 antibody at 1–5 μ g/200 μL/well and react at 4°C overnight (O/N).
6. After three washes with PBS, add anti-rat AlexaFluor 647 at a concentration of 1–5 μg/200 μL/well and react at 4°C for 45 min.
7. After three washes with PBS, add rabbit anti-EGFR antibody at a concentration of 1–5 μg/200 μL/well and react at 4°C for 45 min.
8. After washing three times with PBS, add anti-rabbit AlexaFluor 488 at a concentration of 1–5 μg/200 μL/well and react at 4°C for 45 min (*see* **Note 13**).
9. Prepare a working stock of Acti-stain™ 555 phalloidin, add it at 200 μL/well, and react for 1 h at 4°C.
10. Wash three times with PBS, add DAPI, and react for 5 min at room temperature.
11. Wash the plates three times with PBS, dry well, add a few drops of Fluoromount-G, and seal with a cover glass.

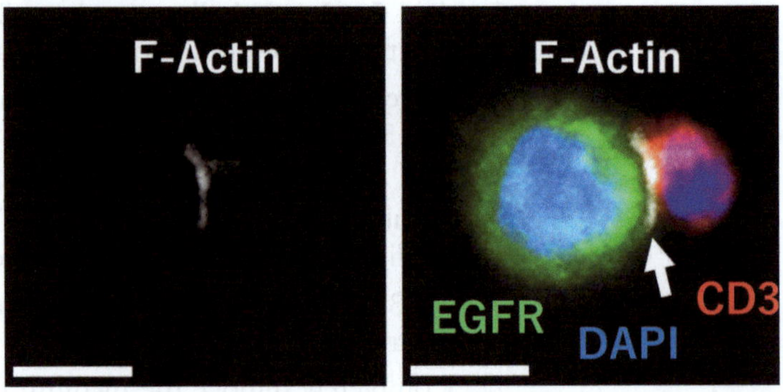

Fig. 1 IS formation by TDB. Staining of Fluorescent immunostaining of co-culture models of cancer cells, T cells, and TDB for F-Actin (white), EGFR (green), CD3 (red), and DAPI (blue). Scale bar: 10 μm

12. The plate is light-shielded with aluminum foil and dried O/N in a draft before measurement under a fluorescence microscope (Fig. 1).
13. After the fluorescence microscopy measurements is completed, store the plate in dark at 4°C.

4 Notes

1. Before preparation, RPMI-1640 is stored at 4°C, FBS and SPF at −20°C. FBS is placed in a refrigerator at 4°C O/N, dissolved and deactivated in a thermostatic bath at 56°C for 30 min before use. FBS is filtered and placed in a thermostatic bath at 37°C before use; SPF is placed in a thermostatic bath at 37°C before use. The prepared RPMI-1640 should be stocked at 4°C and warmed in a 37°C thermostatic bath 30 min before use.
2. Blocking buffer should be prepared each time it is used: PBS in a 15 mL tube, BSA in a 15 mL tube, and mix well by inverting or other means. A 5% skim milk solution diluted in PBS can also be used as a blocking buffer.
3. Use when the cell density is 80% confluent. Any further culture may not only decrease the cell growth rate but may also adversely affect the growth of the cells after passaging.
4. Trypsin should be reacted for a minimum reaction time, as prolonged reaction times of several tens of minutes may damage the cells, resulting in poor viability.
5. Tilt the plate slightly to encourage the cells to detach from the plate by pouring the solution over them. However, be careful not to tilt the plate too much, as the solution can spill easily.

6. To determine how many times medium can be applied to peel off the cells, it is convenient to observe the adherent cells under a microscope each time medium is applied. In most cases, almost all cells can be recovered after 3–4 times of this procedure.
7. Pipet the cells into a uniform suspension by pipetting 5–10 times, using a sharp tip if possible.
8. After mixing the trypan blue with the cell suspension, measure as quickly as possible. In the case of T cells, T cell viability is reduced by 10–20% when the cells are left in trypan blue for about 10 min.
9. T cells reach a viability of 70% in about 4 days and 80–90% in about 10 days. PBMCs contain cells other than T cells. Because they are floating cells, dead cells cannot be removed. Therefore, the survival rate of PBMCs and differentiated CD8+ T cells is never 100%, even if the number of days of culture is appropriate. In addition, since CD8+ T cells that have been cultured for 40 days will be senescent, it is advisable to set the number of days of culture in advance when using them in experiments. The survival rate is relatively good when CD8+ T cells are used on the 8th to 14th day after the start of culture.
10. As T cells are floating cells, trypsin is not necessary, and they can be collected directly.
11. To induce differentiation of PBMCs into CD8+ T cells, IL-2 must be added.
12. TDB is later mixed with the T cell suspension at a ratio of 1:1, so the concentration must be twice the target concentration at the preparation stage. When adding the two, add T cells and TDB in that order and mix gently. Antibody binding to the antigen proceeds more rapidly, and intercellular cross-linking is considered to be the rate-limiting step [19].
13. Phalloidin is dissolved in 500 μL of 100% methanol and stored at 4°C in a light-shielded tube. When used, 1.4 μL of phalloidin is diluted with 200 μL of PBS to make a working stock. Since Pphalloidin is a poison, handle it carefully in a safety cabinet whenever possible.

Acknowledgments

The authors thank the members of the Yasunaga laboratory for their helpful discussions and M. Shimada for secretarial support. This research was supported by AMED under Grant Number JP21ak0101182 to M.Y. and the National Cancer Center Research and Development Fund (Number 23-A-08) to M.Y.

References

1. Suzuki T, Matsuzaki T, Hagiwara H et al (2007) Recent advances in fluorescent labeling techniques for fluorescence microscopy. Acta Histochem Cytochem 40:131–137
2. Im K, Mareninov S, Diaz MFP et al (2019) An introduction to performing immunofluorescence staining. Methods Mol Biol 1897:299–311
3. Cogliati B, Maes M, Pereira IVA et al (2016) Immunohisto- and cytochemistry analysis of connexins. Methods Mol Biol 1437:55–70
4. Coons AH, Creech HJ, Jones RN (1941) Immunological properties of an antibody containing a fluorescent group. Proc Soc Exp Biol Med 47:200–202
5. Hoyt CC (2021) Multiplex immunofluorescence and multispectral imaging: forming the basis of a clinical test platform for immuno-oncology. Front Mol Biosci 8:674747
6. Piña R, Santos-Díaz AI, Orta-Salazar E et al (2022) Ten approaches that improve immunostaining: a review of the latest advances for the optimization of immunofluorescence. Int J Mol Sci 23:1426
7. Fels Elliott DR, Fitzgerald RC (2013) Molecular markers for Barrett's esophagus and its progression to cancer. Curr Opin Gastroenterol 29:437–445
8. Tazawa H, Shigeyasu K, Noma K et al (2022) Tumor-targeted fluorescence labeling systems for cancer diagnosis and treatment. Cancer Sci 113:1919–1929
9. van Dam S, Baars MJD, Vercoulen Y (2022) Multiplex tissue imaging: spatial revelations in the tumor microenvironment. Cancers 14:3170
10. Raskov H, Orhan A, Christensen JP et al (2021) Cytotoxic CD8+ T cells in cancer and cancer immunotherapy. Br J Cancer 124:359–367
11. Liu S, Galat V, Galat Y et al (2021) NK cell-based cancer immunotherapy: from basic biology to clinical development. J Hematol Oncol 14:7
12. Steele NG, Carpenter ES, Kemp SB et al (2020) Multimodal mapping of the tumor and peripheral blood immune landscape in human pancreatic cancer. Nat Cancer 1:1097–1112
13. Kamakura D, Asano R, Kawai H et al (2021) Mechanism of action of a T cell-dependent bispecific antibody as a breakthrough immunotherapy against refractory colorectal cancer with an oncogenic mutation. Cancer Immunol Immunother 70:177–188
14. Trabolsi A, Arumov A, Schatz JH (2024) Bispecific antibodies and CAR-T cells: dueling immunotherapies for large B-cell lymphomas. Blood. Cancer J 14:27
15. Offner S, Hofmeister R, Romaniuk A et al (2006) Induction of regular cytolytic T cell synapses by bispecific single-chain antibody constructs on MHC class I-negative tumor cells. Mol Immunol 43:763–771
16. Trabolsi A, Arumov A, Schatz JH (2019) T cell-activating bispecific antibodies in cancer therapy. J Immunol 203:585–592
17. Bromley SK, Burack WR, Johnson KG et al (2001) The immunological synapse. Annu Rev Immunol 19:375–396
18. Klein C, Brinkmann U, Reichert JM et al (2024) The present and future of bispecific antibodies for cancer therapy. Nat Rev Drug Discov 23:301–319
19. Liu C, Zhou J, Kudlacek S et al (2023) Population dynamics of immunological synapse formation induced by bispecific T cell engagers predict clinical pharmacodynamics and treatment resistance. Elife 12:e83659
20. Moon D, Tae N, Park Y et al (2022) Development of bispecific antibody for cancer immunotherapy: focus on T cell engaging antibody. Immune Netw 22:e4
21. Labrijn AF, Janmaat ML, Reichert JM et al (2019) Bispecific antibodies: a mechanistic review of the pipeline. Nat Rev Drug Discov 18:585–608
22. Kamakura D, Asano R, Yasunaga M (2021) T cell bispecific antibodies: an antibody-based delivery system for inducing antitumor immunity. Pharmaceuticals 14:1172

Chapter 10

Measurement of T Cell Migration by T Cell-Dependent Bispecific Antibody in Cancer Cells and T Cell Cross-Linking

Rikuto Nakamura, Ryo Tsumura, Takahiro Anzai, Ryutaro Asano and Masahiro Yasunaga

Abstract

T cell-dependent bispecific antibodies (TDBs) have revolutionized the treatment of innovative cancer therapies with their potent anti-tumor effects, especially against hematologic malignancies. T cell responses to TDBs induce T cell migration, which is believed to play an important role in drug efficacy. Effective migration of T cells within the tumor microenvironment increases their contact with tumor cells and enhances T cell activation and cytotoxic activity against tumor cells. In this protocol, we present a method using TDB co-culture supernatants and a trans-well to evaluate T cell migration. This will provide a reliable approach to analyze in detail the mechanism of T cell migration in TDB-mediated immunotherapy and contribute to improving its drug efficacy.

Keywords Bispecific antibody, T cell migration, T cell invasion, Trans-well

1 Introduction

The emergence of next-generation antibody drugs, particularly T cell-dependent bispecific antibodies (TDBs), has produced remarkable clinical results in the treatment of hematologic malignancies [1–3].

Strong induction of T cell responses is an important mechanism underlying the clinical efficacy of TDBs, and many studies have confirmed that TDBs promote the release of proinflammatory cytokines [4–6]. It has also been suggested that TDBs not only recruit T cells but also induce T cell migration secondary to enhanced chemokine release [6–9].

To understand the relationship between T cell migration and therapeutic efficacy, it is important to classify "cold tumors" and "hot tumors" in the tumor microenvironment: with cancer immunotherapy, including TDBs, "cold tumors" have poor T cell infiltration

Masahiro Yasunaga (ed.), *Bispecific Antibodies*, Methods in Molecular Biology, vol. 2997,
https://doi.org/10.1007/978-1-0716-5037-0_10,
© The Author(s), under exclusive license to Springer Science+Business Media, LLC, part of Springer Nature 2026

and are resistant to treatment, while "hot tumors" have active T cell infiltration and therapeutic efficacy [10–12].

In this protocol, we introduce a method for quantitative evaluation of T cell migration. Specifically, we will perform a trans-well assay using TDB co-culture supernatants to analyze the effect of factors secreted by tumor cells on T cell directional migration. This method will provide a framework for elucidating the interactions between tumor cells and T cells and for elucidating the mechanisms of T cell migration. Implementation of this protocol will provide insight to further understand the importance of T cell migration in TDB therapy and to develop new therapeutic strategies to convert cold tumors to hot tumors.

2 Materials

2.1 T Cell Proliferation

1. Peripheral Blood Mononuclear Cells (PBMCs)
2. Thermostatic bath (37°C)
3. CO_2 humidified incubator (37°C, 5% CO_2)
4. Culture plate (6 wells, 150 mm dish)
5. RPMI-1640: 10% fetal bovine serum (FBS), 1% penicillin-streptomycin (*see* **Note 1**)
6. Human anti-CD3 antibody (OKT-3)
7. Human interleukin-2 (IL-2)
8. 0.5% Trypan Blue solution

2.2 Co-Culture with TDB

1. Cancer cell
2. TDB
3. 96-well culture plate
4. 0.25% Trypsin-EDTA
5. Phosphate buffered saline (PBS)
6. Reagent reservoir
7. 8-row aspirator (*see* **Note 2**)

2.3 T Cell Migration

1. XenoLight DiR
2. 5.0 μm Trans-well
3. 24-well culture plate
4. 96-well culture plate
5. 2.0 mL tube
6. Tweezers
7. Flow cytometer

3 Methods

3.1 Preparation for CD8+ T Cell

1. OKT-3 diluted to 1 μg/mL in PBS is coated at 1 mL/well in 3 wells of a 6-well culture plate and incubated for 2 h.
2. Suspend PBMCs (1.0×10^7 cells/tube) in RPMI and collect in 15 mL conical centrifuge tubes.
3. Centrifuge (2000 rpm, 4°C, 5 min) to remove the supernatant.
4. Remove the 6-well culture plate coated with OKT-3 from the incubator and remove all the coating liquid.
5. Add 12 mL of medium to a pellet of PBMCs in a 15 mL conical centrifuge tube and suspend well by pipetting to make the solution a uniform cell concentration.
6. Add 4 mL/well of PBMCs suspension to 6 well culture plates.
7. Add further IL-2 to 100 IU/mL and incubate for 3–4 days at 37°C, 5% CO_2 (*see* **Note 3**).
8. T cells are collected in 15 mL conical centrifuge tubes and centrifuged (2000 rpm, 4°C, 5 min).
9. Check that the T cells have formed a pellet and remove the solution except for the cell pellet by aspiration.
10. After this, count the number of cells and viable cells in the same way as in Sect. 3.1.
11. Add medium and IL-2 and pass on.
12. Maintain a cell concentration of 1.0×10^5–1.0×10^7 cells/mL.

3.2 Co-Culture of Cancer Cells, T Cells, and TDB

1. Cancer cells are pre-seeded in 24-well plates at 5.0×10^4 cells/500 μL/well and adhered to the plates overnight.
2. T cells are collected in 15 mL conical centrifuge tubes and centrifuged (2000 rpm, 4°C, 5 min).
3. Check that the T cells have formed a pellet and remove the solution except for the cell pellet by aspiration.
4. Add 5 mL of medium to the cell pellet and suspend well by pipetting to bring the solution to a uniform cell concentration.
5. Place 50 μL of the cell suspension in a well of a 96-well plate, add 50 μL of Trypan blue, and mix by pipetting.
6. The solution is placed in a hemocytometer to determine the number of live and dead cells; T cells are resuspended with medium again to a concentration of 1.0×10^6 cells/mL.
7. TDB is prepared by adding medium to twice the target concentration.
8. Completely remove the medium from the 24-well plate using an aspirator.

9. Add 250 μL (= 2.5 × 10^5 cells/well, E:T=5:1) of T cell suspension to the 24-well plate. Then, add 250 μL of the prepared TDB to bring the total to 500 μL.
10. Incubate for 24 h at 37°C in 5% CO_2.

3.3 Acquisition and Storage of Co-Culture Supernatants

1. After 24 h of incubation, remove the 24-well plate from the incubator. Prepare another 24-well plate.
2. Remove all the supernatant from the co-culture wells with a pipettor and put it into a 2.0 mL tube without mixing with the supernatant from the other wells. Discard the original 24-well plate.
3. Centrifuge the 2.0 mL tubes containing the supernatant (2000 rpm, 5 min, 4°C). While waiting, prepare another 2.0 mL tubes.
4. After centrifugation, take only the cell-free supernatant from the 2.0 mL tubes and place each supernatant in a new 2.0 mL tube (*see* **Note 4**).
5. Store the cell-free supernatant at −20°C.

3.4 T Cell Migration Assay

1. Take the supernatant from −20°C and melt it at 4°C (*see* **Note 5**).
2. Put 400 μL of the supernatant into a 24-well culture plate and store at 4°C.
3. Count the cultured T cells and suspend them at 1.0 × 10^6 cells/mL in medium.
4. Transfer the T cell suspension to a 15 mL tube and add 2.5 μL of XenoLight DiR per 1 mL (*see* **Note 6**).
5. Shield the 15 mL tube from light and incubate for 30 min at 37°C.
6. Remove the 15 mL tube from the incubator, add PBS, and wash 3 times.
7. Resuspend the tube at 1.0×10^6 cells/mL after re-counting the cells.
8. Insert 400 μL of supernatant into a 24-well, add 200 μL of DiR-stained T cell suspension to the top of the trans-well, and insert the trans-well into the 24-well (*see* **Note 7**).
9. Incubate for 2 h at 37°C, shaded from light (*see* **Note 8**).
10. After the reaction, remove the 24-well plate from the incubator and discard the trans-wells by grabbing them with tweezers.
11. Mix the samples contained within 24-well to make them homogeneous, then take 100 μL of each sample and place it into the 96-well plate for measurement.
12. Count the number of cells using a flow cytometer for an arbitrary number of seconds (*see* **Note 9**).
13. Gating is performed using analysis software to calculate the number of migrated T cells (*see* **Note 10**, Fig. 1).

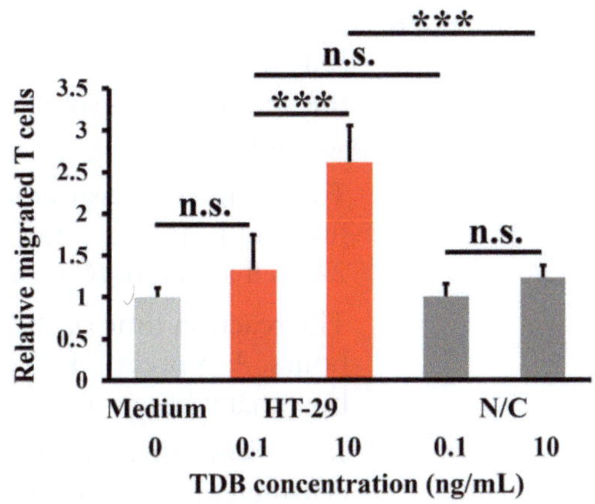

Fig. 1 T cell migration using supernatants co-cultured with TDB. Each relative ratio of migrated T cells was calculated based on the results using medium. HT-29: HT-29 + T cell + TDB, N/C: Negative control (T cell + TDB), Error bars represent the mean ± SD (n≧3), *p<0.05, ***p<0.0005, *n.s.* not significant

4 Notes

1. The proliferation rate of T cells may vary greatly depending on the lot of FBS and other factors. Before conducting this experiment, check the FBS lot and select the one with the fastest T cell proliferation speed and a survival rate that does not deteriorate significantly.
2. Although a standard aspirator can be used, an eight-aspirator can be used to minimize errors. The aspirator can be operated quickly, preventing cells from drying out and damaging after the medium is removed.
3. PBMCs are a mixture of several cells. To expand and culture only CD8+ T cells, it is common to stimulate T cells with OKT-3 and to add IL-2, which is necessary for their survival and differentiation.
4. After centrifugation, the supernatant should be obtained from the top of the 2.0 mL tube to avoid sucking the pelleted cells. For example, in the case of a 2.0 mL tube with 500 μL/well, about 400 μL of supernatant should be obtained from the upper side, never including any cells. If you are not familiar with the experimental procedure, check the supernatant under a microscope to see if it contains any cells.
5. Once frozen at –20°C, the supernatant should be used as soon as possible. When using the supernatant, place it on ice and allow it to thaw gently over a period of 30 min to 1 h. If there is excess supernatant, it should not be refrozen. Repeated freezing may adversely affect cytokines and other substances contained in the supernatant.

6. By staining with DiR, it is possible to distinguish migrating T cells from cells in the supernatant, even if they are mixed in the acquired supernatant. DiR can be replaced with a simple staining reagent.
7. To minimize measurement errors, insert the trans-well carefully and slowly. After inserting the trans-well, carry the 24-well as carefully as possible without shaking it.
8. The reaction time of the T cell migration assay is often several hours. As a preliminary experiment, it is advisable to consider how many hours of reaction time should be measured.
9. If the flow rate and other settings can be changed on the flow cytometer, the conditions should be studied in a preliminary experiment, and the same settings should be used each time thereafter. The assay should be performed with medium, not supernatant, as the negative control.
10. Since DiR can be detected at a wavelength of approximately 750 nm, the number of migrated T cells is determined by gating at a wavelength of 750 nm. After the same procedure is performed on all samples, the relative ratio of migrating T cells in the supernatant is calculated using the number of migrating T cells in the medium negative control as a reference value.

Acknowledgments

The authors thank the members of the Yasunaga laboratory for their helpful discussions and M. Shimada for secretarial support. The authors also thank Dr. R. Asano for providing the material and support. This research was supported by AMED under Grant Number JP21ak0101182 to M.Y. and the National Cancer Center Research and Development Fund (Number 23-A-08) to M.Y.

References

1. Clynes RA, Desjarlais JR (2019) Redirected T cell cytotoxicity in cancer therapy. Annu Rev Med 70:437–450
2. Lejeune M, Köse MC, Duray E et al (2020) Bispecific, T-cell-recruiting antibodies in B-cell malignancies. Front Immunol 11:762
3. Middelburg J, Kemper K, Engelberts P et al (2021) Overcoming challenges for CD3-bispecific antibody therapy in solid tumors. Cancers (Basel) 13:287
4. Golubovskaya V, Sienkiewicz J, Sun J et al (2023) mRNA-lipid nanoparticle (LNP) delivery of humanized EpCAM-CD3 bispecific antibody significantly blocks colorectal cancer tumor growth. Cancers 15
5. Kemper K, Gielen E, Boross P et al (2022) Mechanistic and pharmacodynamic studies of DuoBody-CD3x5T4 in preclinical tumor models. Life Sci Alliance 5
6. Kamakura D, Asano R, Kawai H, Yasunaga M (2021) Mechanism of action of a T cell-dependent bispecific antibody as a breakthrough immunotherapy against refractory colorectal cancer with an oncogenic mutation. Cancer Immunol Immunother 70:177–188

7. Himmels P, Nguyen TTT, Mitzner MC et al (2023) T cell-dependent bispecific antibodies alter organ-specific endothelial cell-T cell interaction. EMBO Rep 24:e55532
8. Liu D, Bao L, Zhu H et al (2023) Microenvironment-responsive anti-PD-L1 × CD3 bispecific T-cell engager for solid tumor immunotherapy. J Control Release 354:606–614
9. Sam J, Colombetti S, Fauti T et al (2020) Combination of T-cell bispecific antibodies with PD-L1 checkpoint inhibition elicits superior anti-tumor activity. Front Oncol 10:575737
10. Cremasco F, Menietti E, Speziale D et al (2021) Cross-linking of T cell to B cell lymphoma by the T cell bispecific antibody CD20-TCB induces IFNγ/CXCL10-dependent peripheral T cell recruitment in humanized murine model. PLoS One 16:e0241091
11. Khazen R, Müller S, Lafouresse F et al (2019) Sequential adjustment of cytotoxic T lymphocyte densities improves efficacy in controlling tumor growth. Sci Rep 9: 12308
12. Nelson N, Lopez-Pelaez M, Palazon A et al (2019) A cell-engineered system to assess tumor cell sensitivity to CD8+ T cell-mediated cytotoxicity. Oncoimmunology 8:1599635

Chapter 11

Optimized In Vitro T-Cell Exhaustion Assay for Evaluating T-Cell-Dependent Bispecific Antibodies

Mayu Oka, Mikiko Itsukaichi and Masahiro Yasunaga

Abstract

T-cell exhaustion is characterized by low proliferative capacity, suppressed effector function, and the persistent expression of co-suppressive receptors such as PD-1 and TIM-3. Chronic activation of T cells, as seen in chronic viral infections or cancer, leads to T-cell exhaustion. In addition, T-cell exhaustion is a major obstacle to the clinical application of T-cell-dependent bispecific antibodies (TDBs). On the other hand, blocking the co-suppressive receptors has improved T-cell function, highlighting the importance of recovery from T-cell exhaustion to enhance therapeutic efficacy. Ongoing research focuses on recovering from exhausted T cells using genetic engineering, particularly in CAR-T cell therapy. Although mouse models infected with LCMV are widely used to generate exhausted T cells, this method raises concerns about animal welfare. We established a method using magnetic beads to induce T-cell exhaustion in vitro, simulating the immune state caused by chronic activation. Moreover, this in vitro exhaustion-induction method is applicable to the evaluation of TDB functional activity and the examination of the T-cell manipulation methods. Collectively, the current in vitro T-cell exhaustion assay system will be helpful for the development of TDBs to overcome T-cell exhaustion.

Keywords T-cell-dependent bispecific antibodies, T-cell exhaustion, In vitro exhaustion induction, Chronic activation, Magnetic beads, Cell killing assay

1 Introduction

Exhausted T cells are defined as cells with low proliferative capacity, suppressed effector function, and persistent expression of co-inhibitory receptors such as PD-1 and TIM-3 [1]. Effector T cells are known to be activated by signals to TCR/CD3 and CD28. Although transient activation of T cells is primarily harmless, chronic activation of T cells due to chronic viral infection or cancer causes T-cell exhaustion. It has been shown that cancer patients' tumor-infiltrating lymphocytes (TILs) express co-inhibitory receptors such as PD-1 and TIM-3 [2, 3]. Besides, it has been reported that cytokines such as IFN-γ decrease in PD-1+ TIM-3+ LAG-3+ CD8+ T cells; these can be exhaustion markers [3]. Furthermore, it has been reported that

Masahiro Yasunaga (ed.), *Bispecific Antibodies*, Methods in Molecular Biology, vol. 2997,
https://doi.org/10.1007/978-1-0716-5037-0_11,
© The Author(s), under exclusive license to Springer Science+Business Media, LLC, part of Springer Nature 2026

concomitant blockade of these co-inhibitory receptors improves T-cell effector function, suggesting that recovery from T-cell exhaustion is the key to improving therapeutic efficacy. Against this background, promoting recovering from the exhausted T cells becomes an active area of research. It is widely known that decreased TCF expression and increased TOX expression occur in exhausted T cells [4]. In addition, recent studies have shown that regulating the expression of transcription factors such as MYC, SOX4, and IRF2 can modulate and even promote recovery from the function of exhausted T cells [5–7]. Such research on exhausted T cells typically involves genetic engineering, particularly active in CAR-T cell therapy.

We hypothesized that impaired T-cell quality is one of the factors contributing to reduced efficacy of TDB and evaluated the effect of exhausted T cells on the TDB-dependent cytotoxicity. We found that the hEx3-dependent cytotoxicity was significantly reduced in exhausted T cells compared to normal T cells (Fig. 1). This suggests that exhausted T cells negatively affect the TDB-dependent cytotoxicity. However, there has been no report on improving the efficacy of TDB from the aspect of host-side factors, namely, exhausted T cells. Based on these results, we considered that improving the quality of exhausted T cells was the key to enhancing the efficacy of TDB, and techniques such as gene overexpression in exhausted T cells could be expected to optimize TDB-dependent cytotoxicity.

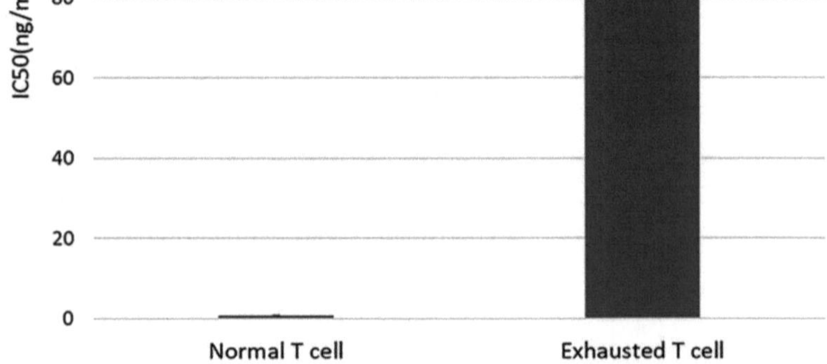

Fig. 1 Verification of the hEx3-dependent cytotoxicity in proliferating and exhausted T cells. This figure shows the IC50 of proliferative T cells (Normal T cell) and T cells exhausted by 5 consecutive days of CD3/CD28 co-stimulation (Exhausted T cell). The hEx3-dependent cytotoxicity was significantly reduced in exhausted T cells compared to normal T cells. ** $p<0.005$, n=3

Upregulation of exhaustion-related genes in exhausted T cells, such as the cytokine receptor, improves the TDB-dependent cytotoxicity (Fig. 2). Further research into T-cell exhaustion, which is an obstacle to the development of TDB, will help improve its efficacy and realize its clinical application.

In general, exhausted T cells are generated by chronic viral infection in mice. Among various viral infections, LCMV (lymphocytic choriomeningitis virus) infection has been widely used to establish experimental models of CD8+ T cell exhaustion [8, 9]. However, this method of producing exhausted T cells requires a large number of experimental animals, an obvious drawback from the perspective pf animal welfare. Furthermore, in vitro, induction of exhausted T cells is considered to be relatively homogeneous and reproducible. Although there was a report of successful in vitro production of exhausted T cells [10], in vitro exhaustion production has not yet to be generalized. Thus, we established an in vitro T-cell exhaustion assay system. We used magnetic beads (Dynabeads) coated with anti-CD3 and anti-CD28 antibodies, which bind to TCR/CD3 and CD28 on the T-cell surface, to deliver co-stimulatory for seven consecutive days to induce T-cell exhaustion. This approach induces chronic activation of T cells similar to observed during chronic viral infection or cancer.

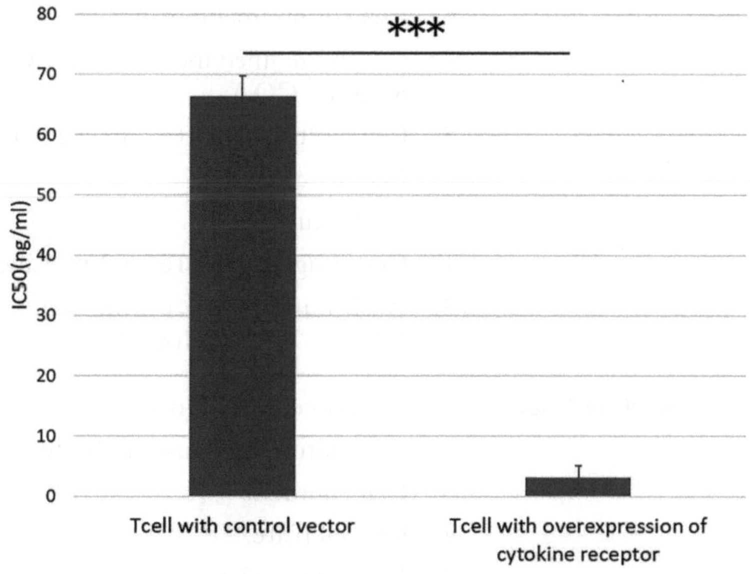

Fig. 2 Verification of the TDB-dependent cytotoxicity after gene overexpression. This figure shows the IC50 of T cells induced to exhaustion by CD3/CD28 co-stimulation after lentiviral infection of controls (T cells with control vector) and T cells overexpressing the cytokine receptor by lentiviral vector followed by CD3/CD28 co-stimulation induced exhaustion (T cell with overexpression of cytokine receptor). T cells with overexpression of the cytokine receptor significantly improved the hEx3-dependent cytotoxicity. *** $p<0.0001$, n=3

2 Materials

2.1 Extended Culture for T Cells

1. OKT-3.
2. D-PBS.
3. 6 well plate.
4. Human Peripheral Blood Mononuclear Cells (PBMC) (e.g., from Cellular Technology Limited, Lonza, or Precision for Medicine).
5. Cell culture medium (*see* **Note 1**).
6. Centrifuge tubes (15-, 50-ml tubes).
7. Aspirator.
8. Recombinant human IL-2 (100 IU/ml).
9. CO_2 humidified incubator (37°C ± 1°C, 90% ± 5% humidity, 5% ± 1% CO_2).

2.2 Generation of Exhausted T Cells

1. Vortex mixer.
2. Dynabeads human T-Activator CD3/CD28.
3. Recombinant human IL-2. (100 IU/ml)
4. Pipetman.
5. T cell.
6. Cell culture medium.
7. CO_2 humidified incubator (37°C ± 1°C, 90% ± 5% humidity, 5% ± 1% CO_2)
8. Cell culture plates (6-, 12-well plates)
9. Magnetic stand
10. 1.5 ml tubes
11. Centrifuge tubes (15-, 50-ml tubes)
12. Wash buffer: 0.1% bovine serum albumin, 2 mM EDTA in PBS pH 7.4 (*see* **Note 2**)

2.3 Cell Killing Assay

1. Cancer cells (*see* **Note 3**).
2. Cell culture medium (*see* **Note 4**).
3. Cell counter.
4. 96-well plate.
5. Pipetting device.
6. Reservoir.
7. CO_2 humidified incubator (37°C ± 1°C, 90% ± 5% humidity, 5% ± 1% CO_2).
8. Centrifuge tubes (15-, 50-ml tubes).
9. Magnetic stand.

10. Normal T cell.
11. Exhausted T cell.
12. Aspirator.
13. T-cell-dependent bispecific antibody (TDB).
14. 1.5 ml tube.
15. Cell counting kit-8.
16. Microplate reader.

3 Methods

3.1 Extended Culture for T Cell

1. Add OKT3 diluted to 1 μg/ml in D-PBS to 3 wells of a 6-well plate and incubate at 37°C for at least 1 h to coat.
2. Suspend PBMC (1×10^7 cells) in cell culture medium and collect in a tube.
3. Centrifuge the previous tube at 2000 rpm, 4°C, 5 min.
4. Remove the supernatant.
5. Suspend the cell pellet in 12 ml of cell culture medium.
6. Remove OKT3 solution for coating from the 6-well plate and add 4 ml of PBMC suspended in cell culture medium to each well.
7. Add 100 IU/ml recombinant human IL-2.
8. Incubate at 37°C for 3–4 days.

3.2 Generation of Exhausted T Cell In Vitro

1. Resuspend the Dynabeads Human T-Activator CD3/CD28 in the vial with a vortex mixer (*see* **Notes 5 and 6**).
2. Transfer the desired volume of Dynabeads to a 1.5 ml tube (*see* **Note 7**).
3. Add 1 ml Wash Buffer or Culture Medium to the tube, and vortex for 5 s (*see* **Notes 2 and 6**).
4. Place the tube on a magnetic stand for 1 min and remove the supernatant.
5. Take the tube from the magnetic stand and resuspend the washed Dynabeads in the same volume of Culture Medium as the initial volume of Dynabeads taken from the vial.
6. Count cultured T cells.
7. Prepare T-cell suspension at 1×10^6 cells/ml in a suitable cell culture plate according to our previous report or methods described in other chapters [11] (*see* **Note 8**).
8. Add beads at 25 μl of Dynabeads per 1×10^6 cells to the plate (*see* **Note 9**)

9. Add 100 IU/ml recombinant human IL-2.
10. Incubate in a humidified CO_2 incubator at 37°C for 3 or 4 days.
11. After 3 or 4 days, collect cell suspension with Dynabeads to a suitable centrifuge tube.
12. Centrifuge the tubes at 2000 rpm for 5 min at 4°C.
13. Remove the supernatant and resuspend the cell pellet in 1 ml of fresh medium.
14. Transfer the 1 ml cell suspension to a 1.5 ml tube.
15. Place the tube on a magnetic stand for 1 min and collect cell suspension only to another tube.
16. Count the number of T cells (*see* **Note 10**).
17. Wash the fresh Dynabeads following steps 1–5.
18. Subculture the T cells at 1×10^6 cells/ml in a suitable tissue culture plate (*see* **Notes 8 and 11**).
19. Add 25 μl of Dynabeads per 1×10^6 cells to the plate (*see* **Note 9**).
20. Add 100 IU/ml recombinant human IL-2.
21. Incubate in a humidified CO_2 incubator at 37°C for 3 or 4 days (*see* **Note 12**)

3.3 Cell Killing Assay

1. Count cultured cancer cells (*see* **Note 13**).
2. Prepare approximately 10 ml (more than 9.6 ml per 96-well plate) of cancer cell suspension at 1×10^5 cells/ml.
3. Pour the cell suspension into the reservoir.
4. Culture the cancer cells in 96-well plates at 100 μl/well (*see* **Notes 14, 15, and 16**)
5. Incubate in a humidified CO_2 incubator at 37°C for 24 h.
6. On day 2, place the tube with the prepared exhausted T cells on a magnetic stand for 1 min and collect the cell suspension only to another tube to remove the Dynabeads (*see* **Note 17**)
7. Count normal T cells and exhausted T cells, respectively (*see* **Note 13**)
8. Prepare normal T-cell suspension and exhausted T-cell suspension at 1×10^6 cells/ml, respectively (*see* **Notes 8, 18, and 19**).
9. Pour the normal T-cell suspension and exhausted T-cell suspension into the reservoir, respectively.
10. Aspirate the culture supernatant from the 96-well plate prepared 24 h previously with an aspirator.

11. Culture the T cells and exhausted T cells in the appropriate lanes in the 96 well at 50 μl/well and temporarily incubate the plate in a humidified CO_2 incubator at 37°C while preparing concentrations of TDB (*see* **Notes 16, 20, and 21**).
12. Prepare TDB solution diluted with the culture medium for T cell in each suitable tube for each concentration (*see* **Note 22**).
13. Pour the TDB solution into the reservoir of each lane for each concentration (*see* **Note 23**).
14. Add TDB to the suitable lanes of the 96-well plate for each concentration at 50 μl/well (*see* **Notes 16 and 24**).
15. Incubate in a humidified CO_2 incubator at 37°C for suitable hours (more than 24 h) (*see* **Notes 25 and 26**).
16. On day 3, dilute the Cell Counting Kit-8 10-fold with medium and prepare approximately 10 ml of diluted solution (more than 9.6 ml per 96-well plate).
17. Aspirate the culture supernatant from the 96-well plate prepared 48 h previously with an aspirator.
18. Wash 3 times with 200 μl of the culture medium.
19. Add THE diluted solution of Cell Counting Kit-8 to every well of the 96-well plate at 100 μl/well (*see* **Notes 16 and 27**).
20. Incubate in a humidified CO_2 incubator at 37°C for 1–2 h (*see* **Note 28**).
21. Measure absorbance at 450 and 600 nm with a microplate reader.
22. Subtract the absorbance value at 600 nm from that at 450 nm (*see* **Note 29**).
23. Calculate the cell viability and IC50 (*see* **Note 30**).

4 Notes

1. We used RPMI-1640 supplemented with 10% heat-inactivated for cancer cells and T cells.
2. A culture medium can replace the wash buffer.
3. Here, the pancreatic cancer cell line BxPC3 was used as an example.
4. For BxPC3, we used RPMI-1640 supplemented with 10% heat-inactivated.
5. Separate stimulation of CD3 and CD28 does not work well for inducing exhaustion; simultaneous stimulation of CD3

and CD28 is necessary for exhaustion induction. When plates were coated with OKT-3, anti-CD28 antibodies were added, and T cells were stimulated by respective antibodies, PD-1 and TIM-3 expression did not increase. The method using Dynabeads is successful based on this result. Dynabeads are magnetic beads with a diameter of 4.5 μm, similar in size to antigen-presenting cells, and with anti-CD3 and anti-CD28 antibodies covalently attached to their surface, which send co-stimulatory signals to TCR/CD3 and CD28. Stimulation for seven consecutive days induces T cell exhaustion caused by chronic activation of T cells. The use of Dynabeads allows efficient contact with T cells, promoting activation and thereby inducing exhaustion. This approach induces chronic activation of T cells similar to that observed during chronic viral infection or cancer.

6. Vortex gently to avoid breaking the structure of the Dynabeads. Use the lowest vibration setting when vortexing. Also, make sure that the Dynabeads are homogeneously distributed.
7. The following steps should be carried out in a clean bench to prevent the solution from being exposed to any foreign substances.
8. T-cell suspension at 1×10^6 cells/ml indicates the presence of live cells in the cell suspension at 1×10^6 cells/ml. Make sure that this is the concentration of the number of live cells, not the number of all cells, including dead cells.
9. Make sure that Dynabeads are spread evenly throughout the dish, not in any particular place.
10. When counting cells using trypan blue, always use the average of at least three counts, as the number of cells can vary as apoptosis progresses.
11. Use new plates to prevent contamination.
12. Although this protocol describes the procedure for 7 days of continuous stimulation, the number of days required for exhaustion depends on the PBMC lot and the state of the cells. Therefore, it is necessary to consider the number of days of continuous stimulation, such as 5 days, 7 days, or even longer. The expression of PD-1 and TIM-3, which are well-known markers of exhaustion, should be examined, and if there are no signs of exhaustion, the number of days of continuous stimulation should be extended (Fig. 3).
13. To confirm reproducibility, carefully count cells and use the average of at least three counts, as changes in the number of cells cultured can affect the results.

14. Use a 96-well plate with a flat bottom when using adherent cancer cells.
15. Take care to avoid producing air bubbles in the solution. To prevent air bubbles, dispense the solution with the tip of the pipette along the inner wall of the well.
16. If the solution adheres to the inner walls, lightly tap the sides of the plate, taking care not to splash the solution.
17. Make sure to remove the Dynabeads before counting.
18. The addition of the TDB solution will doubly dilute the final cell concentration.
19. Effector cells(E) to Targeted cells(T) ratio of 1:5 is desired.
20. Do not use the outer rows (rows A and H) and outer columns (columns 1 and 12) of the 96-well plate co-culture with T cells, as edge effects-caused by reduced medium volume in the outer wells of the plate during cell culture-can affect cell condition and growth. When seeding cancer cells, it is acceptable to seed them in outer wells, but these wells should not be used for experiments. PBS may also be added to the outer wells to minimize edge effects.
21. After culturing, temporarily incubate the plate not to make T cells impaired.
22. The range of concentrations of the TDB solution will vary, as the effective concentration of TDB will depend on the amount of cancer antigen expressed by the cancer cells and the condition of the antibodies. For example, we prepare hEx3 solution

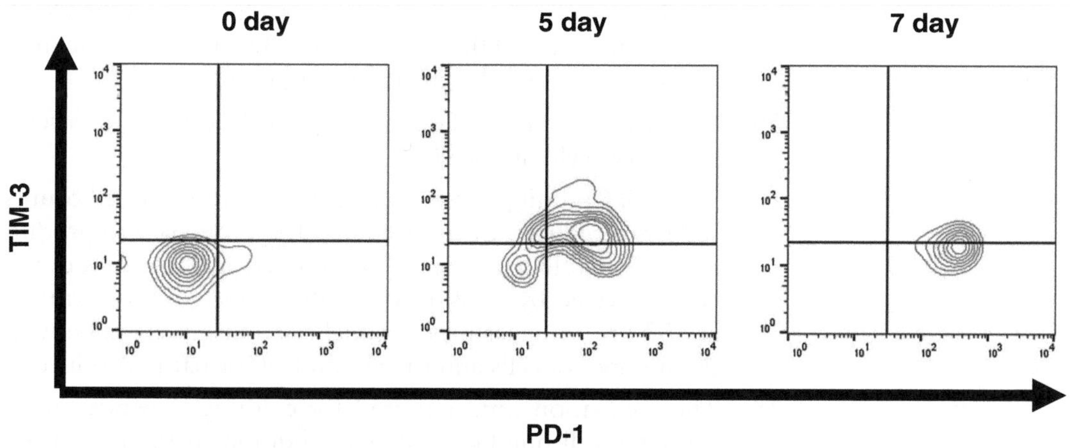

Fig. 3 Expression levels of PD-1 and TIM-3 in T cells after CD3/C28 co-stimulation. The number in the upper right-hand corner of each graph shows the percentage of cells with double-positive exhaustion markers when the total number of cells is set to 100%; at 0 days there are almost no cells with double-positive exhaustion markers, but at 5 and 7 days of continuous stimulation with CD3/CD28, the percentage of cells with double-positive exhaustion marker double-positive cells is markedly increased

at 0.02, 0.2, 2, 20, 200 ng/ml. (You must consider that the final TDB concentration will be doubly diluted because you have already added 50 µl of T-cell solution to the wells.) Because our repeated trials have shown that normal T cells have an IC50 (the value at which the cancer cell survival rate is calculated to be 50%, setting the cancer cell survival rate at 0 ng/ml as 100%) between 0.1 and 10 ng/ml, and exhausted T cells have an IC50 between 10 and 100 ng/ml. In other examples, with solitomab, we prepare solitomab solution at 0.002, 0.02, 0.2, 2, 20 ng/ml. Of course, other arrangements can be experimented, and this is only one example. You need to consider that the range of concentration depends on the kind of TDB and the cancer cell type.

23. When pipetting a large reservoir is difficult, a reservoir liner with a smaller volume and a line shape is easier to handle.

24. If you would like to test n=3 and use one 96 well plate, here is an example of the placement of T-cell samples. Culture a sample of normal T cells in rows B–D from columns 4–11 and a sample of exhausted T cells in rows E–G after columns 4–11. In the column 2, add only culture medium as a blank. Culture only cancer cells in column 3. Add 0 ng/ml of TDB solution, i.e., just culture medium, in column 3. From column 4–11, add TDB solution sequentially to increase concentration. Then, the results for n=3 can be calculated on one 96-well plate. Of course, other arrangements can be experimented with; this is only one example. It does not matter if some columns are not used.

25. The incubation time also depends on the cancer cell types and the kind of TDB. Sometimes, 24 h may be not sufficient for the cancer cells to die, and the results may not be clearly distinct for each TDB concentration. In such cases, the incubation time should be changed, e.g., to 48 h.

26. Under a microscope, floating cancer cells can be observed if the cancer cells are dead.

27. The Cell Counting Kit-8 is a kit for measuring live cell counts. WST-8 is reduced by intracellular dehydrogenase to produce a water-soluble formazan. The number of viable cells can be easily assessed by measuring the absorbance of this formazan at 450 nm. A linear proportional relationship exists between the number of cells and the amount of formazan produced.

28. The incubation time required for coloring depends on the cell type, cell number, and cell condition. If the plate is left too long for incubation, the formazan reaches saturation, and accurate cell viability cannot be measured. You have to consider incubation times that are not saturated but are colored enough to show differences in coloration for each TDB concentration. In our repeated trials, 1 hour was adequate

incubation time. However, this is only an example, and as mentioned earlier, the appropriate time depends on the situation, so the optimum time must be determined.

29. This step is to eliminate background due to turbidity in the cell culture medium.

30. The value for the sample without TDB added, in which only cancer cells and T cells are co-cultured, is used as the value of negative control. Determine the cancer cell viability of the sample with TDB added when the cancer cell viability in the negative control sample is set to 100%.

Acknowledgments

The authors thank the members of the Yasunaga laboratory for their helpful discussions and M. Shimada for secretarial support. The authors also thank Dr. R. Asano for providing the material and support. This research was supported by AMED under Grant Number JP21ak0101182 to M.Y. and the National Cancer Center Research and Development Fund (Number 23-A-08) to M.Y.

References

1. Wherry EJ (2011) T cell exhaustion. Nat Immunol 12(6):492–499. https://doi.org/10.1038/ni.2035
2. Datar I et al (2019) Expression analysis and significance of PD-1, LAG-3, and TIM-3 in human non-small cell lung cancer using spatially resolved and multiparametric single-cell analysis. Clin Cancer Res 25(15):4663–4673. https://doi.org/10.1158/1078-0432.CCR-18-4142
3. Woroniecka K et al (2018) T-cell exhaustion signatures vary with tumor type and are severe in glioblastoma. Clin Cancer Res 24(17):4175–4186. https://doi.org/10.1158/1078-0432.CCR-17-1846
4. Chow A et al (2022) Clinical implications of T cell exhaustion for cancer immunotherapy. Nat Rev Clin Oncol 19(12):775–790. https://doi.org/10.1038/s41571-022-00689-z
5. Wang R et al (2011) The transcription factor Myc controls metabolic reprogramming upon T lymphocyte activation. Immunity 35(6):871–882. https://doi.org/10.1016/j.immuni.2011.09.021
6. Zheng L et al (2021) Pan-cancer single-cell landscape of tumor-infiltrating T cells. Science (New York, N.Y.) 374(6574):abe6474. https://doi.org/10.1126/science.abe6474
7. Lukhele S et al (2022) The transcription factor IRF2 drives interferon-mediated CD8+ T cell exhaustion to restrict anti-tumor immunity. Immunity 55(12):2369–2385. https://doi.org/10.1016/j.immuni.2022.10.020e10
8. Zhou S et al (2004) Differential tissue-specific regulation of antiviral CD8+ T-cell immune responses during chronic viral infection. J Virol 78(7):3578–3600. https://doi.org/10.1128/jvi.78.7.3578-3600.2004
9. Wherry EJ et al (2003) Viral persistence alters CD8 T-cell immunodominance and tissue distribution and results in distinct stages of functional impairment. J Virol 77(8):4911–4927. https://doi.org/10.1128/jvi.77.8.4911-4927.2003
10. Corselli M et al (2022) Single cell multiomic analysis of T cell exhaustion in vitro. Cytometry A 101(1):27–44. https://doi.org/10.1002/cyto.a.24496
11. Kamakura D et al (2021) Mechanism of action of a T cell-dependent bispecific antibody as a breakthrough immunotherapy against refractory colorectal cancer with an oncogenic mutation. Cancer Immunol Immunother 70(1):177–188. https://doi.org/10.1007/s00262-020-02667-9

Chapter 12

Advanced Evaluation of T-Cell-Dependent Bispecific Antibodies Using Fluorescence Imaging and In Vivo Models

Mikiko Itsukaichi and Masahiro Yasunaga

Abstract

In this chapter, advanced methods for evaluating T-cell-dependent bispecific antibodies (TDBs) using in vivo fluorescence imaging and animal models are presented. TDBs, which target tumor-associated antigens on cancer cells and the TCR/CD3 complex on T cells, have shown success in hematologic malignancies but face challenges in solid tumors due to factors like reduced T-cell quality and low tumor-infiltrating T cells. Comprehensive protocols for cancer and T-cell preparation, flow cytometry, xenograft model generation, and imaging techniques are outlined, aiming to enhance the understanding of TDB dynamics and improve their efficacy in solid tumor treatment.

Keywords T-cell-dependent bispecific antibodies (TDBs), Fluorescence imaging, In vivo models, Cancer immunotherapy, Xenograft models, PDX models (patient-derived xenograft), CDX models (cell line-derived xenograft), Solid tumors

1 Introduction

Antibody drugs are often used in cancer immunotherapy, one of the four major cancer therapies [1]. Typical examples are immune checkpoint inhibitors such as anti-PD-1 antibodies represented by nivolumab and ipilimumab and anti-CTLA-4 antibodies [2, 3]. There are also chimeric antigen receptor (CAR)-T therapies and T-cell-dependent bispecific antibody (TDB) therapies [4].

TDB is an antibody that targets tumor-associated antigens (TAAs) and membrane proteins on the surface of T cells to bridge cancer cells and T cells.

The TAA targets on the cancer cell side vary depending on the type of TDB, while the T-cell targets are often designed to bind specifically to the TCR/CD3 complex expressed on the T-cell surface [5, 6]. During this redirection of T cells to cancer cells, the binding of TDB to the TCR/CD3 complex activates the T cells to produce cytotoxicity against the cancer cells. The mechanism of action of

Masahiro Yasunaga (ed.), *Bispecific Antibodies*, Methods in Molecular Biology, vol. 2997,
https://doi.org/10.1007/978-1-0716-5037-0_12,
© The Author(s), under exclusive license to Springer Science+Business Media, LLC, part of Springer Nature 2026

TDB is cell–cell contact-dependent and cell–cell contact-independent cytotoxic activity mediated by immune synapses, which is induced by granzyme perforin, interferon γ (IFN-γ) released from activated T cells, and tumor necrosis factor α (TNFα) released from activated T cell [7].

Although clinical applications of T-cell-dependent bispecific antibodies (TDBs) are underway, exemplified by the approval of blinatumomab targeting CD19 and CD3 and the partial conditional approval of TDBs targeting CD3 and CD20, all approved TDBs target hematologic malignancies, and no TDBs targeting solid tumors have been approved [5, 6, 8, 9].

Thus, compared to TDB targeting hematologic malignancies, the clinical application of TDB in solid tumors has been challenging. In addition, while TDB targeting solid tumors has shown efficacy in pre-clinical studies, its efficacy in clinical trials has been limited.

It has been discussed that the challenges to the clinical application of TDB targeting solid tumors are a decrease in the quality of the patient's T cells and a decline in the number of tumor-infiltrating T cells [10, 11].

The low number of tumor-infiltrating T cells in cancer patients' tumors is because cancer cell lines have different characteristics from cancer cells and tumors in cancer patients [12, 13]. In vitro experiments using cancer cell lines and mouse tumor models derived from cancer cell lines may not fully mimic the actual tumors [12]. Therefore, it is considered necessary to update experimental methods, such as conducting experiments using patient cancer cells or tumors.

The causes of low tumor-infiltrating T cells in cancer patients are considered immune desert and excluded T cells [11, 14]. Immune desert is a tumor with no or few T cells in it, whereas T-cell exclusion occurs when T cells are present around the tumor but cannot infiltrate into it [15]. In contrast, tumors with abundant T-cell infiltration show an inflamed phenotype, which allows T cells to enter the tumor and attack the cancer cells.

Changing the immune desert and T-cell exclusion to the state of Inflammation is recognized as the key to successful cancer immunotherapy [15]. Still, the dynamics of TDB and its effect on T-cell dynamics during TDB administration are unknown. A sufficient amount of T cells must also accumulate in the tumor for TDB to be effective [14].

In this chapter, we will focus on in vivo fluorescence simultaneous imaging of T cells and TDB and describe the experimental protocols required for in vivo imaging and its periphery, including methods for culturing cancer cells and T cells, flow cytometry to confirm antigen expression levels, HE staining to help understand the tumor microenvironment, and methods for producing CDX and PDX models. The lecture will describe the experimental protocols required for in vivo imaging and related fields.

2 Materials

2.1 In Vitro Culture

1. Cell line or human peripheral blood mononuclear cells (PBMCs). Here, the pancreatic cancer cell line BxPC3 is used as an example. PBMC is commercially available (e.g., from Cellular Technology Limited, Lonza, or Precision for Medicine).
2. Cell culture medium and supplements (10% fetal bovine serum (FBS) and 1% penicillin–streptomycin solution). We use RPMI-1640 supplemented with 10% heat-inactivated FBS and 1% penicillin–streptomycin solution for BxPC3.
3. Fetal bovine serum (FBS) (*see* **Note 5**).
4. Culture dish (10 cm dish).
5. CO_2 incubator (37°C, 5% CO_2).
6. Trypsin EDTA: 0.025% trypsin and 0.01% EDTA in PBS.
7. Cell freezing medium.
8. 0.4% Trypan Blue solution.
9. Cell counter.
10. Flow cytometer.

2.2 Mouse Models

1. Expanded cancer cells for cell line-derived xenograft (CDX) model (1×10^7 per mouse).
2. Primary tumor samples for patient-derived xenograft (PDX) model (tumor block size 5–20 mm^3 per mouse, F3~F5 PDX model, usually takes 6–12 months).
3. Immunodeficient mouse (Nude mouse) aged 5–8 weeks.

2.3 Histological Analysis

1. Optimal cutting temperature (OCT) compound
2. Cryomold
3. Liquid nitrogen
4. Cryostat
5. Microscope adhesive glass slides (76 × 26 mm)
6. 4% paraformaldehyde
7. Monoclonal or polyclonal antibodies (goat anti-hEGFR antibody)
8. Alexa Fluor® 647 goat-anti human IgG
9. PBS (phosphate-buffered saline)
10. Nuclear dyes, 1 μg/ml 4′, 6-diamidino-2′-phenylindole, dihydrochloride (DAPI) in phosphate-buffered saline (PBS)
11. Fluoromount-G (Mounting medium)
12. Fluorescence microscope

2.4 In Vivo Imaging of TDB

1. Mouse models, CDX or PDX
2. T-cell-dependent bispecific antibodies (TDBs)
3. Protein labeling kit
4. T cells expanded from PBMC (2×10^7 per mouse)
5. Fluorescent T-cell-dependent bispecific antibodies (TDBs)
6. Cell Labeling Kit
7. Injection syringe (27-G)
8. Fluorescent in vivo imaging system (IVIS in vivo imaging system)
9. Living Image Software (software for analysis)

3 Methods

3.1 Expansion of Cancer Cells (see Note 1)

3.1.1 Thaw Frozen Cells

1. Add 10% fetal bovine serum (FBS) and 1% penicillin–streptomycin solution to a suitable medium (*see* **Note 2**).
2. Add the medium prepared in Step 1 to a 15 ml tube.
3. Dissolve the frozen cells at –80°C in a 37°C chamber and quickly transfer them to the medium prepared in Step 2 (*see* **Note 3**).
4. Centrifuge at 1500 rpm for 3 min (*see* **Note 4**).
5. Absorb supernatant.
6. Add an optimal amount of medium prepared in Step 1.
7. Scatter on a dish to make the density uniform (*see* **Note 5**).
8. All cells are cultured under 37°C and 5% CO_2 conditions.

3.1.2 Passage

1. Add Trypsin EDTA to the culture dish.
2. Incubate for 5–7 min under 37°C and 5% CO_2 condition (*see* **Note 6**).
3. Add medium to the dish when the cancer cells begin to float.
4. Transfer the cell suspension of Step 3 to a tube and centrifuge at 1500 rpm for 30 min.
5. Discard the supernatant (*see* **Note** 7).
6. Suspend the precipitated cancer cells after adding the culture medium.
7. Add medium and the optimal volume of the cell suspension of Step 6 to the dish (*see* **Note 8**).
8. Incubate under 37°C and 5% CO_2 condition.

3.1.3 Stock

1. Follow the same steps as Method 3.1.2, Steps 1–5. Add cell freezing medium and transfer the cells to a tube for freezing (*see* **Note 9**).
2. Freeze at −80°C.

3.1.4 Cell Count

1. Follow the same steps as Method 3.1.2, Steps 1–5. Remove a portion of the cell suspension, add Trypan Blue, and pipet well (*see* **Note 10**).
2. Count the cells on a cell counter.

3.2 Flow Cytometry

1. Add 0.5% BSA and 2 mM EDTA to PBS to prepare wash solution (B.E. PBS).
2. Add dissociation buffer to the dish where cells are cultured and incubate at 37°C for 15–20 min to collect cancer cells (*see* **Note 11**).
3. After washing with B.E. PBS, add the primary antibody and incubate at 4°C for 30 min.
4. Add fluorescent secondary antibody and incubate at 4°C for 30 min.
5. After washing with B.E. PBS again, resuspend in 1 μg/ml of propidium iodide.
6. Measure samples on a flow cytometer and analyze data using the software for analysis (*see* **Note 12**).

3.3 Expansion of T Cells

Human peripheral blood mononuclear cells (PBMCs) derived from healthy young adults or cancer patients are prepared and cultured.

1. Prepare 10 μg/ml of anti-CD3 antibody coated plates (*see* **Note 13**).
2. Add PBMC-derived T cells suspended with the medium on the plate above and culture for 3 days in the presence of 100 IU/ml IL-2 under 37°C and 5% CO_2 conditions (*see* **Note 14**).
3. After 3 days, collect T-cell culture medium and centrifuge at 2000 rpm for 5 min (*see* **Note 15**).
4. Measure the number of cells after staining cells with trypan blue.
5. Change medium and 100 IU/ml IL-2 is added every 3–5 days.

3.4 How to Prepare Cell Line-Derived Xenograft Model (CDX)

1. Cancer-bearing mice are generated by transplanting 1×10^7 cells of cancer cells of the cell line into the dorsal skin of BALB/c-nu mice (4-week-old female) (*see* **Note 16**).
2. When tumor volume grew to around 300 mm^3, the mice were randomly grouped.
3. Tumor volume is measured according to the following formula: [tumor volume = (diameter) × (short diameter) 2 × 1/2].

3.5 How to Prepare Patient-Derived Xenograft Model (PDX)

1. Cancer-bearing mice are generated by transplanting 1×10^7 cells of patient-derived cancer cells into the dorsal skin of BALB/c-nu mice (*see* **Note 16**).
2. When tumor volume grew to around 300 mm^3, the mice were randomly grouped.
3. Tumor volume is measured according to the following formula: [tumor volume = (diameter) × (short diameter) 2 × 1/2].

3.6 Immunohistochemistry of Targeted Antigens

1. Remove the tumor from the BALB/c-nu mouse and freeze it in the O.C.T. compound (*see* **Note 17**).
2. Using a cryostat, prepare frozen sections of the tumors from Step 1 and place them on glass slides.
3. Air dry the frozen sections on glass slides for 30 min (*see* **Note 18**).
4. Treat with 4% paraformaldehyde for 15 min at room temperature (*see* **Note 19**).
5. Wash three times with PBS.
6. Mix 15 ml of 30% H_2O_2 and 135 ml of methanol to prepare a 3% H_2O_2 endogenous peroxidase inactivation solution.
7. Add the solution prepared in Step 6 to the sample and incubate at room temperature for 30 min.
8. Wash the samples three times with PBS, add 5% skim milk, and incubate for 30 min at room temperature (*see* **Note 20**).
9. Wash again with PBS three times.
10. Rewash the samples with PBS three times, dilute the primary antibody with 5% skim milk, add to the samples, and incubate at room temperature for 1 h.
11. Rewash the samples with PBS three times (*see* **Note 21**), add fluorescent-labeled secondary antibody, and incubate at room temperature for 1 h.
12. Wash again with PBS three times (*see* **Note 22**), add Fluoromount-G, and cover with cover glass to seal.
13. Take images using a fluorescence microscope.

3.7 HE Staining

1. Prepare frozen samples and fix samples with 4% paraformaldehyde as described in Steps 1–4 of Method 3.5.
2. Wash three times with PBS and perform nuclear staining with hematoxylin for 10 min.
3. Remove color from the samples by running tap water for 10 min (*see* **Note 23**).
4. Dye with eosin for 3–5 min.
5. Wash with tap water and drain off the water (*see* **Note 23**).

6. Soak in ethanol for 1 min three times for dehydration and permeability.
7. Soak in xylene solution for 1 min twice and seal with a cover glass.
8. Take images under a fluorescence microscope.

4 In Vivo Fluorescent Imaging of TDB (*see* Note 24)

4.1 TDB and Control Antibody Labeling

1. Antibody-reactive dye is added to an antibody solution of TDB, and the control antibody is prepared at 2 mg/ml and allowed to react for 1 h at room temperature with agitation.
2. After labeling, the labeled antibody and free dye are separated by gel filtration chromatography (*see* **Note 25**).

4.2 T-Cell Labeling

1. Prepare XenoLight DiR solution (hereafter referred to as DiR) by adding 1 μl of DiR to 500 μl of Roswell Park Memorial Institute medium-1640 (RPMI 1640).
2. Suspend T cells in DiR solution to a cell concentration of 1 × 10^7 cells/ml and incubate at 37°C for 45 min.
3. Wash the cells twice with PBS and collect the labeled cells (DiR-T cells).

4.3 Injection of Labeling T Cell and Antibody to the Mice

1. Suspend 2 × 10^7 cells of DiR-T cells in 100 μl of PBS and transplant them intraperitoneally into carcinoma-bearing mice.
2. After 24 h, 5 mg/kg of Alexa Fluor 647-labeled or control IgG is injected into each mouse (Fig. 1).
3. After injection, mice are photographed on fluorescent in vivo imaging system (Fig. 2).
4. The images are analyzed by Living Image Software.

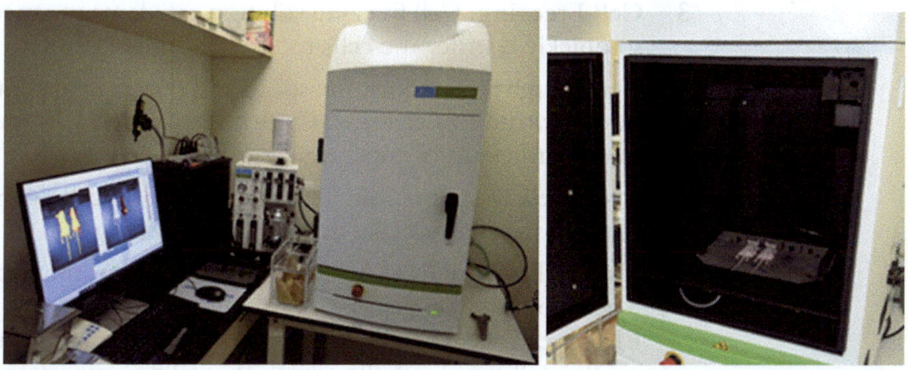

Fig. 1 The IVIS Lumina S5 (PerkinElmer) machine used to capture fluorescence in vivo imaging and the image being captured

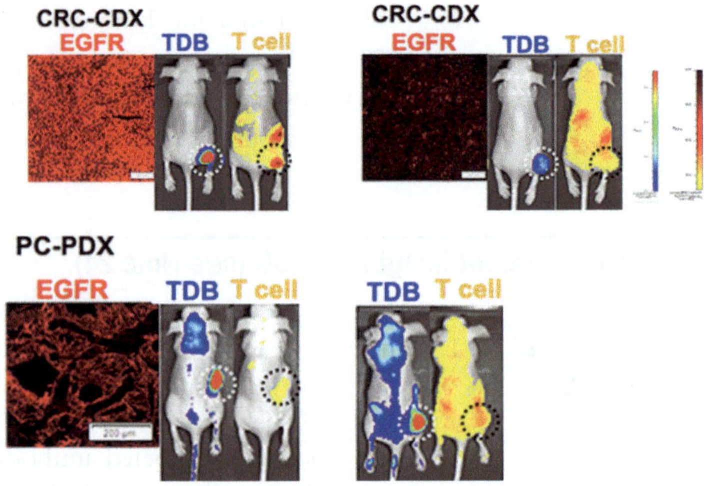

Fig. 2 The simultaneous detection of T-BsAb and T cells, with the upper figure representing the experiment using CDX and the lower figure using PDX. Fluorescent immunostaining shows the amount of EGFR expression in each tumor; the high red levels are more abundant. The bars on the right side of the bottom row indicate the amount of T-BsAb and T-cell accumulation individually. Experiments using the protocol described above obtained reproducible detection

5 Notes

1. **Pausing Experiments**: When pausing an experiment, keep the cells at 4°C (e.g., on ice) to prevent them from weakening.
2. **FBS Lot Variation**: Cell culture efficiency can vary depending on the lot number of FBS. It's recommended to compare multiple lot numbers (e.g., purchase about five different lots) and select the one with the highest cell growth rate. For consistency and reproducibility, use the same lot number whenever possible.
3. **Cell Freezing**: Cell freezing solution can damage cancer cells, so quickly transfer the cells to the medium after thawing.
4. **Centrifugation**: Adjust the speed and duration of centrifugation according to cell size. Centrifugation at 1500 rpm for 3 min is sufficient for most cancer cells.
5. **Even Cell Distribution**: To achieve uniform cell distribution in the dish, gently shake the dish after seeding.
6. **Using Trypsin-EDTA**: Remove cells from the incubator when about 80% floating. Prolonged exposure to trypsin-EDTA can weaken the cells. The incubation time depends on the cell type; initially, check the cells 5 min after adding Trypsin-EDTA. If the cells are floating but not completely

separated, gently shake the dish or pipette the cells to help with detachment.

7. **Removing Trypsin-EDTA**: Centrifuge and discard the supernatant to remove Trypsin-EDTA.
8. **Cell Counting and Seeding**: After counting the cells as described in Section 3.1.4, adjust the cell suspension volume before transferring it to the culture dish.
9. **Freezing Cells**: To avoid cell damage, freeze cells immediately after being in the freezing solution.
10. **Accurate Cell Counting**: Ensure cells are fully separated into individual cells before counting, achieved by thorough pipetting.
11. **Plasma Membrane Protein Analysis**: Since trypsin-EDTA can temporarily disrupt the plasma membrane, use a dissociation buffer to check membrane protein expression levels.
12. **Flow Cytometry Wavelengths**: Save the analyzed wavelengths when repeating experiments to maintain consistency and ease of comparison. This helps check reproducibility and quantify differences in protein expression across experiments.
13. **Preparing Anti-CD3-Coated Plates**: You can prepare anti-CD3 antibodies by diluting anti-CD3 antibodies in PBS at 10 μg/ml or purchase pre-coated plates.
14. **Adding IL-2**: Always add IL-2 to the culture, as T cells will not proliferate and die without it.
15. **Centrifuging T Cells**: T cells, smaller than cancer cells, precipitate more effectively at higher centrifugation speeds.
16. **Cell Transplantation Accuracy**: Variations in transplanted cell numbers can lead to differences in tumor sizes among mice. Carefully calculate cell counts using the average of multiple measurements. When transplanting, ensure cells do not spill from the syringe. All animal experiments and human sample usage comply with ethical guidelines and be approved by the Institutional Animal Care and Use Committee (IACUC) and the Institutional Review Board (IRB).
17. **Embedding in OCT Compound**: Embed the OCT compound carefully to avoid air bubbles, as these can complicate sectioning.
18. **Frozen Section Preparation**: Avoid high temperatures during frozen section preparation to prevent tissue damage.
19. **Sample Fixation**: Ensure proper sample fixation to maintain structural integrity.
20. **Overnight Incubation**: It is possible to incubate samples overnight at 4°C during specific steps.

21. **Washing After Primary Antibody Incubation**: Wash samples three times to remove any excess primary antibody.
22. **Washing After Secondary Antibody Incubation**: Wash samples three times to ensure no residual secondary antibody remains.
23. **Washing with Tap Water**: When rinsing with tap water, ensure the water does not directly hit the sample to prevent damage.
24. **Fluorescent Labeling for In Vivo**: Label each cell with a different fluorescent dye to observe the dynamics of TDBs and T cells simultaneously in vivo.
25. **Labeling Kit**: For antibody labeling, we used a commercially available protein labeling kit utilizing Alexa Fluor 647 (e.g., from Thermo Fisher Scientific).

Acknowledgments

The authors thank the members of the Yasunaga laboratory for their helpful discussions and M. Shimada for secretarial support. The authors also thank Dr. R. Asano for providing the materials and support. This research was supported by AMED under Grant Number JP21ak0101182 to M.Y. and the National Cancer Center Research and Development Fund (Number 23-A-08) to M.Y.

References

1. Goydel RS, Rader C (2021) Antibody-based cancer therapy. Oncogene 40(21):3655–3664
2. Topalian SL, Hodi FS, Brahmer JR, Gettinger SN, Smith DC, McDermott DF, Powderly JD, Carvajal RD, Sosman JA, Atkins MB et al (2012) Safety, activity, and immune correlates of anti-PD-1 antibody in cancer. N Engl J Med 366(26):2443–2454
3. Hodi FS, O'Day SJ, McDermott DF, Weber RW, Sosman JA, Haanen JB, Gonzalez R, Robert C, Schadendorf D, Hassel JC et al (2010) Improved survival with ipilimumab in patients with metastatic melanoma. N Engl J Med 363(8):711–723
4. Holstein SA, Grant SJ, Wildes TM (2023) Chimeric antigen receptor t-cell and bispecific antibody therapy in multiple myeloma: moving into the future. J Clin Oncol: Off J Am Soc Clin Oncol 41(27):4416–4429
5. Wei AH, Ribera JM, Larson RA, Ritchie D, Ghobadi A, Chen Y, Anderson A, Dos Santos CE, Franklin J, Kantarjian H (2021) Biomarkers associated with blinatumomab outcomes in acute lymphoblastic leukemia. Leukemia 35(8):2220–2231
6. Hutchings M, Morschhauser F, Iacoboni G, Carlo-Stella C, Offner FC, Sureda A, Salles G, Martínez-Lopez J, Crump M, Thomas DN et al (2021) Glofitamab, a novel, bivalent CD20-targeting T-cell-engaging bispecific antibody, induces durable complete remissions in relapsed or refractory B-cell lymphoma: a phase I trial. J Clin Oncol: Off J Am Soc Clin Oncol 39(18):1959–1970
7. Kamakura D, Asano R, Kawai H, Yasunaga M (2021) Mechanism of action of a T cell-dependent bispecific antibody as a breakthrough immunotherapy against refractory colorectal cancer with an oncogenic mutation. Cancer Immunol Immunother 70(1):177–188
8. Budde LE, Sehn LH, Matasar M, Schuster SJ, Assouline S, Giri P, Kuruvilla J, Canales M, Dietrich S, Fay K et al (2022) Safety and efficacy of mosunetuzumab, a bispecific antibody,

in patients with relapsed or refractory follicular lymphoma: a single-arm, multicentre, phase 2 study. Lancet Oncol 23(8):1055–1065
9. Hutchings M, Mous R, Clausen MR, Johnson P, Linton KM, Chamuleau MED, Lewis DJ, Sureda Balari A, Cunningham D, Oliveri RS et al (2021) Dose escalation of subcutaneous epcoritamab in patients with relapsed or refractory B-cell non-Hodgkin lymphoma: an open-label, phase 1/2 study. Lancet (London, Engl) 398(10306):1157–1169
10. Middelburg J, Kemper K, Engelberts P, Labrijn AF, Schuurman J, van Hall T (2021) Overcoming challenges for CD3-bispecific antibody therapy in solid tumors. Cancers 13(2):287
11. Bonaventura P, Shekarian T, Alcazer V, Valladeau-Guilemond J, Valsesia-Wittmann S, Amigorena S, Caux C, Depil S (2019) Cold tumors: a therapeutic challenge for immunotherapy. Front Immunol 10:168
12. Hegde PS, Chen DS (2020) Top 10 challenges in cancer immunotherapy. Immunity 52(1):17–35
13. Dranoff G (2011) Experimental mouse tumour models: what can be learnt about human cancer immunology? Nat Rev Immunol 12(1):61–66
14. Kamakura D, Asano R, Yasunaga M (2021) T cell bispecific antibodies: an antibody-based delivery system for inducing antitumor immunity. Pharm (Basel, Switz) 14:11
15. Zheng S, Wang W, Shen L, Yao Y, Xia W, Ni C (2024) Tumor battlefield within inflamed, excluded or desert immune phenotypes: the mechanisms and strategies. Exp Hematol Oncol 13(1):80

Check for updates

Chapter 13

Fluorescent-Based Analysis of T Cells and T-Cell-Dependent Bispecific Antibody in a Brain Tumor Mouse Model

Yuri Nakamura, Teppei Azuma and Masahiro Yasunaga

Abstract

T-cell-dependent bispecific antibody (TDB) is a novel therapeutic antibody considered one of the promising immunotherapies for treating brain tumors. TDB activates T cells via CD3-mediated signaling by crosslinking CD3 expressed on T cells and tumor-associated antigens expressed on cancer cells. In preclinical studies, the efficacy of TDB targeting EGFRvIII, one of tumor-associated antigens, has been validated against malignant brain tumors. Despite its therapeutic potential, the in vivo dynamics of TDB and the T cells remain poorly understood. In this chapter, we describe a method to analyze the in vivo dynamics of TDB and T cells in brain tumor-bearing mice using fluorescence imaging.

Keywords T-cell-dependent bispecific antibodies, Brain tumor, Cell line-derived xenograft model, Brain tumor mouse model, Fluorescent imaging, Ex vivo imaging, Cancer immunotherapy

1 Introduction

From 1999 to 2020, the number of brain tumor cases in the United States reached about 440 000 [1]. Brain tumors seriously affect patients' neurological functions and quality of life and are considered a major global concern [2]. The standard treatment for brain tumors includes surgical resection, radiation therapy, and chemo therapy [3]. However, surgical approaches are often limited due to the highly localized functions of the brain, the invasive nature of malignant tumors, and their proximity to critical structures [3]. Radiation therapy is also constrained by the risk of radiation-induced damage and subsequent brain dysfunction [4]. Furthermore, conventional systemic chemotherapy has demonstrated limited efficacy against brain tumors due to the restricted drug delivery imposed by the blood–brain barrier (BBB) and the blood–brain tumor barrier (BBTB) [5, 6].

Recent studies have challenged the long-standing notion that the central nervous system (CNS) is an immune-privileged site devoid of

Masahiro Yasunaga (ed.), *Bispecific Antibodies*, Methods in Molecular Biology, vol. 2997,
https://doi.org/10.1007/978-1-0716-5037-0_13,

© The Author(s), under exclusive license to Springer Science+Business Media, LLC, part of Springer Nature 2026

immune surveillance [7–9]. As a result, immunotherapy has emerged as a promising alternative to overcome therapeutic limitations in brain tumors [10].

One of the immunotherapies with significant potential for clinical application in brain tumors is T-cell-dependent bispecific antibody (TDB) [11]. TDB is a novel therapeutic antibody binding both CD3 on T cells and tumor-associated antigens on cancer cells [11], activating T cells via CD3-mediated signaling [12, 13].

Activated T cells release cytotoxic molecules such as perforin and granzyme, as well as cytokines like interferon-gamma (IFN-γ) and tumor necrosis factor-alpha (TNFα), exerting cytotoxic activity [12]. Additionally, activated T cells proliferate in response to intracellular signaling within the tumor [14, 15] and secrete T-cell-attracting chemokines, recruiting peripheral T cells to the tumor site [16, 17]. Furthermore, the cytotoxic molecules and cytokines released by activated T cells induce cytotoxicity in adjacent cells, regardless of the target antigen expression on these cells [18]. This feature makes TDB effective against refractory tumors with heterogeneous antigen expression. Since cytokine-mediated cytotoxicity is independent of the effector-to-target cell (E:T) ratio, TDB can also be effective in noninflammatory tumors with low T-cell infiltration [19]. Moreover, immunosuppressive regulatory T cells (Tregs) can also be redirected by TDB to cancer cells, inducing granzyme-mediated cytotoxicity [20].

Brain tumors are noninflammatory in nature, with low T-cell infiltration [21, 22], causing them to be poorly responsive to conventional immunotherapies. TDB, which promotes T-cell infiltration, proliferation, and activation at tumor sites, emerged as a promising approach for treating malignant brain tumors [11, 23, 24].

In preclinical studies, a TDB targeting epidermal growth factor receptor variant III (EGFRvIII) on tumor cells induced tumor regression in patient-derived xenograft (PDX) models of glioblastoma multiforme (GBM), demonstrating the effectiveness of TDB for malignant brain tumors [11].

Despite these promising findings, the pharmacokinetics and biodistribution of TDBs in brain tumors remain poorly understood. Key questions remain unanswered, such as whether TDB effectively penetrates the brain tumor microenvironment, how T cells distribute and proliferate postadministration, and whether TDB-driven T-cell activation sustains long-term anti-tumor immunity. Conventional ex vivo analyses provide only endpoint measurements, limiting our ability to capture real-time therapeutic dynamics.

To address these gaps, this chapter describes a fluorescence imaging protocol to investigate the ex vivo dynamics of TDB and T cells in a brain tumor-bearing mouse model. By utilizing near-infrared fluorescent labeling, this method enables longitudinal and monitoring of antibody distribution and T-cell infiltration in the tumor microenvironment. Understanding these spatiotemporal dynamics will be crucial for optimizing TDB-based immunotherapy for brain tumors.

2 Materials

2.1 Cancer Cell Culture

1. Water bath
2. Cell culture medium and supplements (10% fetal bovine serum (FBS) and 1% Penicillin-Streptomycin Solution)
3. Cancer cells. Here, the breast cancer cell line HCC1954 is used as an example.
4. Culture dish (25 cm)
5. Aspirator
6. Micropipette
7. CO_2 incubator (37°C, 5% CO_2)
8. Centrifuge

2.2 Brain Tumor Model Mouse (see Note 7)

1. Immunodeficient mouse (Nude mouse)
2. Medetomidine Hydrochloride
3. Midazolam
4. Butorphanol Tartrate
5. Atipamezole Hydrochloride
6. Iodine swab
7. Ethanol
8. Cotton soaked in 70% ethanol
9. Cotton (4 × 4 cm)
10. Disposable scalpel
11. Cotton swab
12. Drill
13. Hamilton syringe (10 μL)
14. Dulbecco's phosphate buffered saline (D-PBS), Ca2+-/ Mg2+ -free pH 7.4
15. Ultrapure water
16. Bone wax
17. Suture with needle, 11 mm, 135°C, reverse cutting needle
18. Scissor
19. Cell line expressing the bioluminescent enzyme luciferase
20. Transplant instrument

21. Heating pad
22. Pencil
23. Needle holder
24. Tweezer

2.3 Monitoring of Cancer Cell Bioluminescence

1. Cell line expressing the bioluminescent enzyme luciferase
2. D-PBS
3. Saline
4. 1.5 mL light-shielding tubes
5. Injection syringe (27-G)
6. Micropipette
7. IVIS Fluorescence and Luminescence imaging system
8. Software for Fluorescence and Luminescence Analysis

2.4 T-cell Culture

1. 6-well plate
2. Anti-CD3 antibody
3. D-PBS
4. Human Peripheral blood mononuclear cells (PBMC) (e.g., from Cellular Technology Limited, Lonza, or Precision for Medicine)
5. Cell culture medium and supplements (10% fetal bovine serum (FBS) and 1% Penicillin-Streptomycin Solution). We use RPMI-1640 supplemented with 10% heat-inactivated FBS and 1% Penicillin-Streptomycin Solution for T cells.
6. Interleukin-2 (IL-2)
7. Culture dish (25 cm)
8. Aspirator
9. Micropipette
10. CO_2 incubator (37°C, 5% CO_2)
11. Centrifuge

2.5 Fluorescent Labeling of TDB and T Cells

1. T-cell-dependent bispecific antibody (TDB)
2. T cells
3. Fetal bovine serum (FBS)
4. Penicillin-Streptomycin Solution
5. Roswell Park Memorial Institute medium-1640 (RPMI1640)
6. Cell Labeling Kit. Here, we use XenoLight DiR for T-cell labeling.
7. Antibody labeling Kit
8. 25 mL tube

9. Aspirator
10. Micropipette
11. CO_2 incubator (37°C, 5% CO_2)
12. Centrifuge

2.6 Fluorescence Imaging of T Cell and TDB Kinetics in Brain Tumor Mouse Model

1. Brain Tumor Model Mouse
2. Labeled TDB
3. Labeled T cells
4. Injection syringe
5. Cotton soaked in 70% ethanol (4 × 4 cm)
6. Cotton (4 × 4 cm)
7. Scissor
8. Tweezer
9. 24-well plate
10. Saline
11. Kimwipe
12. Cork board
13. Kimtowel
14. Fixation apparatus
15. Anesthetic
16. IVIS Fluorescence and Luminescence Imaging System
17. Software for Fluorescence and Luminescence Analysis

3 Methods

3.1 Preparation of Brain Tumor Mouse Model

3.1.1 Culture Cancer Cells

1. Prepare the culture medium by adding 10% fetal bovine serum (FBS) and 1% Penicillin-Streptomycin Solution to a suitable medium (*see* **Note 1**).
2. Thaw the frozen cancer cells in a 37°C water bath and dissolve them in the prepared medium (*see* **Note 2**).
3. Centrifuge at 1200 rpm for 3 min (*see* **Note 3**).
4. Remove the supernatant and add an appropriate volume of culture medium.
5. Seed the cells into dishes and culture them in an incubator set at 37°C with 5% CO_2 (*see* **Note 4**).

3.1.2 Create CDX Mouse Model (see Note 5)

1. Mix Medetomidine Hydrochloride, Midazolam, and Butorphanol Tartrate (M/M/B) for anesthesia in a ratio of M/M/B = 0.75/4/5 mg/kg.

2. Prepare a cancer cell suspension at a concentration of 5.0×10^6 cells/150 μL D-PBS) (*see* **Note 6**).
3. Measure the body weight of each mouse (*see* **Note 7**).
4. Inject the anesthetic intraperitoneally at 100 μL/10 g body weight (*see* **Note 8**).
5. Secure the mouse in the transplant instrument (*see* **Note 9**).
6. Disinfect the head area with iodine cotton, then wipe with ethanol.
7. Make an incision on the head with a scalpel (*see* **Note 10**).
8. Expose the skull and wipe off the blood with a cotton swab (*see* **Note 11**).
9. Install the Hamilton syringe on the transplantation device, and mark a position 1 mm anterior and 1.8 mm right from Bregma (the point on the skull where two seams meet: the coronal suture (running from side to side near the forehead) and the sagittal suture (running from front to back along the top of the head)).
10. Move the Hamilton syringe to a noninterfering position and drill a burr hole at the marked site (*see* **Note 12**).
11. Wipe off the blood on the skull using a cotton swab.
12. Wash the Hamilton syringe by rinsing it with ethanol (three times) and D-PBS (three times).
13. Insert the Hamilton syringe to a depth of 3 mm from the brain surface through the burr hole (*see* **Note 13**).
14. Inject the prepared cancer cell suspension at a rate of 0.25 μL/30 s for 5.5 min (*see* **Note 14**).
15. Let the syringe remain in place for 5 min.
16. Detach the Hamilton syringe from the transplant instrument and wash it with ethanol (three times) and D-PBS (three times) (*see* **Note 15**).
17. Wipe off the leaked cells and blood on the surface of the skull with a cotton swab (*see* **Note 16**).
18. Seal the burr hole with bone wax (*see* **Note 17**).
19. Suture the skin to close the incision (*see* **Note 18**).
20. Inject the Atipamezole Hydrochloride as antagonist intraperitoneally at 0.75 mg/kg.
21. Place the mouse on a heating pad and monitor for recovery and awakening (*see* **Note 19**).

3.1.3 Monitoring Brain Tumors

1. Dissolve 1 g of D-Luciferin, the substrate for the luciferase enzyme, in 40 mL of D-PBS to prepare a 25 mg/mL solution.
2. Aliquot 500 μL into 1.5 mL light-shielding tubes and store at −20°C.

3. Before use, thaw the D-Luciferin solution and add 500 μL of saline to each tube (final concentration: 12.5 mg/mL).
4. Administer D-Luciferin intraperitoneally to brain tumor model mice at a dose of 125 mg/kg.
5. Ten minutes after administration, image the tumor-bearing mice using the IVIS Fluorescence and Luminescence Imaging System.
6. Measure the luminescence intensity of cancer cells using software for fluorescence and luminescence analysis (Fig. 1). All animal experiments comply with ethical guidelines and are approved by the Institutional Animal Care and Use Committee (IACUC).

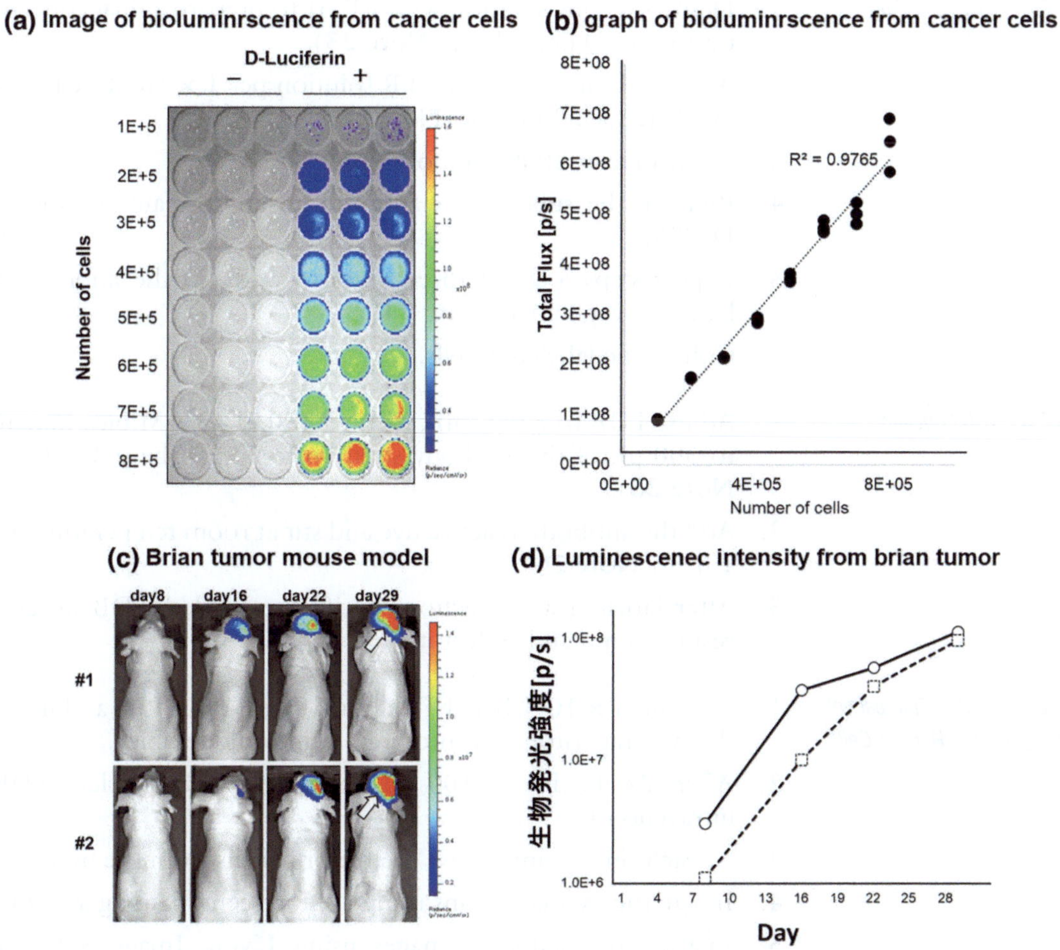

Fig. 1 The increase in luminescence intensity corresponds to the number of luciferase-expressing cancer cells after adding luciferin (**a**). An image (**b**) is a line graph. Tumor luminescence in brain tumor model mice (**c**) images (**d**) a line graph

3.2 Fluorescence Imaging of T Cell and TDBT Kinetics in Brain Tumor Mouse Model

3.2.1 Expansion of T Cell

1. Add 1 mL/well of anti-CD3 antibody diluted to 10 μg /mL in D-PBS to a 6-well plate and incubate for 1 h (*see* **Note 20**).
2. Remove the coating solution using an aspirator.
3. Prepare the culture medium by adding 10% fetal bovine serum (FBS) and 1% Penicillin-Streptomycin Solution to Roswell Park Memorial Institute medium-1640 (RPMI1640) (*see* **Note 1**).
4. Thaw the frozen PBMC in a 37°C water bath and dissolve them in the prepared medium.
5. Centrifuge at 2000 rpm for 5 min.
6. Remove the supernatant and add 12 mL of RPMI.
7. Seed 4 mL/well into the prepared 6-well plate.
8. Expand T cells by adding 100 IU/mL of IL-2 and passaging them three to four times (*see* **Notes 21** and **22**).

3.2.2 T Cell Labeling

1. Dilute the DiR solution with RPMI to prepare a final concentration of 50 μg/mL (*see* **Note 23**).
2. Add 1 mL of the diluted DiR solution per 1 × 10^7 T cells and incubate for 20 min (*see* **Noe 23**).
3. Centrifuge at 2000 rpm for 5 min.
4. Remove the supernatant and add an appropriate volume of D-PBS.
5. Repeat steps 3 and 4 three times to wash off the fluorescent labeling reagent (*see* **Note 24**).
6. Collect the labeled T cells.

3.2.3 TDB Labeling

1. Adjust TDB to 2 mg/mL, then add 50 μL of 1 M bicarbonate to 500 μL of the antibody solution to adjust the pH 7.4 (*see* **Note 25**).
2. Add the antibody-reactive dye and stir at room temperature for 1 h (*see* **Note 26**).
3. After labeling and reaction, purify the labeled TDB using a resin column and collect it.

3.2.4 Ex Vivo Fluorescent Imaging of TDB and T Cell

1. Suspend 4 × 10^7 labeled T cells in 100 μL of D-PBS and inject them intraperitoneally into brain tumor model mice.
2. After 24 h, inject 100 μL of 1 mg/mL labeled TDB intravenously.
3. At each time point, excise the major organs from the mouse.
4. Image the excised organs using a fluorescent imaging system.
5. Analyze the acquired images using Living Image Software (Fig. 2).

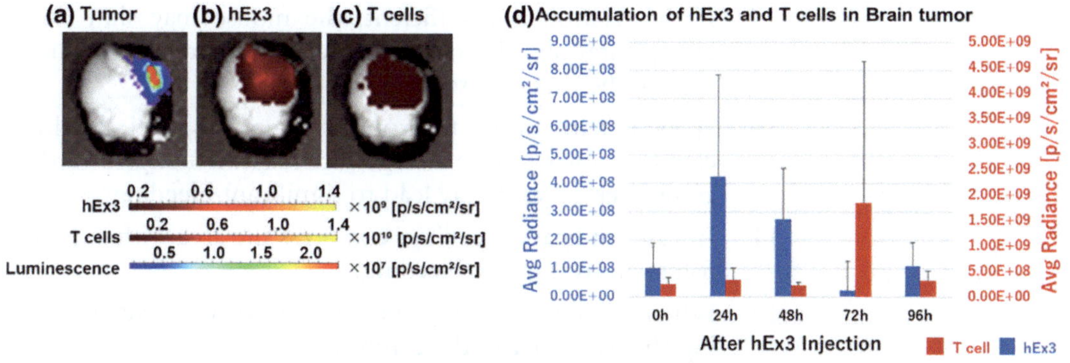

Fig. 2 Ex vivo imaging at 72 h after hEx3 administration showing (**a**) brain tumor luminescence, (**b**) fluorescence of hEx3, (**c**) fluorescence of T cells. (**d**) fluorescence of hEx3 (blue) and T cells (red) in brain tumors at each time point

4 Notes

1. The efficiency of cell culture may vary depending on the Lot number of FBS. It is recommended to compare about five different lots and select the one that provides the highest cell growth rate. To ensure consistency and reproducibility, always use the same Lot number.
2. Cell freezing solution can damage cancer cells, so quickly transfer the cells to the medium after thawing.
3. Adjust the speed and duration of centrifugation according to cell size. Centrifugation at 1500 rpm for 3 min is sufficient for most cancer cells.
4. Gently tap to ensure uniform cell distribution in the dish.
5. Older mice tend to have a lower cell engraftment rate. It is recommended to use 4–6-week-old mice for transplantation. All animal experiments comply with ethical guidelines and be approved by the Institutional Animal Care and Use Committee (IACUC).
6. The cell density of the suspension affects tumor engraftment rate and duration. Since the optimal cell density varies depending on the cell line, it is necessary to compare and evaluate at least three different cell densities.
7. Transplantation should not be performed on immature or underweight mice, as it imposes a significant burden on them. For nude mice, transplantation should not be performed on individuals weighing less than 18 g.
8. Place the mouse on a heating pad for approximately 10 min to maintain body temperature while waiting for the anesthesia to take full effect.

9. If the fixation is insufficient, the mouse may shift during transplantation, placing a significant burden on the animal. Ensure the mouse is securely fixed.
10. While it is important to properly secure the surgical field, be cautious not to make a large incision extending toward the cranial vertex, as it may lead to significant bleeding.
11. In cases of heavy bleeding, compress the area with a dry cotton swab to stop the bleeding.
12. When drilling a burr hole, be cautious to avoid excessive drilling that could harm the cerebrum.
13. If the Hamilton syringe is inserted too shallowly, cells may leak onto the brain surface. Conversely, if the insertion is too deep, it may lead to neurological damage. Therefore, take the utmost care when inserting the syringe.
14. When injecting, carefully ensure that cells are not leaking onto the brain surface.
15. To prevent blood or cells from clogging the syringe, wash it with ethanol and D-PBS before the blood coagulates on the syringe.
16. If blood adheres to the skull, it may hinder the proper sealing of the burr hole in subsequent steps. Additionally, if cells adhere to the skull, it may lead to metastasis to regions outside the brain.
17. Using too much bone wax may exert pressure on the brain, so apply an appropriate amount.
18. Suture at least two points.
19. Warming the mouse on a heating pad shortens the time it takes to awaken and reduces its stress. Prolonged hypothermia can place a significant burden on the mouse, so ensure the transplantation procedure is performed carefully and smoothly. Leave the mouse on the heating pad for at least 20 min to allow it to recover its body temperature.
20. Gently tap to ensure the antibody dilution covers the bottom of the well.
21. Always add IL-2 to the culture, as T cells will not proliferate and die without it.
22. T cells, smaller than cancer cells, precipitate more effectively at higher centrifugation speeds.
23. The higher the concentration of the DiR solution, the shorter the staining time, which helps reduce damage to the cells. However, an excessively high DiR concentration can also cause cellular damage. Therefore, at least three different concentrations should be tested to determine the optimal concentration. In this experiment, a relatively high concentration

of 50 μg/mL was used to improve T cell viability by keeping the staining time relatively short at 20 min. If unbound labeling reagents are not thoroughly removed, their fluorescence may interfere with the accurate evaluation of T cell distribution in each organ during ex vivo imaging. Therefore, ensure thorough washing.

24. If the antibody concentration is not appropriate, labeling may fail. Check the optimal concentration for each kit used.
25. Cover the container with aluminum foil to protect it from light during the reaction.
26. Fill a 24-well plate with saline and immerse the excised organs to wash them. For the brain, do not immerse it in saline; instead, gently use a Kimwipe to remove blood.

Acknowledgments

The authors thank the members of the Yasunaga laboratory for their helpful discussions and M. Shimada for secretarial support. The authors also thank Dr. R. Asano for providing the material and support. This research was supported by AMED under Grant Number JP21ak0101182 to M.Y. and the National Cancer Center Research and Development Fund (Number 23-A-08) to M.Y.

References

1. Siegel RL, Giaquinto AN, Jemal A (2024) Cancer statistics. CA Cancer J Clin 74(1):12–49
2. Manake R, Phillips V, Gangi A, Ravikumar J (2024) Trends in the incidence of brain cancer: an observational study. Cureus 16(10):e72805
3. 近藤聡英 (2024) 脳腫瘍の最新治療. 神経治療 41(4):479–484
4. Owonikoko TK, Arbiser J, Zelnak A, Shu H-KG, Shim H, Robin AM et al (2014) Current approaches to the treatment of metastatic brain tumours. Nat Rev Clin Oncol 11(4):203–222
5. Eichler AF, Chung E, Kodack DP, Loeffler JS, Fukumura D, Jain RK (2011) The biology of brain metastases-translation to new therapies. Nat Rev Clin Oncol 8(6):344–356
6. Fortin D (2012) The blood-brain barrier: its influence in the treatment of brain tumors metastases. Curr Cancer Drug Targets 12(3):247–259
7. Louveau A, Smirnov I, Keyes TJ, Eccles JD, Rouhani SJ, Peske JD et al (2015) Structural and functional features of central nervous system lymphatic vessels. Nature 523(7560):337–341
8. Kipnis J (2016) Multifaceted interactions between adaptive immunity and the central nervous system. Science 353(6301):766–771
9. Laman JD, Weller RO (2013) Drainage of cells and soluble antigen from the CNS to regional lymph nodes. J Neuroimmune Pharmacol 8(4):840–856
10. McGranahan T, Therkelsen KE, Ahmad S, Nagpal S (2019) Current state of immunotherapy for treatment of glioblastoma. Curr Treat Options Oncol 20(3):24
11. Iurlaro R, Waldhauer I, Planas-Rigol E, Bonfill-Teixidor E, Arias A, Nicolini V et al (2022) A novel EGFRvIII T-cell bispecific antibody for the treatment of glioblastoma. Mol Cancer Ther 21(10):1499–1509
12. Bacac M, Fauti T, Sam J, Colombetti S, Weinzierl T, Ouaret D et al (2016) A novel carcinoembryonic antigen T-cell bispecific antibody (CEA TCB) for the treatment of solid tumors. Clin Cancer Res 22(13):3286–3297
13. Staerz UD, Kanagawa O, Bevan MJ (1985) Hybrid antibodies can target sites for attack by T cells. Nature 314(6012):628–631. https://doi.org/10.1038/314628a0
14. Chiu D, Tavaré R, Haber L, Aina OH, Vazzana K, Ram P et al (2020) A PSMA-targeting CD3 bispecific antibody induces antitumor responses

that are enhanced by 4-1BB costimulation. Cancer Immunol Res 8(5):596–608
15. Sam J, Colombetti S, Fauti T, Roller A, Biehl M, Fahrni L et al (2020) Combination of T-cell bispecific antibodies with PD-L1 checkpoint inhibition elicits superior anti-tumor activity. Front Oncol 10:575737
16. Franciszkiewicz K, Boissonnas A, Boutet M, Combadière C, Mami-Chouaib F (2012) Role of chemokines and chemokine receptors in shaping the effector phase of the antitumor immune response. Cancer Res 72(24):6325–6332
17. Uppaluri R, Sheehan KCF, Wang L, Bui JD, Brotman JJ, Lu B (2008) Prolongation of cardiac and islet allograft survival by a blocking hamster anti-mouse CXCR3 monoclonal antibody. Transplantation. 86(1):137–147
18. Isaaz S, Baetz K, Olsen K, Podack E, Griffiths GM (1995) Serial killing by cytotoxic T lymphocytes: T cell receptor triggers degranulation, re-filling of the lytic granules and secretion of lytic proteins via a non-granule pathway. Eur J Immunol 25(4):1071–1079
19. Kamakura D, Asano R, Kawai H, Yasunaga M (2021) Mechanism of action of a T cell-dependent bispecific antibody as a breakthrough immunotherapy against refractory colorectal cancer with an oncogenic mutation. Cancer Immunol Immunother 70(1):177–188
20. Choi BD, Gedeon PC, Herndon 2nd JE, Archer GE, Reap EA, Sanchez-Perez L et al (2013) Human regulatory T cells kill tumor cells through granzyme-dependent cytotoxicity upon retargeting with a bispecific antibody. Cancer Immunol Res 1(3):163
21. Chongsathidkiet P, Jackson C, Koyama S, Loebel F, Cui X, Farber SH et al (2018) Sequestration of T cells in bone marrow in the setting of glioblastoma and other intracranial tumors. Nat Med 24(9):1459–1468
22. Thorsson V, Gibbs DL, Brown SD, Wolf D, Bortone DS, Ou Yang T-H et al (2018) The immune landscape of cancer. Immunity 48(4):812–830
23. Razpotnik R, Novak N, Čurin Šerbec V, Rajcevic U (2017) Targeting malignant brain tumors with antibodies. Front Immunol 8:1181
24. Faresjö R, Bonvicini G, Fang XT, Aguilar X, Sehlin D, Syvänen S (2021) Brain pharmacokinetics of two BBB penetrating bispecific antibodies of different size. Fluids Barriers CNS 18(1):26

Chapter 14

Peripheral Blood Mononuclear Cell (PBMC) Humanized Mice as a Platform for a Preclinical In Vivo Evaluation of T-Cell-Dependent Bispecific Antibodies

Ryo Tsumura, Aya Uchida and Masahiro Yasunaga

Abstract

Amid advancements in antibody-based cancer immunotherapy, T-cell-dependent bispecific antibodies (TDBs) hold great promise in cancer treatment. Peripheral blood mononuclear cell (PBMC) humanized mice serve as a platform for preclinical characterization of TDBs, although its induction of xenogeneic graft-versus-host disease (GvHD) limits the experimental time window and introduces confounding factors. NSG-MHC I/II DKO mouse is a highly immunodeficient mouse knocked out for mouse MHC class I and II molecules, which delay GvHD onset following human PBMC engraftment. Here, we describe the materials and methods for generating PBMC humanized mouse models with NSG-MHC I/II DKO mice. Using this model, we demonstrate the feasibility of testing the *in vivo* efficacy of TDBs in NSG-MHC I/II DKO mice engrafted with the colorectal cancer cell line HT29.

Keywords T-cell-dependent bispecific antibody, Humanized mice, PBMC, NSG, NSG-MHC I/II DKO

1 Introduction

Along with the advancement of antibody engineering technologies, T-cell-dependent bispecific antibodies (TDBs) have emerged as one of the promising antibody-based cancer immunotherapies over the past decades. TDBs simultaneously bind to two different antigens: one on a membrane of the T cell (*e.g.*, CD3) and another on the surface of target cells. Cross-linking the T cell and target cancer cell via TDBs facilitates the T-cell activation and its target cell killing [1]. After the approval of blinatumomab (anti-CD3/CD19 antibody) for the treatment of relapsed or refractory B-cell acute lymphoblastic leukemia (B-ALL) in 2014 [2], the FDA has successively approved TDBs targeting different tumor-associated antigens (TAAs) including CD20 [3, 4], B-cell mutation antigen (BCMA) [5], G protein-coupled receptor class C group 5 member D (GPRC5D) [6], and Delta-like

Masahiro Yasunaga (ed.), *Bispecific Antibodies*, Methods in Molecular Biology, vol. 2997,
https://doi.org/10.1007/978-1-0716-5037-0_14,
© The Author(s), under exclusive license to Springer Science+Business Media, LLC, part of Springer Nature 2026

ligand 3 (DLL3) [7]. TDBs targeting a wider array of TAAs have been explored in various stages of clinical development, and these novel TDBs are expected to improve the treatment outcomes of future cancer patients. Despite the promising potential of TDBs, several limitations and challenges remain to be addressed in its preclinical studies. Syngeneic murine models expressing human TAAs and non-human primates (NHPs) have been commonly utilized as preclinical models of TDBs [8, 9]. Although these immunocompetent syngeneic models provide insights into the mode of action and help us select TDB candidates, their species differences can impact the interpretation of its *in vivo* efficacy and their potential side effects represented as off-target effects and cytokine release syndrome (CRS).

Highly immunodeficient mice engrafted with human immune cells (hereinafter referred to as "humanized mice") serve as a powerful tool to overcome these issues. Humanized mice transplanted with human patient-derived cancer cells or tumor tissue recapitulate the human immune response upon TDB administration, hence, it allows us to acquire clinically relevant data. There are two major methods for immune humanization of the mouse: (1) the engraftment of human CD34$^+$ hematopoietic progenitor cells (HPCs) and (2) human peripheral blood mononuclear cells (PBMCs). HPC humanization allows the recipient mice to possess various immune cell subsets, which enable us to monitor the complex immune response [10]. However, the engraftment of HPCs consumes time and cost. On the other hand, PBMC humanization leads to the reconstruction of the human immune system predominated by T cells, which takes a relatively short time window [10]. Since PBMC contains mature T cells that went through thymic selection in humans, this model exerts robust immune responses, which makes it suited for the *in vivo* evaluation of TDB candidates. Because PBMC humanization takes a relatively short time and PBMCs are easier to obtain than CD34$^+$ HPCs, it allows us to perform preclinical studies of TDBs within a specified time frame at an affordable cost. As this model recapitulates the variability of individual PBMC donors, it enables simulating patient-specific treatment responses *in vivo*, thereby offering a potent tool to advance personalized medicine.

One of the important considerations about PBMC humanization is the development of acute xenogeneic graft-versus-host disease (GvHD) [11]. The transplanted mature human T cells are predominantly stimulated through their recognition of mouse major histocompatibility complex (MHC) molecules, leading to the immune response attacking the host to cause GvHD [12]. This limits the time window of experiments while adding a confounding factor that complicates the interpretation of data. Using highly immunodeficient mice knocked out for the genes encoding murine MHC class I and class II (MHC I/II DKO) serves as a favorable solution to overcome this issue, as the onset of GvHD delayed significantly in this strain after the PBMC engraftment [12, 13]. Hereafter, we introduce materials and methods for PBMC humanization of NSG mice and NSG-MHC I/II DKO mice and further illustrate the administration of solitomab [14] (anti-human CD3 and anti-human epithelial cellular adhesion molecule [EpCAM] TDB) in the colorectal cancer cell-line-based xenograft (CDX)-engrafted PBMC humanized mice.

2 Materials

2.1 Mouse (see Notes 1 and 2)

1. 5–6 weeks-old NSG (NOD.Cg-$Prkdc^{scid}$ $Il2rg^{tm1Wjl}$/SzJ, JAX Stock Number #005557) mice (*see* **Notes 3** and **4**).
2. 5–6 weeks-old NSG-MHC I/II DKO (NOD.Cg-$Prkdc^{scid}$ $H2\text{-}K1^{b\text{-}tm1Bpe}$ $H2\text{-}Ab1^{g7\text{-}em1Mvw}$ $H2\text{-}D1^{b\text{-}tm1Bpe}$ $Il2rg^{tm1Wjl}$/SzJ, JAX Stock Number #025216) mice (*see* **Notes 4** and **5**).

2.2 Cells

1. Human PBMC sourced from a healthy donor (*see* **Note 6**).
2. HT29, human colorectal adenocarcinoma cell line (*see* **Note 7**), or any suitable cell line.
3. Complete cell culture medium for HT29 cells; McCoy's 5A medium supplemented with 10% heat-inactivated and sterile filtered fetal bovine serum (FBS), Penicillin (100 units/mL)-Streptomycin (100 μg/mL)-Amphotericin B (0.25 μg/mL) suspension.
4. Sterile cell culture dish (150 mm × 25 mm, tissue culture [TC]-treated).
5. Dulbecco's phosphate-buffered saline (DPBS), no calcium, no magnesium.
6. 0.25 w/v% trypsin-1mmol/L EDTA-4Na solution.
7. 0.4% Trypan blue stain solution.
8. 70 μm cell strainer.
9. Sterilized 50 mL conical centrifuge tubes.
10. Cell culture incubator (37°C, 5% CO_2 atmosphere).
11. 1.5 mL conical tubes.
12. Cell counter: In this study, we use TC20 Automated cell counter.

2.3 Cell Transplantation and TDB Administration

1. Injection needle and syringe: 1 mL syringe with 27-gauge (G) needle (*see* **Note 8**).
2. Isoflurane inhalation solution.
3. Anesthesia vaporizer for isoflurane with oxygen supply and anesthetic gas absorbent.
4. Inhalation mask and chamber.

5. Small animal heating pad (*see* **Note 9**).
6. Mouse restrainer for tail vein injection.
7. Cotton moistened with 70% ethanol.
8. Solitomab; a TDB targeting human EpCAM and human CD3. Other TDBs targeting different antigens can be used to test efficacy, depending on the specific needs of your experiment.
9. Weight scale to measure the body weight of mice.
10. Digital caliper to measure width and length of the tumor engrafted in mice, for calculating tumor volume.

2.4 Blood Collection

1. Heparin sodium (10 000 Units/10 mL).
2. Razor blade to cut tail vein for blood collection.
3. Protein low binding 1.5 mL tube.
4. Mini centrifuge.
5. Cotton moistened with 70% ethanol.
6. Sterile gauze.

2.5 Red Blood Cell Lysis

1. Red blood cell (RBC) lysis solution (10×).
2. Distilled water.
3. 5 mL polystyrene round-bottom tube.

2.6 Flow Cytometry

1. B.E.DPBS: DPBS with 0.5% bovine serum albumin and 2 mM EDTA (pH 8.0).
2. Human Fc receptor (FcR) blocking reagent.
3. Mouse BD Fc blocking reagent, or similar reagent.
4. Fluorescently labeled isotype control antibody: In this study, we use REAfinity REA control antibody, human IgG1.
5. Fluorescently labeled anti-mouse CD45 antibody: In this study, we use REAfinity anti-mouse CD45 antibody (human IgG1).
6. Fluorescently labeled mouse IgG1, kappa isotype control antibody.
7. Fluorescently labeled anti-human CD45 antibody (mouse IgG1 kappa).
8. BD Horizon Fixable Viability Stain 520, or similar reagent.
9. 5 mL polystyrene round-bottom tube with a cell strainer snap cap (35 μm mesh).
10. Corning 96-well Clear V-Bottom TC-treated microplate.
11. Flow cytometer: In this study, we use a Guava easyCyte 8HT.

12. Quality control fluorescent microbeads for flow cytometer: In this study, we use a Guava easyCheck kit.
13. Cleaning solution of flow cytometry: In this study, we use an ICF instrument cleaning fluid.

3 Methods

A thorough preliminary experiment is the key to success for the efficacy testing of TDBs using PBMC humanized mice (*see* **Notes 10–12**). It is recommended to assign a minimal $n = 5$–8 animals per group for each study.

3.1 Cell Culture of HT29

1. Place the frozen vial of HT29 cells in a 37°C water bath to thaw the cells.
2. Resuspend the cells in a 50 mL conical centrifuge tube with 20–50 mL of complete culture medium and centrifuge at 300 × *g* for 3 min at 4°C.
3. Discard the supernatant and resuspend the cells with a complete culture medium.
4. Seed HT29 cells on a sterile cell culture dish at the seeding density of 4–8 × 10^3 cells/cm^2 and culture them at 37°C in an incubator (5% CO_2 atmosphere) until it reaches 70–80% confluency.

3.2 Preparation of HT29 Cells

1. Discard the culture medium and wash the dish with DPBS.
2. Add 2–3 mL of 0.25 w/v% trypsin-1mmol/L EDTA-4Na solution and incubate for 3 min at 37°C in the incubator.
3. Add 20–30 mL of complete culture medium to stop the reaction.
4. Collect cell suspension into a 50 mL tube and centrifuge at 300 × *g* for 3 min 4°C.
5. Discard the supernatant and resuspend the cells with DPBS.
6. Take 50 μL of the cell suspension to a 1.5 mL tube with 50 μL 0.4% Trypan blue solution for counting cells and assess cell viability using TC20 Automated Cell Counter.
7. Centrifuge them at 300 × *g* for 3 min at 4°C.
8. Discard the supernatant and resuspend the cells with DPBS to achieve a final cell concentration of 1 × 10^7 cells/mL (*see* **Note 13**).
9. Keep the tube containing cells on ice until injection.

3.3 Subcutaneous Transplantation of Tumor Cells

1. Place the mouse in an anesthetic chamber and administrate isoflurane with oxygen. Make sure that the animal is anesthetized by monitoring the respiratory rate.
2. Shave the hair from the area where you will transplant the cells subcutaneously. Normally, it is the flank area.
3. Clean the injection site with cotton moistened with 70% ethanol.
4. Gently mix the HT29 cells and draw 100 μL of HT29 cells (1×10^6 cells/mouse) into a syringe equipped with a 27G needle, taking care to avoid any air bubbles.
5. Inject the cell suspension of HT29 subcutaneously into the flank. Confirm that there is no leakage from the injection site.
6. Place the mouse in a recovery area and allow it to recover from anesthesia naturally.

3.4 Preparation of Human PBMC

1. Take the PBMC vial from liquid nitrogen and thaw the cells in the water bath at 37°C. During this process, watch the vial closely and remove it from the water bath as soon as only a small ice crystal remains. This minimizes exposure of cells to cryoprotectants (*i.e.,* DMSO) at 37°C, helping to preserve cell viability.
2. Immediately after thawing, transfer the vial to a biosafety cabinet or clean bench. Pipette the cell suspension gently into the 50 mL tube containing pre-warmed 45 mL DPBS with 10% FBS. This step helps dilute the cryoprotectant and prevent PBMCs from adhering to the tube walls. Rinse the original tube to collect any remaining cells.
3. Centrifuge cells at 300 × *g* for 10 min at room temperature.
4. Discard the supernatant and tap the bottom of the tube a few times. Resuspend the cell pellet in 40 mL DPBS and mix gently by pipetting.
5. Centrifuge cells at 300 × *g* for 10 min at room temperature.
6. Discard the supernatant and tap the bottom of the tube a few times. Resuspend the cell pellet in 40 mL DPBS and gently mix by pipetting.
7. To avoid cell aggregates, filter the cell suspension through a 70 μm cell strainer.
8. Take 50 μL of the cell suspension to a 1.5 mL tube with 50 μL 0.4% Trypan blue solution for counting cells and assess viability using TC20 Automated Cell Counter.
9. Centrifuge cells at 300 × *g* for 10 min at room temperature.
10. Discard the supernatant and tap the bottom of the tube a few times. Resuspend the cell pellet in DPBS to achieve a final cell concentration of 5×10^7 cells/mL.

11. Keep the tube containing cells on ice until injection. Once the cells are prepared, proceed with PBMC administration as quickly as possible.

3.5 Intravenous PBMC Transplantation (see Note 14)

1. Place the mice on a heating pad to dilate the tail veins before the injection. Note that it is not recommended to anesthetize the animal for an intravenous injection.
2. Hold the animal with a mouse restrainer and keep the tail out for the injection.
3. Disinfect the tail with cotton moistened with 70% ethanol. Scrubbing the tail with cotton helps increase the visibility of the tail vein.
4. Gently tap the tube containing the cell suspension of PBMCs by inverting the 1.5 mL tube. If needed, spin down the cell suspension on the tube cap using a mini centrifuge for a few seconds.
5. Draw 200 μL of PBMCs (1 × 107 cells/mouse) into a syringe with a 27G needle, ensuring no bubbles are present (*see* **Note 15**).
6. Inject the PBMCs slowly in the tail vein.

3.6 Monitoring the Mice

1. Monitor the general health conditions of mice engrafted with PBMC and/or tumor cells according to the approved protocol of animal care and use in each institution (*see* **Note 1**). PBMC engraftment can lead to the occurrence of GvHD, while excessive growth of the engrafted tumor impacts the survival of mice. Throughout the experiment, mice should be monitored carefully to check their body weight and general health condition after the PBMC and/or tumor cell engraftment (*see* **Note 16**). In our study, significant body weight loss is observed in all NSG mice engrafted with human PBMC in a month, indicating their occurrence of GvHD (Fig. 1).
2. Measure body weight of mice using a weight scale and the tumor size using a digital caliper at least twice a week. Tumor volume is calculated by the formula: volume = length × $(\text{width})^2 \times 1/2$.
3. Mice should be euthanized at the humane endpoint, which in this study is defined as a tumor volume of 2000 mm^3, body weight loss exceeding 20%, or the presence of abnormal signs such as changes in physical appearance or hypothermia.

3.7 Blood Sample Collection to Check the PBMC Engraftment (see Note 17)

1. Prepare protein low binding 1.5 mL tubes, each containing 4 μL of heparin sodium as an anticoagulant.
2. Hold a mouse in a restrainer, ensuring the tail is accessible for puncture.

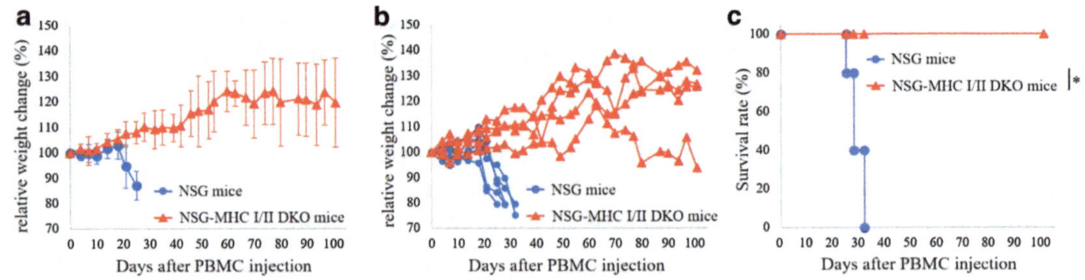

Fig. 1 Three-panel figure showing X-Y charts of relative weight change and survival rate over time for NSG and NSG-MHC I/II DKO mice after PBMC injection. Panels (**a**) and (**b**) display relative weight change (%) on the *y*-axis against days after injection on the *x*-axis, with NSG mice in blue and NSG-MHC I/II DKO mice in red (NSG; n = 5, NSG-MHC I/II DKO; n = 4). Error bars indicate SD. Panel (**c**) shows survival rate (%) on the *y*-axis over the same time period, with a significant difference indicated between the two groups

3. Wipe the tail with cotton moistened with 70% ethanol to disinfect the area where the blood will be collected.
4. By using a sterile blade, make a shallow puncture near the tip of the mouse tail vein, targeting the tail vein.
5. Gently stroke the tail to stimulate blood flow, then collect up to 100 μL of blood into the heparinized collection tube. For flow cytometry analysis, collecting 50 μL per mouse is sufficient.
6. Immediately tap the tube a few times to ensure that the blood is thoroughly mixed with heparin. This step is crucial for preventing blood coagulation.
7. Spin down the tube using a mini centrifuge for a few seconds to ensure that blood adhered to the tube wall is collected at the bottom.
8. Press the puncture site with a sterile gauze until the bleeding stops.
9. Keep the collection tubes on ice until the next steps.

3.8 Red Blood Cell Lysis

1. Prepare 1× RBC lysis buffer by diluting 10× stock with distilled water.
2. Add 1 mL of 1× RBC to the collection tube containing 50 L of the blood sample. Gently mix by inverting the tube several times, then briefly spin down the tube using a mini centrifuge.
3. Transfer the sample to a 5 mL polystyrene round-bottom tube and add 2 mL of 1× RBC lysis buffer.
4. Centrifuge cells at 300 × *g* for 5 min at 4°C.
5. Discard the supernatant and resuspend the cell pellet in 3 mL of 1× RBC lysis buffer and mix gently by pipetting.

6. Repeat steps 4 and 5 if the cell pellet still contains visible RBC contamination (*see* **Note 18**).
7. Resuspend the cell pellet in 1 mL of B.E. PBS and mix gently by pipetting. Keep the tube on ice until the next steps.

3.9 Flow Cytometry Analysis to Monitor Immune Humanization

1. Count the cells and assess cell viability in each sample using a cell counter.
2. Transfer approximately 2 × 10^4 cells per sample to a 5 mL polystyrene round-bottom tube.
3. Centrifuge cells at 300 × *g* for 5 min at 4°C and carefully remove the supernatant.
4. Add 25 μL of mouse Fc blocking reagent (0.01 mg/mL, diluted with B.E. PBS) and 20 μL of human FcR blocking reagent to each tube.
5. Mix well by pipetting and incubate for 10 min on ice.
6. Add 2 μL of anti-mouse CD45 antibody and 5 μL of anti-human CD45 antibody to each tube.
7. Mix well by pipetting and incubate on ice in the dark for 20 min.
8. Add 3 mL of DPBS in each tube to wash the cells.
9. Centrifuge the tube at 300 × *g* for 5 min, then carefully discard the supernatant.
10. Add 200 μL of Fixable Viability Stain 520 (1000-fold diluted in DPBS) to each tube.
11. Mix well by pipetting and incubate on ice in the dark for 30 min.
12. Add 3 mL of B.E. PBS to wash the cells.
13. Centrifuge the tube at 300 × *g* for 5 min and discard the supernatant.
14. Resuspend the cell pellet in 200 μL of B.E. PBS and transfer the cells into a 5 mL polystyrene round-bottom tube with a cell strainer cap (35 μm mesh).
15. Transfer the sample to a 96-well V-bottom plate.
16. Use the Guava easyCheck kit to verify the performance of the Guava eacyCyte system.
17. Measure the fluorescence intensity of the sample using the Guava eacyCyte cytometer.
18. Analyze the data by using flow cytometry analysis software. For this study, FlowJo is used. Representative flow cytometry plots are shown in Fig. 2a, b, while the proportion of human CD45-positive cells within the total (mouse and human) CD45-positive population is shown in Fig. 2c, d (*see* **Note 19**).

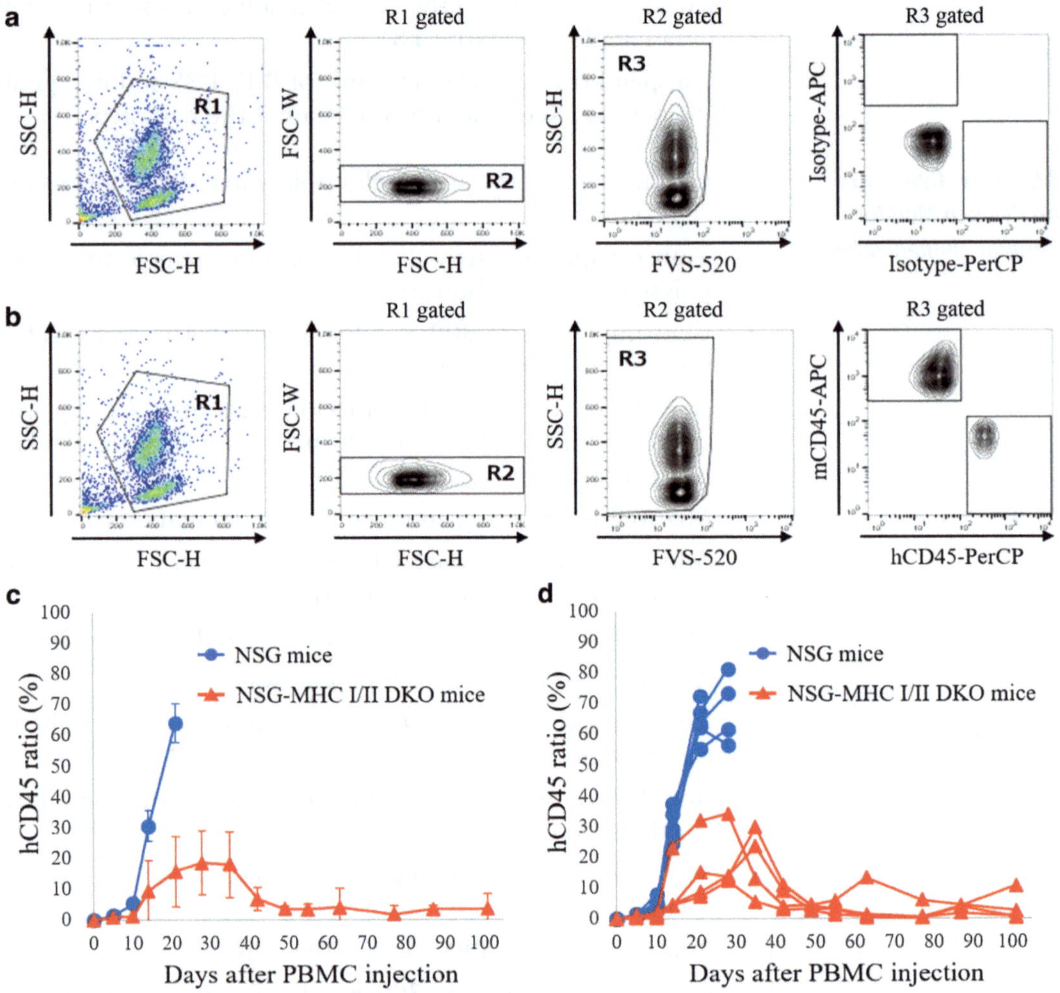

Fig. 2 Human immune cell reconstitution in PBMC engrafted NSG and NSG-MHC I/II DKO mice, assessed by the proportion of human CD45-positive cells among total CD45-positive cells in peripheral blood. Firstly, the cell population is gated based on FSC/SSC to exclude debris and dead cells. Secondly, doublets can be removed by gating singlets on the FSC-H vs. FSC-W plot. Thirdly, dead cells should be excluded based on Fixable Viability Stain 520 staining. Fourthly, the proportion of human immune cells can be analyzed based on human and mouse CD45 expression. Representative flow cytometry plots staining with (**a**) isotype control antibodies and (**b**) anti-human CD45 and anti-mouse CD45 antibodies. (**c** and **d**) Temporal change of human CD45-positive cell proportions in NSG and NSG-MHC I/II DKO mice. (**c**) The average and (**d**) the individual proportion of human CD45-positive cells (NSG; $n = 5$, NSG-MHC I/II DKO; $n = 4$). Error bars indicate SD

3.10 Efficacy Testing of Solitomab in PBMC Humanized NSG-MHC I/II DKO Mice Engrafted with HT29

1. Engraft NSG-MHC I/II DKO mice subcutaneously with 1×10^6 HT29 cells. The timeline of the TDB treatment in the present study is shown in Fig. 3a (*see* **Note 20**).
2. 7 days after HT29 engraftment, administrate 1×10^7 PBMCs intravenously.

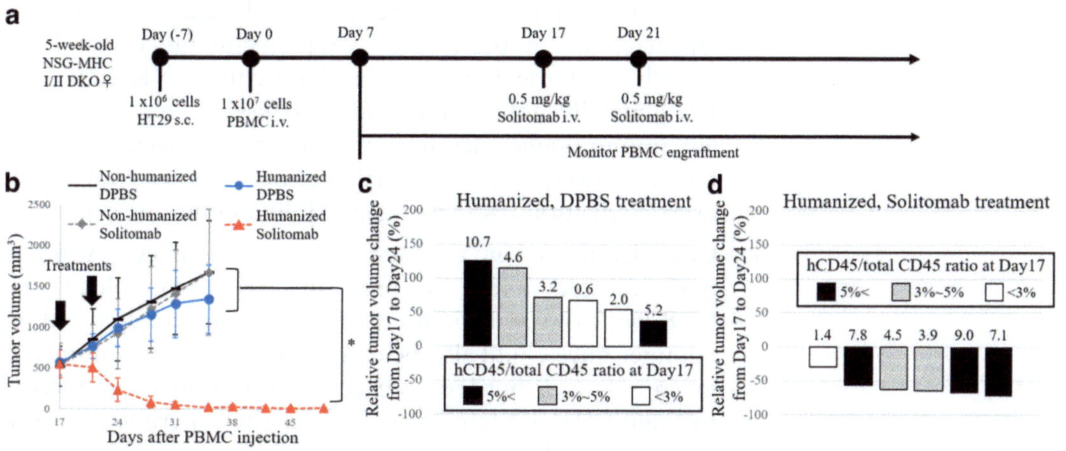

Fig. 3 *In vivo* efficacy testing of TDB using PBMC humanized NSG-MHC I/II DKO mice engrafted with HT29. (**a**) The timeline of the present study. (**b**) Tumor growth suppression of HT29 following solitomab treatment. The black line represents tumor growth in non-humanized mice treated with DPBS ($n = 4$), while the gray line shows tumor growth in non-humanized mice treated with solitomab ($n = 3$). The blue line indicates tumor growth in PBMC humanized mice treated with DPBS ($n = 6$), and the red line represents tumor growth in PBMC humanized mice treated with solitomab ($n = 6$). Arrows denote the time points of treatment. Repeated-measures ANOVA is used for statistical analysis. Error bars indicate SD. $^*p < 0.001$. Relative tumor volume changes from Day 17 to Day 24 in humanized mice treated with (**c**) DPBS and (**d**) solitomab. Values indicated on the individual bar plot represent the proportion of human CD45 positive cells at Day 17 post-PBMC engraftment

3. At 17 and 21 days post-PBMC engraftment, intravenously administrate DPBS or solitomab (0.5 mg/kg) (*see* **Note 21**).
4. Monitor mice until they reach the endpoint; Measure body weight and tumor volume twice per week from the start of the study, while collecting blood samples to assess the engraftment rate of human immune cells once per week after PBMC engraftment (Fig. 3).

4 Notes

1. Ethical considerations: All animal experiments should follow the institution's animal welfare and ethical use guidelines, with all procedures approved by the Institutional Animal Care and Use Committee.
2. Housing and husbandry of mice: As recipient mice used for PBMC humanization are severely immunocompromised, they are sensitive to opportunistic pathogens that are non-pathogenic in immunocompetent mice [15]. It is important to maintain them in the highest health status room as possible and autoclave or sterilize all husbandry supplies. See the guidebook for the detailed instructions [16].

3. NSG mice: This is a severe immunodeficient mouse strain characterized by the lack of complement, functional T and B cells, and the impaired development of NK cells [17]. It's generally not recommended to use mouse strains with less immunodeficiency than this strain for effective immune humanization, as the residual murine immune components exert a rejection response against human immune cells.
4. Considerations on the age and sex of mice: Although 5–6-week-old mice are used in this study, older mice (7–8-week-old) can be used without major concerns. While younger mice could also be employed, consideration must be given to tumor burden, as excessive tumor load can increase mortality. We use female animals for the experiment, as they are relatively docile and easy to handle.
5. NSG-MHC I/II DKO mouse: This strain is an NSG mouse knocked out for MHC class I (H2-Ab1) and II (H2-K1 and H2-D1) molecules, which are the major contributors to the GvHD. Using this strain as a recipient delays GvHD upon PBMC engraftment, but it does not completely prevent the occurrence of GvHD [12]. Some PBMC lots are prone to induce the early onset of GvHD even in this strain, thus pre-validation of the PBMC lot *in vivo* is crucial.
6. Source of PBMC: In this study, we use human PBMCs obtained from a vendor. It is also feasible to use PBMCs obtained from a patient, although it is essential to confirm its cell count and viability before use.
7. Selection of human tumor cells to engraft: Not all human-derived cells or tissues engraft successfully in NSG mice, despite their high level of immunodeficiency. If you are using CDX, prioritize selecting cell lines for which successful engraftment in NSG mice has been documented in previous studies. Nevertheless, be sure to perform preliminary engraftment experiments using either NSG mice or NSG-MHC I/II DKO mice to plot the growth curve.
8. Consideration for syringes for cell transplantation: Using thinner needles above 27G needles is not recommended because they can cause cell damage due to high sheer stress while they get clogged easily as cells likely get aggregated by being pushed through narrow spaces. Using a combined syringe and needle allows for efficient use of cell suspension, as it minimizes the dead space.
9. Tail vein injection: Warming the animal with a heating pad causes vasodilation of the vein, making it easier to visualize and access the injection. Alternatively, stroking the tail with 70% ethanol-soaked gauze can help encourage blood flow and make the tail vein prominent. Before injection, it is

recommended to verify that the needle is correctly positioned within the vein by checking for the blood flashback through negative pressure. In this case, you must change the needle each time you inject a different animal as coagulated blood can clog the needle.

10. Drawing tumor growth curve with different cell numbers engrafted: A preliminary experiment to determine the optimal number of cancer cells for engraftment is crucial, as cell quantity directly impacts tumor growth rate. This should be done using either NSG or NSG-MHC I/II DKO mice, as tumor growth varies by strain and immune status. In this study, 5–6-week-old female NSG-MHC I/II DKO mice are used to draw tumor growth curves with HT29 at cell numbers (1) 1×10^6, (2) 2×10^6, (3) 5×10^6, (4) 10×10^6 cells/mouse. Careful consideration of cell number is crucial, as an excess number of tumor cells can increase lethality and trigger a stronger allogeneic immune response.
11. PBMC lot profiling and selection: The success of PBMC humanization largely depends on the PBMC lots used. It is recommended that at least three different PBMC lots be tested prior to the main experiment to identify the optimal lot. Some PBMC lots can trigger the early onset of GvHD, even in NSG-MHC I/II DKO mice. Additionally, the number of PBMCs to engraft should be carefully determined in the preliminary experiment. Ensure that the same mouse strain (*e.g.*, NSG or NSG-MHC I/II DKO) is used for preliminary studies and the main experiments, as the rate of human cell engraftment can vary depending on the strain and immune profiles.
12. Optimizing the engraftment timeline for tumor cells and PBMCs: The engraftment timeline needs to be carefully optimized based on the tumor growth curve and the time-course of PBMC engraftment, as the optimal timing largely depends on the characteristics of cancer cells. For fast-growing cancer cells, PBMC injection should precede cancer cell transplantation, while for slow-growing tumors, the order should be reversed. Potential immune allogeneic responses between PBMCs and tumor cells should be considered. In this experiment, HT29 cells were engrafted one week before PBMC transplantation to avoid an allogeneic response that would have hindered HT29 engraftment if PBMCs had been injected first. Reducing the number of tumor cells for engraftment may further minimize the allogeneic response. The injection route or the tumor cells also influences the allogeneic response; subcutaneous injection generally induces a weaker allogeneic response than intravenous injection.

13. Matrigel to facilitate tumor engraftment: If the cancer cell does not engraft well in the recipient mice, using Matrigel can often improve engraftment efficiency. Although it is not used in this study, our laboratory typically mixes the cell suspension with Matrigel in a 1:1 ratio for subcutaneous transplantation in mice.
14. Irradiation of mice before PBMC humanization: While this study does not perform irradiation of mice, either irradiating the recipient or treating the recipient with busulfan before PBMC injection would affect the PBMC engraftment rate and subsequent occurrence of GvHD [18].
15. Number of PBMCs to engraft: Injected PBMC cell numbers would affect the PBMC engraftment rate and time required for achieving a sufficient humanization rate in the recipient mice. In general, we recommend engrafting a minimum of 1×10^7 PBMCs for the non-irradiated recipient.
16. Monitoring GvHD of mice: Body weight is the most significant indicator of GvHD, while other symptoms of GvHD include ruffled fur, hair loss, redness of the skin, severe hunching, ulcerated eyes, diarrhea, and ear swelling [19]. When the animals die suddenly, be sure to perform a necropsy to determine the cause of death.
17. To monitor immune humanization in the PBMC-engrafted mice, blood needs to be collected from individual mice regularly, typically once per week. The frequency of blood collection should be determined based on the volume of blood to be collected and the specific needs of the experiment under the institution's animal welfare and ethical use guidelines.
18. RBC lysis: Incomplete RBC lysis leads to contamination with residual RBCs and cellular debris. This can interfere with subsequent flow cytometry analysis, hindering the accurate identification of cell populations. Be careful to avoid overexposure to the lysis buffer, as this can lead to the loss of viable lymphocytes.
19. Immune humanization dynamics in NSG and NSG-MHC I/II DKO mice: In NSG mice, the proportion of human CD45-positive cells increases rapidly after PBMC engraftment, whereas in NSG-MHC I/II DKO mice, this increase is more gradual (Fig. 2c, d). Since the rapid immune humanization in NSG mice can lead to the early onset of GvHD, the choice of mouse strain should be tailored to the specific needs of your experiment.
20. Study design: The engraftment order and cell numbers of PBMCs and tumor cells should be determined based on the data obtained through preliminary experiments (*see* **Notes 10** and **11**). These steps are crucial for designing the main study

successfully. In this study, we employ two approaches: (1) Engraft PBMC first, followed by HT29 cells and (2) Engraft HT29 cells first, followed by PBMCs. For at least 4 weeks after PBMC engraftment, measure body weight and tumor volume twice per week and assess the proportion of human CD45-positive cells in the peripheral blood once per week to characterize humanization dynamics.

21. Humanization rate required for TDB efficacy testing: The proportion of human CD45 positive cells required to define "humanized" mice varies depending on the study and context. Notably, in the present study, less than a 5% engraftment rate in the peripheral blood is sufficient to inhibit HT29 tumor growth with solitomab treatment.

Acknowledgments

This work was supported by the National Cancer Center Research and Development Fund under Grant Number 2023-A-8 and AMED under Grant Number JP21ak0101182. We would like to thank Mr. T. Komatsu and Ms. Y. Shino for their experimental support.

References

1. Staerz UD, Kanagawa O, Bevan MJ (1985) Hybrid antibodies can target sites for attack by T cells. Nature 314(6012):628–631
2. Topp MS, Kufer P, Gökbuget N et al (2011) Targeted therapy with the T-cell-engaging antibody blinatumomab of chemotherapy-refractory minimal residual disease in B-lineage acute lymphoblastic leukemia patients results in high response rate and prolonged leukemia-free survival. J Clin Oncol 29(18):2493–2498
3. Budde LE, Sehn LH, Matasar M et al (2022) Safety and efficacy of mosunetuzumab, a bispecific antibody, in patients with relapsed or refractory follicular lymphoma: a single-arm, multicentre, phase 2 study. Lancet Oncol 23(8):1055–1065
4. Dickinson MJ, Carlo-Stella C, Morschhauser F et al (2022) Glofitamab for relapsed or refractory diffuse large B-cell lymphoma. N Engl J Med 387(24):2220–2231
5. Usmani SZ, Garfall AL, van de Donk N et al (2021) Teclistamab, a B-cell maturation antigen × CD3 bispecific antibody, in patients with relapsed or refractory multiple myeloma (MajesTEC-1): a multicentre, open-label, single-arm, phase 1 study. Lancet (London, England) 398(10301):665–674
6. Chari A, Minnema MC, Berdeja JG et al (2022) Talquetamab, a T-cell-redirecting GPRC5D bispecific antibody for multiple myeloma. N Engl J Med 387(24):2232–2244
7. Paz-Ares L, Champiat S, Lai WV et al (2023) Tarlatamab, a first-in-class DLL3-targeted bispecific T-cell engager, in recurrent small-cell lung cancer: an open-label, phase I study. J Clin Oncol 41(16):2893–2903
8. Belmontes B, Sawant DV, Zhong W et al (2021) Immunotherapy combinations overcome resistance to bispecific T cell engager treatment in T cell-cold solid tumors. Sci Transl Med 13(608):eabd1524. https://doi.org/10.1126/scitranslmed.abd1524
9. Wang L, Leach V, Muthusamy N et al (2024) A CD3 humanized mouse model unmasked unique features of T-cell responses to bispecific antibody treatment. Blood Adv 8(2):470–481
10. Chuprin J, Buettner H, Seedhom MO et al (2023) Humanized mouse models for immuno-oncology research. Nat Rev Clin Oncol 20(3):192–206
11. Covassin L, Laning J, Abdi R et al (2011) Human peripheral blood CD4 T cell-engrafted non-obese diabetic-scid IL2rγ(null) H2-Ab1 (tm1Gru) Tg (human leucocyte antigen D-related 4) mice: a mouse model of human allogeneic graft-versus-host disease. Clin Exp Immunol 166(2):269–280

12. Brehm MA, Kenney LL, Wiles MV et al (2019) Lack of acute xenogeneic graft- versus-host disease, but retention of T-cell function following engraftment of human peripheral blood mononuclear cells in NSG mice deficient in MHC class I and II expression. FASEB J 33(3):3137–3151
13. Yaguchi T, Kobayashi A, Inozume T et al (2018) Human PBMC-transferred murine MHC class I/II-deficient NOG mice enable long-term evaluation of human immune responses. Cell Mol Immunol 15(11):953–962
14. Kebenko M, Goebeler ME, Wolf M et al (2018) A multicenter phase 1 study of solitomab (MT110, AMG 110), a bispecific EpCAM/CD3 T-cell engager (BiTE®) antibody construct, in patients with refractory solid tumors. Oncoimmunology 7(8):e1450710
15. Foreman O, Kavirayani AM, Griffey SM et al (2011) Opportunistic bacterial infections in breeding colonies of the NSG mouse strain. Vet Pathol 48(2):495–499
16. NRCUCoIC (1989) Immunodeficient rodents: a guide to their immunobiology, husbandry, and use. National Academy of Sciences, Washington, DC
17. Shultz LD, Ishikawa F, Greiner DL (2007) Humanized mice in translational biomedical research. Nat Rev Immunol 7(2):118–130
18. Ye C, Yang H, Cheng M et al (2020) A rapid, sensitive, and reproducible in vivo PBMC humanized murine model for determining therapeutic-related cytokine release syndrome. FASEB J 34(9):12963–12975
19. Elhage A, Sligar C, Cuthbertson P et al (2022) Insights into mechanisms of graft-versus-host disease through humanised mouse models. Biosci Rep 42(9):BSR20211986. https://doi.org/10.1042/BSR20211986

Part III

Quantitative Pharmacology

Chapter 15

Overview of Quantitative Clinical Pharmacology for T-Cell Engaging Bispecific Antibodies

Rong Deng and Chi-Chung Li

Abstract

T-cell bispecific antibodies (TCBs) are an exciting advancement in cancer immunotherapy, enabling potent and durable tumor responses by bridging CD3-positive T cells with tumor-specific antigens. By forming artificial immune synapses, TCBs activate T cells in an MHC-independent manner, initiating robust anti-tumor responses. However, these unique mechanisms present challenges, including complex pharmacokinetics (PK), target-mediated drug disposition (TMDD), and on-target toxicities such as cytokine release syndrome (CRS) and off-tumor effects. This chapter offers a comprehensive review of the clinical pharmacology considerations involved in TCB development, covering their mechanisms of action, PK and pharmacodynamic (PD) characteristics, and bioanalytical strategies. It also summarizes approaches to mitigate risks and optimize therapeutic outcomes.

Advanced modeling techniques, such as quantitative systems pharmacology (QSP) and physiologically based pharmacokinetics (PBPK) modeling, are highlighted as valuable tools for addressing translational challenges. When combined with emerging applications of artificial intelligence and machine learning, these methods can provide further predictability and insights for optimizing dosing regimens, mitigating toxicities, and predicting clinical outcomes. The chapter also discusses strategies for overcoming resistance mechanisms through combination therapies and improving TCB constructs to enhance patient benefit. By integrating these insights, the chapter emphasizes the role of pharmacology and modeling advances in maximizing the therapeutic potential of TCBs for cancer treatment.

Keywords Bispecific antibodies, T-cell engagers, Quantitative clinical pharmacology, Modeling and simulations, Dose optimization, Model-informed drug development (MIDD), Step-up dosing, Cytokine release syndrome (CRS), PK/PD

1 Introduction

T-cell engaging bispecific antibodies (TCB) present a promising new advancement within cancer immunotherapies. This therapeutic approach harnesses the body's immune system in the battle against cancer. TCBs work through dual targeting of CD3, expressed on effector T cells, and target antigens/receptors (targets) on malignant cells. This results in the formation of artificial immune synapses, acting as a trigger for T-cell activation, importantly in an MHC-independent

Masahiro Yasunaga (ed.), *Bispecific Antibodies*, Methods in Molecular Biology, vol. 2997,
https://doi.org/10.1007/978-1-0716-5037-0_15,
© The Author(s), under exclusive license to Springer Science+Business Media, LLC, part of Springer Nature 2026

manner [1]. This activation sets off an immune response cascade, leading to the release of cytokines and cytotoxic granules (e.g., perforin and granzymes), which effectively destroy malignant cells [1, 2].

Key benefits of TCBs include their high potency and durable response rates, as highlighted in recent regulatory approvals for use in relapsed/refractory (r/r) Non-Hodgkin's Lymphoma (NHL) and r/r Multiple Myeloma (MM) [3–13]. TCB treatment has resulted in impressive objective and complete response rates, demonstrating high efficacy. However, the development of TCBs is not without its challenges. Mechanistically, the intended process requires simultaneous binding to T cells and target cells/antigens, leading to unique pharmacokinetics (PK) and distribution properties that contrast with monoclonal antibodies (mAbs). As immune agonists, TCBs function without saturating receptor binding [14], in contrast to checkpoint inhibitors such as anti-PD-1/PD-L1 monoclonal antibodies. Consequently, the PK at clinically active doses may be shaped by target-mediated drug disposition (TMDD), which is modulated by the binding kinetics of each target. The PK and distribution of TCBs are thus determined by the dynamic interplay of their dual binding kinetics and TMDD, influenced by factors such as relative binding affinities, target expression levels, and target turnover rates [15]. Such complex disposition properties were described by Mandekian et al., suggesting that the higher CD3 affinity and distribution to T-cell-rich tissues shift the systemic exposure and tumor exposure to T-cell-containing tissues in HER2 tumor expressing mice [16].

Additionally, approved mAbs targeting the same tumor target as the TCB often constitute standard-of-care or prior lines of therapies, particularly in r/r settings. For example, rituximab and obinutuzumab are both approved CD20-target mAbs, and their respective concentrations have been used in the characterization of exposure–response (E–R) relationships for mosunetuzumab (CD20/CD3 TCB) for the treatment of r/r NHL [14, 17]. This potential binding competition for the same targets can affect the target engagement levels of TCBs, making it a critical factor in understanding clinical dose–response relationships. This is particularly significant in dosing regimens and disease contexts sensitive to acute, temporal fluctuations in target engagement. Examples include scenarios with aggressive disease progression, where rapid tumor control is vital for achieving long-term benefit, as well as dose ranges or molecules characterized by steeper exposure–response relationships.

Lastly, like other T-cell therapies such as chimeric antigen receptor (CAR) T-cell therapies [18, 19], the clinical development of TCBs is often challenged by on-target acute toxicities, including cytokine release syndrome (CRS) and neurological adverse events (NAEs), due to the mechanism of T-cell activation. Therefore, early-stage development of TCBs must prioritize characterizing the pharmacological drivers of both efficacy and safety to uncover critical insights into the therapeutic index (TI). This understanding is essential for devising strategies to maximize clinical benefit while addressing on-target safety concerns. Optimizing dosing regimens and exploring modifications to PK and E–R relationships through alternative routes of administration have proven effective in expanding the TI. These advancements have contributed to achieving a favorable benefit-risk profile for TCBs, particularly in r/r NHL [20].

In this chapter, we provide an overview of key clinical pharmacology considerations in the development of TCBs, highlighting quantitative modeling approaches and their applications in generating insights, informing dose regimens, determining therapeutic windows, and maximizing the benefit-risk profile to enable the clinical development of TCBs for oncology indications.

2 Mechanism of Action

Bispecific antibodies (BsAb) can achieve their therapeutic effect through two distinct mechanisms of action (MOAs). One category functions by bridging the patient's own immune cells (e.g., through CD3+ T cells or CD16+ NK cells) with target-positive tumor cells to facilitate their attack. These types of bispecific constructs have one arm directed against an antigen on the immune effector cells and the other arm directed against a target cell surface antigen overexpressed by tumor cells [16]. Simultaneous engagement of both arms results in formation of an immunologic synapse between the target tumor cell and effector cell, which leads to killing of the target tumor cells (e.g., through granzyme- and perforin-induced cell lysis). In contrast, the other category of the bispecific antibodies does not require to bridge the two target cells simultaneously for efficacy, but acts through engaging multiple targets. Bispecific antibodies in this category may target two soluble cytokines, bind different epitopes of the same tumor cell or viral antigen, or engage different targets to mimic the function of an endogenous protein. Such an approach may have advantages over combination therapy with multiple antibodies [20].

3 Bioanalytical Strategy for Pharmacokinetics and Immunogenicity

This provides a summary and overview of the bioanalytical strategy for BsAbs [20, 21], with similar considerations applying to the subcategory of TCB molecules. In general, the regulatory guidance and requirements for PK and anti-drug antibody (ADA) method validation in supporting monospecific antibody (mAb) development also apply to BsAb programs. Additional bioanalytical work may be required for BsAbs by taking into consideration the MOA, target biology and clinical study design. For clinical studies, the PK assays should be designed to detect the intact format of the BsAb. Assay sensitivity requirements for BsAbs may be different from the conventional monospecific antibodies. For example, higher sensitivity is required for immune cell engager BsAbs. Some oncology-focused BsAbs are being evaluated as a later line of treatment initially or as a combination treatment with other therapeutics. These BsAbs may share the same target(s) with therapeutics used in the earlier lines of treatment. Therefore, verification of the assay specificity/selectivity is essential to ensure that the PK assay measures the intended analyte by design. Orthogonal assays may be required to confirm structural stability/biotransformation as needed. For nonclinical studies, a similar strategy often applies for PK and ADA measurements in the monospecific antibody and BsAb programs.

Compared to a monospecificantibody therapeutic, it takes more engineering modifications and complex manufacturing processes to generate a BsAb, which may contain neoepitopes and domains with more immunogenic potential. While a similar tiered approach (screening, confirmatory, and titering) is used for ADA measurements in BsAb clinical studies, additional characterization work should be considered if applicable, such as domain characterization. The critical reagent considerations may be different in a BsAb ADA assay development, for example, the selection of a surrogate positive control using antibodies against the whole molecule or against a certain domain of the molecule. In terms of neutralizing antibody (NAb) assay development, if the MOA requires a BsAb binding to both targets, the NAb assay should be developed using two relevant cell lines.

4 Pharmacokinetics

4.1 Unique PK Properties of TCBs Dependent on Constructs

The characterization of the PK properties of a BsAb, including a TCB, needs to take consideration of its format and could be governed by molecular weight, physicochemical properties, interaction with Fc receptors, and binding to target, etc., as described in the literature [22]. There are around 100 different formats of BsAbs, ranging from fusion proteins to modified IgG-type monoclonal antibodies, with molecular weight (MW) of the molecule ranging from a few kilodaltons (kD) to hundreds kD [23]. Different MW will have a significant impact on the clearance of the bispecific molecules, similar as described for mAbs.

4.2 Target-Mediated Drug Disposition (TMDD) and Time-Dependent PK

For TCBs, the mechanism of action as a conditional agonist tends to be highly potent and active at lower target engagement levels. Thus, the dose range can be at the lower end, and the elimination of the drug can be impacted by both nonspecific clearance and target-mediated drug disposition (TMDD). Several TCBs, such as teclistamab, talquetamab, and elranatamab, have time-dependent CL in humans, where CL decreases with time. Teclistamab clearance decreases over time, with a clearance of 0.472 L/day at the 13th treatment dose and a mean maximal reduction from baseline to the 13th treatment dose of 40.8%. The unique PK consideration for a TCB is that it binds to two different targets, so that the TMDD can be influenced by binding affinity and the turnover rate of each individual target. It has been observed that there was a strong correlation between CD3 affinity and distribution of an anti-HER2/

CD3 BsAb to T-cell-rich tissues, with higher CD3 affinity reducing its systemic exposure and shifting its distribution away from tumor to T-cell-containing tissues in a solid tumor mouse model [16]. The binding avidity of typical mAbs would be higher than that for BsAbs as two arms of the typical mAbs can bind to the specific target vs for a bispecific antibody, only one arm would bind to the specific target. The avidity difference could theoretically lead to the PK difference at the low dose level between mAbs and TCBs, that share the same targets. Additionally, time-dependent CL can reflect changes in disease state and cachexia, which can evolve with therapeutic intervention and over time [24].

4.3 Comparison with Monoclonal Antibodies

PK characteristics comparisons between mAbs and bispecific Abs are summarized in Table 1. The overall of approved bispecific Abs are summarized in Table 2.

Table 1
Summary of PK characteristics for monoclonal antibodies and bispecific antibodies

Properties	Monoclonal antibodies	Bispecific antibodies
Molecular weight	Few KDa to around 150 KDa	Few kDa to 1000 kDa
Potency and binding specificity	Low potency and high specificity. Binding very specific for target protein or antigen	Low to high potency and high specificity Binding very specific for target protein or antigen
PK	Nonlinear PK at low doses; linear PK at high doses after saturation of target (i.e., target mediated disposition (TMDD))	Nonlinear PK at low doses; linear PK at high doses after saturation of target (i.e., target mediated disposition (TMDD))
PKPD relationship	PK usually dependent of PD	PK usually dependent of PD
Half-life	Long $t_{1/2}$ (days or weeks)	Varies from hours to weeks due to different molecular weight (MW)
Administration route	Need parenteral dosing. Subcutaneous (SC) or intramuscular (IM) is possible	Need parenteral dosing. Subcutaneous (SC) or intramuscular (IM) is possible
Metabolism	Catabolism by proteolytic degradation	Catabolism by proteolytic degradation
Renal clearance	No renal clearance of intact antibody. May be eliminated by damaged kidneys. Antibody fragment might be eliminated by renal clearance	Major route if Mol wt is <69 kDa
Distribution	Distribution usually limited to blood and extra-cellular space	Distribution usually limited to blood and extra-cellular space
Immunogenicity	Immunogenicity may be seen	Immunogenicity may be seen
Therapeutic window	Large therapeutic window	Narrow to large therapeutic window depending on the targets and dosing regimens

Table 2
Therapeutic bispecific antibody approved to date (Up to 2024)[a]

			Target		PK				
Drug name	Therapeutic area/ indication[a]	Antibody composition	Receptor/ antigen/ activity	Type	Behavioral[b]	Elimination/ terminal half-life[c]	Clearance	Vss	Route/ dosing highlights
Emicizumab-kxwh	Hemophilia A/ Bleeding due to hemophilia A	Humanized bi-specific IgG4	Factor IXa and factor X	Soluble	Linear withing dose range of 0.3–3.0 mg/ kg	27.8±8.1 days	CL/F: 0.24 L/day (95% CI: 0.22-0.26)	V/F: 11.4 L (95% CI: 10.6–12.1)	SC injection, mg/kg dose
Amivantamab	Oncology/Non-small cell lung cancer	Human IgG1 bispecific	EGFR, cMet	Cell-bound	Linear	11.3 (± 4.53) days	360 (± 144) mL/day	5.13 (± 1.78) L	IV infusion, flat mg dose
Blinatumumab	Oncology/ Relapsed or refractory precursor B-cell acute lymphoblastic leukemia (ALL)	Antibody Fragment	CD19 and CD3	Cell-bound	Linear	2.1 h	3.11 L/h	4.35 L	IV infusion flat mcg dose
Catumaxomab (EU only)	Oncology/ Malignant ascites	Chimeric:rat-mouse trifunctional MAB	EpCAM and CD3	Cell bound	NR	2.5 days (range 0.7 to 17 days)	NR	NR	IP infusion flat mg dose
Elranatamab	Oncology/ RRMM	Humanized bi-specific IgG2-alanine (IgG2Δa) kappa	BCMA/CD3	Cell-bound	Linear PK (6 to 76 mg)	22 days at the 76 mg dosage following 24 weeks dosing	Clearance of 0.324 L/ day at 76 mg dosage following 24 weeks dosing	7.76 L	SC injection, flat dose
Epcoritamab	Oncology/ DLBCL	Humanized bi-specific IgG1	CD3/CD20	Cell-bound	More than proportionally over a full dosage range from 1.5 to 60 mg	22 days at 48 mg at the end of cycle 3	0.53 L/day after cycle 3	25.6 L	SC injection, flat dose

Glofitamab	Oncology/ DLBCL	Humanized bi-specific IgG1	C20 and CD3	Cell-bound	Linear (0.005 to 30 mg)	7.6 days	CL at steady sate 0.617 L/day	5.6 L	IV infusion, flat dose
Mosunetuzumab	Oncology/ relapsed or refractory follicular lymphoma	Humanized bi-specific IgG1	C20 and CD3	Cell-bound	Linear (0.2 mg to 60 mg)	16.1 days	CL at baseline 1.08 L/ day; CL at steady stae 0.548 L/ day	5.49 L	IV infusion, flat dose
Talquetamab-tgvs	Oncology/ relapsed or refractory multiple myeloma	Humanized bi-specific IgG4-PAA	GPRC5D/ CD3	Cell-bound	Linear PK (0.005 to 0.8 mg/kg weekly and 0.8 to 1.2 mg/kg every 2 weeks)	8.4 days after the first treatment dose and 12.2 days at 16 weeks after the first treatment dose	Clearance decreases over time, CL is 0.90 L/day (63%) at 16 weeks after the first treatment dose	10.1 L	SC injection, mg/kg dose
Tarlatamab-dlle	Oncology/ extensive stage small cell lung cancer	Humanized bi-specific IgG	DLL3/CD3	Cell-bound	Linear PK (1 to 100 mg)	11.2 days (4.3 to 26.5)	0.65 L/day	8.6 L	IV infusion, flat dose
Tebentafusp-tebn	Oncology/ unresectable or metastatic uveal melanoma	Bispecific fusion protein	gp100 peptide presented by HLA/CD3	Cell-bound	Linear	7.5 h (range: 6.8–7.5 h)	16.4 L/day	7.56 L	IV infusion, flat mg dose
Teclistamab	Oncology/ relapsed and refractory multiple myeloma	Humanized bi-specific IgG4	BCMA/CD3	Cell-bound	Linear (SC 0.08 mg/kg to 3 mg/kg)	3.8±1.7 days (after first IV treatment)	CL decreased with time, CL after 13th treatment is 0.472 L/day	5.63 L	SC injection, mg/kg dose
Faricimab	Ophthalmology/ Wet AMD and DME	Humanized bi-specific IgG1	Ang-2/ VEGF-A	Soluble	Linear	Apparent systemic half-life 7.5 days	Plasma CL/F 2.33 mL/ day	Vc/F: 1.48 L	IVT, flat dose

CL clearance, F bioavailability, IV intravenous, IVT intravitreal, SC subcutaneous, Vc volume of central compartment, Vss volume of distribution at steady state

[a]First approved indication

[b]Where PK are nonlinear, parameters are reported at usual clinical dose

[c]Terminal t1/2 reported for linear PK only

5 Pharmacodynamics

Pharmacodynamic (PD) biomarkers are useful to inform key questions in early drug development, including to demonstrate evidence of target engagement, the biological proof-of-concept (POC) of MOA, making inferences to clinical efficacy and safety, and to inform dose/regimen selection. For TCBs, PD biomarkers focus on assessing T-cell activation and the subsequent immune response. This includes measuring the number and percentage of activated T cells, cytokine production (such as increased levels of IL-6, IL-1, IL-2, IL-8, IL-10, IFN-γ, and TNF-alpha), CRP levels, and immune cell subsets to understand functional changes after TCB treatment [25]. Since clinical CRS is a common toxicity associated with TCBs, characterized by elevated secretion of multiple pro-inflammatory cytokines [18, 19, 26], PD biomarker assessments are frequently used in TCB clinical development to predict the risk of CRS and inform clinical dosing regimen selection. Examples of using IL-6 through integrated assessment of quantitative systems pharmacology (QSP) modeling framework were used to inform step–up dose regimen selection [27]. In the recently approved regulatory packages, cytokines are consistently reported as part of the clinical pharmacology package submitted for regulatory filing and approval purposes [6–13]. See Section "Cytokine Release Syndrome (CRS) Management."

6 Cytokine Release Syndrome (CRS) Management

6.1 On-Target Acute Toxicities

CRS is an on-target toxicity caused by T-cell activation, which triggers a cascade of immune responses involving macrophages, monocytes, and endothelial cells. Although a recent high-level guideline for CRS assessment and mitigation has been developed [28], standard practices are still evolving and remain an active area of research. CRS is highly context-dependent and multi-factorial, influenced by patient-specific factors (such as disease and immune status) and molecule-specific factors (such as binding affinity, construct, and dosing regimen). As a result, CRS risk varies significantly across different molecules and diseases, depending on exposure and timing.

6.2 Step-up (Priming) Dosing Regimens

Quantitative pharmacological modeling has been widely applied to inform strategies for mitigating, managing, and preventing CRS following TCB treatments. One effective approach is step-up dosing (SUD), which involves administering smaller initial doses before reaching the full or maximally effective dose. This method helps expand the therapeutic index by reducing cytokine peaks and desensitizing T cells to cytokine production with repeated dosing [14, 25]. By lowering immediate target activation in peripheral blood, SUD decouples the factors driving efficacy from those causing CRS [29]. As a result, SUD regimens have been adopted in all recently approved TCB therapies. In addition, SC administration, by slowing the time to define results in a slower and less severe CRS onset, potentially enabling higher clinical doses compared to IV administration [30–33]. The selection of SUD and SC regimens remains an area of active research [34]. For mosunetuzumab, SUD dosing was chosen based on exposure–response curves for CRS and model-derived maximal receptor occupancy, aiming to minimize CRS risk in relapsed/refractory follicular lymphoma patients [14]. Some studies found no clear exposure-response link between Cmax and CRS [35], but observed significant differences in CRS risk between dosing intervals.

6.3 Use of Tocilizumab for CRS Management

Tocilizumab, which binds to both soluble and membrane-bound IL-6 receptors, is a widely used agent for mitigating CRS [36]. The tocilizumab dosing regimen can influence the exposure–response (E–R) relationship for safety following TCB administration, making it a critical consideration when characterizing E–R relationships for CRS. Furthermore, tocilizumab impacts the pharmacological profile of IL-6 and its signaling pathway, altering the relationship between measured IL-6 concentrations and CRS risk [37]. Higher IL-6 levels have been observed in tocilizumab-treated patients compared to untreated ones, likely due to reduced IL-6 receptor bound to tocilizumab. Thus, caution is needed when modeling the relationship between PK, IL-6, and CRS, with proper consideration of tocilizumab use [38].

Although tocilizumab is approved for CRS management in CAR T-cell therapies, there is no formal guideline for its use with TCB molecules. From a pharmacological perspective, the clinical dose of tocilizumab may vary depending on the disease context, dosing regimens, and specific TCB molecules, which could result in different extents and patterns of CRS in clinical settings. Mechanistic PK-PD modeling has been used to characterize the binding kinetics and effects of tocilizumab in patients treated with teclistamab for CRS management or prevention [39]. This modeling is crucial for removing the confounding effect of tocilizumab on CRS exposure–response relationships and for determining the appropriate dosing regimen for CRS management.

7 Quantitative Pharmacology Modeling Approaches

7.1 Model-Informed Drug Development (MIDD)

Modeling and simulation in TCB development follow principles of model-informed drug development (MIDD) and are applied similarly to other therapeutic modalities to support various stages of development [40, 41]. As outlined in the following sections, modeling techniques address diverse questions such as dose and regimen selection, prediction of dose–response relationships, dose adjustments for special populations (e.g., organ impairment, elderly, pediatric), and assessment of drug interaction risks to inform product labeling. These model-driven insights, integrated with preclinical and clinical data, contribute to the totality of evidence that guides decision-making in drug development.

7.2 Human PK Prediction

Over the years, many theories and approaches have been proposed and used for scaling preclinical PK data to humans. Allometric scaling, based on a power–law relationship between size of the body and physiological and anatomical parameters, is the simplest and most widely used approach [42–44]. While no single scaling method has been shown to definitively predict human PK in all cases, especially for small-molecule drugs [45], the PK for mAbs can be predicted reasonably well, especially for mAbs at doses where the dominant clearance route is likely to be independent of concentration. Most therapeutic mAbs bind to nonhuman primate antigens more often than to rodent antigens, due to the greater sequence homology observed between nonhuman primates and humans. The binding epitope, in vitro binding affinity to antigen, binding affinity to FcRn, tissue cross-reactivity profiles, and disposition and elimination pathways of mAbs are often comparable in nonhuman primates and humans. It has been demonstrated that clearance and distribution volume of mAbs with linear PK in humans can be reasonably projected based on data from nonhuman primates alone. Predicting the PK of TCBs generally follows the same paradigm as mAbs; i.e., allometrically scaling preclinical PK parameters to predict human PK.

Importantly, TCBs often exhibit activity at low dose ranges, where TMDD significantly influences their PK. To accurately describe and translate TMDD, a general TMDD model incorporating two distinct binding targets has been proposed [46]. Additionally, the quasi-equilibrium (QE) approximation of the TMDD model has been suggested to streamline characterization and support translational PK efforts [47].

7.3 First-in-Human Dose Selection

There are two methods for first-in-human (FIH) dose selection: the toxicology-based approach and pharmacology-based approach. The pharmacology-based approach, such as pharmacologically active dose (PAD), anticipated therapeutic dose (ATD), and minimal anticipated biological effect level (MABEL), is widely used for BsAbs due to their unique MOA [48, 49]. Retrospective analysis of FIH studies with T-cell engaging bispecific antibodies have been demonstrated that doses corresponding to ~10–30% pharmacological activity (PA) in relevant models have resulted in minimal toxicities and an FIH dose higher than 30% PA may be acceptable depending on the totality of the data [50]. Recently, Liao et al. [51] proposed a quantitative systems pharmacology (QSP) model to address the challenge of translating in vitro drug potency across diverse assay conditions and contexts. This model focuses on immunological synapse (i.e., trimer) formation and provides a unified framework for making E–R predictions across various disease settings and immunological contexts.

7.4 Trimer Modeling

Due to its complex MoA, TCBs exhibite a bell-shaped concentration–response relationship in vitro. This behavior is likely attributed to the formation of the trimer complex (target–TCB–CD3), which drives both efficacy and safety/toxicity. The formation of the trimer complex is concentration dependent. At low TCB concentrations, trimer formation is favored due to efficient bivalent binding. As the TCB concentration increases, trimer formation reaches its peak at an optimal concentration. However, when TCB concentrations exceed this optimal level, excess TCB shifts the binding preference toward monovalent interactions, forming dimers between TCB and T cells or TCB and tumor cells [51, 52]. This shift results in a bell-shaped concentration–response relationship. Exposure–efficacy analysis of epcoritamab indicated that full doses of >48 mg were not expected to lead to significantly higher efficacy [9–13], which could be due to not enough trimer formation when dose >48 mg.

The response window of TCBs depends on several factors that influence trimer formation, including target receptor expression, the effector-to-target (E:T) ratio, and the binding affinities of the TCB for CD3 and the targeted tumor antigen. While the bell-shaped concentration–response relationship has been confirmed preclinically for CD3 TCBs, it is not commonly observed in patients. This discrepancy is likely due to the relatively low TCB concentration ranges tested in clinical settings compared to the theoretical concentrations at which the bell-shaped response occurs. Additionally, the complexity of the in vivo system—including the distribution of target cells, TCB PK across tissues, receptor turnover rates, and the substantial heterogeneity in patient responses—may obscure the emergence of clear bell-shaped efficacy patterns in clinical scenarios.

7.5 Quantitative Systems Pharmacology (QSP)

Due to its complex MOA, mechanistic modeling approaches, including QSP, are a useful tool to understand the complexities and aid in clinical translation, trial design, and prediction of regimens and strategies to reduce dose-limiting toxicities of TCBs. A mosunetuzumab QSP model was successfully developed and used to support the identification of an appropriate dosing regimen to mitigate the acute safety risk of CRS associated with mosunetuzumab [27]. By leveraging this QSP model with a digital twin approach, the dose/exposure–response relationships during dose escalation for mosunetuzumab were characterized and potential pretreatment biomarkers predictive of response were identified [53].

7.6 Physiologically-Based PK (PBPK) Modeling

The potential risk of drug–drug interactions with cytochrome P450 (CYP450) inactivation by the transient systemic elevation of cytokines was evaluated in silico using a physiologically-based PK (PBPK) model for TDB development [54], such as blinatumomab [55] and mosunetuzumab [56]. Transient elevation of cytokines (interleukin (IL)-6, IL-10, interferon-gamma (IFN-γ)) has been observed within the first 48 hours of continuous intravenous blinatumomab infusion. The PBPK work predicted suppression of hepatic CYP450 activities was <30%, and IL-6–mediated changes in exposure to sensitive substrates of CYP3A4, CYP1A2, and CYP2C9 were <two-fold and lasted <1 week [55]. Transient elevation of interleukin-6 (IL-6) occurs in early treatment cycles of mosunetuzumab. At clinically approved doses/regimens, the predicted DDI risk for mosunetuzumab remained low, with exposure changes for midazolam less than two-fold.

A semi-mechanistic, physiologically-based PK/PD model was developed for epcoritamab, a CD3×CD20 BsAb, to quantitatively describe biodistribution, trimer formation, and tumor response using preclinical, clinical PK, biomarker, tumor, and response data from the dose-escalation part of the phase I/II trial to guide selection of 48 mg as the RP2D [35].

8 Future Perspectives

8.1 Pharmacological Drivers for Efficacy and Safety: A Continued Learn-and-Confirm Cycle

TCB molecules have revolutionized cancer immunotherapy, delivering potent and durable responses in oncology. However, their robust mechanism of action also carries the risk of on-target toxicities, such as CRS and neurological AEs, resulting from T-cell activation. To mitigate these risks, novel constructs aim to localize TCB activation to tumor sites, thereby expanding the therapeutic index [57]. Developing successful TCB therapies requires a deep understanding of the pharmacological drivers underpinning both efficacy and safety—across tissues and patient populations—to devise strategies that expand the therapeutic index and optimize the benefit-risk profile.

Key factors influencing efficacy and safety include:

- Molecular design and construct: Affinity, structural design, and selective activation (e.g., prodrug designs).
- Target engagement kinetics: Behavior in blood, normal tissue, and target tissue.
- Disease settings: Target expression, disease dynamics, and prior therapies.
- Patient-specific contexts: Effector-to-target (E:T) ratios, immune status, and disease burden.
- Dosing administration routes and dosing regimen.

Navigating this complex and multifactorial pharmacological landscape requires innovative clinical pharmacology approaches. Central to this is the implementation of a "clinic-in-a-loop" paradigm, which emphasizes iterative learning and confirming cycles:

experiments leading to data generation, data-driven insights driving predictive modeling and simulations, and hypothesis testing informing subsequent experiments. Such intentional cycles are imperative to accelerate translational research and refine strategies in TCB therapy development.

8.2 Role of Artificial Intelligence/Machine Learning (AI/ML)-Based Modeling

Advanced mechanistic modeling, such as QSP and PBPK, has been valuable in guiding clinical design and dose selection. AI and machine learning (ML) algorithms also show promise in predicting patient-specific risks like CRS and supporting personalized treatments [58]. Integrating neural networks with mechanistic models [59] can enhance both interpretability and predictability. The synergy of pharmacological modeling, statistical modeling, and AI/ML-based algorithms can potentially usher in a new era of data-driven insights. This approach could help tackle complex translational questions by integrating large datasets across various molecules, such as predicting CRS risk across different diseases and therapies.

8.3 Combination Therapies and Resistance Mechanisms

Future TCB development is expected to focus on combination therapies to enhance response rates and durability by addressing resistance mechanisms. Resistance to TCB therapies can arise from factors such as T-cell dysfunction within the immunosuppressive tumor microenvironment and antigen loss, both of which significantly impact therapeutic efficacy and durability [57, 60]. Liu et al. [52] also highlighted the bone marrow as a therapeutic "sanctuary site," where tumor evolution and antigen escape occur due to the scarcity of effector cells.

To counter antigen loss, combination therapies targeting alternative antigens within the same disease context are being explored. Similarly, combining TCBs with co-stimulatory molecules offers a promising strategy to address T-cell dysfunction and improve activation [61]. Furthermore, optimizing TCB constructs continues to play a critical role in enhancing the quality of the immune synapse [62]. Fine-tuning the balance between CD3 and target affinity is key to driving tumor or disease response while minimizing systemic activation and associated adverse effects.

Advancements in these areas are crucial for the successful discovery and development of effective TCB therapies. As research progresses, quantitative clinical pharmacology will remain pivotal by:

- Linking tissue and blood data to better understand pharmacokinetics and pharmacodynamics.
- Refining dosing and sequencing strategies for combination regimens.
- Evaluating synergistic or additive effects of combination therapies.
- Identifying disease- and patient-specific factors to inform development decisions and regulatory milestones.

By integrating these insights into the drug development process, combination strategies can be effectively designed to overcome resistance mechanisms and maximize therapeutic potential.

References

1. Zhou S, Liu M, Ren F et al (2021) The landscape of bispecific T cell engager in cancer treatment. Biomark Res 9:38. https://doi.org/10.1186/s40364-021-00294-9
2. Baeuerle PA, Wesche H (2022) T-cell-engaging antibodies for the treatment of solid tumors: challenges and opportunities. Curr Opin Oncol 34(5):552–558. https://doi.org/10.1097/CCO.0000000000000869
3. Dickinson MJ, Carlo-Stella C, Morschhauser F et al (2022) Glofitamab for relapsed or refractory diffuse large B-cell lymphoma. New England J Med 387(24):2220–2231. https://doi.org/10.1056/NEJMoa2206913
4. Sun Y, Yu X, Wang X et al (2023) Bispecific antibodies in cancer therapy: target selection and regulatory requirements. Acta Pharm Sin B 13(9):3583–3597. https://doi.org/10.1016/j.apsb.2023.06.006
5. Klein C, Brinkmann U, Reichert JM et al (2024) The present and future of bispecific antibodies for cancer therapy. Nat Rev Drug Discov 23:301–319. https://doi.org/10.1038/s41573-024-00896-6
6. Center for Drug Evaluation and Research, FDA (2022) Multi-disciplinary review and evaluation BLA 125557: blinatumomab
7. Center for Drug Evaluation and Research, FDA (2021) Multi-disciplinary review and evaluation BLA 761228: Kimmtrak (tebentafusp)
8. Center for Drug Evaluation and Research, FDA (2021) Multi-disciplinary review and evaluation BLA 761291: teclistamab
9. Center for Drug Evaluation and Research, FDA (2022) Multi-disciplinary review and evaluation BLA 761263: Lunsumio (Mosunetuzumab)
10. Center for Drug Evaluation and Research, FDA (2022) Multi-disciplinary review and evaluation BLA 761342: Talquetamab-tgvs
11. Center for Drug Evaluation and Research, FDA (2022) Multi-disciplinary review and evaluation BLA 761324: epcoritamab
12. Center for Drug Evaluation and Research, FDA (2022) Multi-disciplinary review and evaluation BLA 761345: Elranatamab
13. Center for Drug Evaluation and Research, FDA (2022) Multi-disciplinary review and evaluation BLA 761309: Glofitamab-gxbm (Columvi)
14. Li CC, Bender B, Wilkins J et al (2024) A Novel Step-Up Dosage Regimen for Enhancing the Benefit-to-Risk Ratio of Mosunetuzumab in Relapsed or Refractory Follicular Lymphoma. Clin Pharmacol Ther https://doi.org/10.1002/cpt.3445
15. Morcos PN, Li J, Hosseini I, Li CC (2021) Quantitative clinical pharmacology of T-cell engaging bispecifics: current perspectives and Opportunities. Clin Transl Sci 14(1):75–85. https://doi.org/10.1111/cts.12877
16. Mandikian D, Takahashi N, Lo AA et al (2018) Relative target affinities of T-cell-dependent bispecific antibodies determine biodistribution in a solid tumor mouse model. Mol Cancer Ther 17(4):776–785. https://doi.org/10.1158/1535-7163.MCT-17-0657
17. Bender B, Li CC, Marchand M et al (2024) Population pharmacokinetics and CD20 binding dynamics for mosunetuzumab in relapsed/refractory B-cell non-Hodgkin lymphoma. Clin Transl Sci 17(6):e13825. https://doi.org/10.1111/cts.13825
18. Shimabukuro-Vornhagen A, Gödel P, Subklewe M et al (2018) Cytokine release syndrome. J Immunotherapy Cancer 6:56
19. Lee DW, Santomasso BD, Locke FL et al (2019) ASTCT consensus grading for cytokine release syndrome and neurologic toxicity associated with immune effector cells. Biol Blood Marrow Transplant 25:625–638
20. FDA Guidance: bispecific antibody development programs, guidance for industry, May 2021, https://www.fda.gov/media/123313/download
21. Register AC, Tarighat SS, Lee HY (2021) Bioassay development for bispecific antibodies-challenges and opportunities. Int J Mol Sci 22(10):5350. https://doi.org/10.3390/ijms22105350
22. Chen Y, Xu Y (2017) Pharmacokinetics of bispecific antibody. Curr Pharmacol Rep 3:126–137. https://doi.org/10.1007/s40495-017-0090-5
23. Brinkmann U, Kontermann RE (2017) The making of bispecific antibodies. MAbs 9(2):182–212. https://doi.org/10.1080/19420862.2016.1268307
24. Proctor JR, Wong H (2024) Time-dependent clearance can confound exposure-response analysis of therapeutic antibodies: A comprehensive review of the current literature. Clin Transl Sci 17:e13676. https://doi.org/10.1111/cts.13676
25. Radtke KK, Bender BC, Li Z et al (2024) Clinical pharmacology of cytokine release

syndrome with T-cell engaging bispecific antibodies: current insights and drug development strategies. Clin Cancer Res https://doi.org/10.1158/1078-0432.CCR-24-2247

26. Cosenza M, Sacchi S, Pozzi S (2021) Cytokine release syndrome associated with T-cell-based therapies for hematological malignancies: pathophysiology, clinical presentation, and treatment. Int J Mol Sci 22:7652. https://doi.org/10.3390/ijms22147652
27. Hosseini I, Gadkar K, Stefanich E et al (2020) Mitigating the risk of cytokine release syndrome in a phase I trial of CD20/CD3 bispecific antibody mosunetuzumab in NHL: impact of translational system modeling. NPJ Syst Biol Appl 6:1–11
28. Crombie J, Graff T, Falchi L et al (2024) Consensus recommendations on the management of toxicity associated with CD3× CD20 bispecific antibody therapy. Blood 143(16):1565–1575
29. Li J, Piskol R, Ybarra R et al (2019) CD3 bispecific antibody-induced cytokine release is dispensable for cytotoxic T cell activity. Sci Transl Med 11(508):eaax8861. https://doi.org/10.1126/scitranslmed.aax8861
30. Ball K, Dovedi SJ, Vajjah P, Phipps A (2023) Strategies for clinical dose optimization of T cell-engaging therapies in oncology. MAbs 15(1):2181016. https://doi.org/10.1080/19420862.2023.2181016
31. Budde EL, Bartlett NL, Giri P et al (2022) Subcutaneous mosunetuzumab is active with a manageable safety profile in patients with relapsed/refractory B-cell non-Hodgkin lymphomas: updated results from a phase I/II study. Blood 140(Suppl 1):3753–3755. https://doi.org/10.1182/blood-2022-157729
32. Nooka AK, Moreau P, Usmani SZ et al (2022) Teclistamab, a B cell maturation antigen (BCMA) x CD3 bispecific antibody, in patients with relapsed/refractory multiple myeloma (RRMM): updated efficacy and safety results from MajesTEC-1. J Clin Oncol 40(16_suppl):8007–8007. https://doi.org/10.1200/JCO.2022.40.16_suppl.8007
33. Minnema MC, Krishnan AY, Berdeja JG et al (2022) Efficacy and safety of talquetamab, a G protein-coupled receptor family C group 5 member D x CD3 bispecific antibody, in patients with relapsed/refractory multiple myeloma (RRMM): updated results from MonumenTAL-1. J Clin Oncol 40(16_suppl):8015–8015. https://doi.org/10.1200/JCO.2022.40.16_suppl.8015
34. Li T, Polhamus D, Thalhauser C et al (2023) Abstract 2798: simplifying selection and optimization of step-up dosing of subcutaneous Epcoritamab to mitigate CRS risk using repeated time-to-event modeling. Cancer Res 83:2798
35. Li T, Hiemstra IH, Chiu C et al (2022) Semimechanistic physiologically-based pharmacokinetic/pharmacodynamic model informing epcoritamab dose selection for patients with B-cell lymphomas. Clin Pharmacol Ther 112(5):1108–1119. https://doi.org/10.1002/cpt.2729
36. Genentech, Inc. (2025) Actemra (tocilizumab) injection, for intravenous or subcutaneous use [Prescribing information]. U.S. Food and Drug Administration. https://www.gene.com/download/pdf/actemra_prescribing.pdf
37. Nishimoto N, Terao K, Mima T et al (2008) Mechanisms and pathologic significances in increase in serum interleukin-6 (IL-6) and soluble IL-6 receptor after administration of an anti–IL-6 receptor antibody, tocilizumab, in patients with rheumatoid arthritis and Castleman disease. Blood 112:3959–3964
38. Bender B, Schoemaker R, Li Z, Li F, Vadhavkar S, Lu J, Susilo M, Turner D, Li CC, Chanu P (2024) A multi-stage modeling approach towards predicting cytokine release syndrome (CRS) in patients receiving Mosunetuzumab. ACoP Conference 2024, Poster T-011
39. Zhou J, Vishwamitra D, Guo Y et al (2023) Model-based exploration of the impact of prophylactic tocilizumab on IL-6 dynamics in multiple myeloma patients receiving teclistamab treatment. Blood 142:4670
40. Madabushi R, Seo P, Zhao L, Tegenge M, Zhu H (2022) Review: role of model-informed drug development approaches in the lifecycle of drug development and regulatory decision-making. Pharm Res 39(8):1669–1680. https://doi.org/10.1007/s11095-022-03288-w
41. Lim K, Zhu XS, Zhou D, Ren S, Phipps A (2024) Clinical pharmacology strategies for bispecific antibody development: learnings from FDA-approved bispecific antibodies in oncology. Clin Pharmacol Ther 116(2):315–327. https://doi.org/10.1002/cpt.3308
42. Dedrick RL (1973) Animal scale-up. J Pharmacokinet Biopharm 1(5):435–461
43. Mahmood I (2005) Prediction of concentration-time profiles in humans. Pine House Publisher, Rockville
44. Mahmood I (2009) Pharmacokinetic allometric scaling of antibodies: application to the first-in-human dose estimation. J Pharm Sci 98(10):3850–3861. https://doi.org/10.1002/jps.21682
45. Tang H, Mayersohn M (2005) A novel model for prediction of human drug clearance by allometric scaling. Drug Metab Dispos 33(9):1297–1303. https://doi.org/10.1124/dmd.105.004143

46. Trivedi A, Stienen S, Zhu M et al (2017) Clinical pharmacology and translational aspects of bispecific antibodies. Clin Transl Sci 10(3):147–162. https://doi.org/10.1111/cts.12459
47. Schropp J, Khot A, Shah DK, Koch G (2019) Target-mediated drug disposition model for bispecific antibodies: properties, approximation, and optimal dosing strategy. CPT Pharmacometrics Syst Pharmacol 8(3):177–187. https://doi.org/10.1002/psp4.12369
48. Leach MW, Clarke DO, Dudal S, Han C, Li C, Yang Z, Brennan FR, Bailey WJ, Chen Y, Deslandes A, Loberg LI, Mayawala K, Rogge MC, Todd M, Chemuturi NV (2021) Strategies and recommendations for using a data-driven and risk-based approach in the selection of first-in-human starting dose: an International Consortium for Innovation and Quality in Pharmaceutical Development (IQ) assessment. Clin Pharmacol Ther 109(6):1395–1415. https://doi.org/10.1002/cpt.2009
49. Nagaraja Shastri P, Shah N, Lechmann M, Mody H, Retter MW, Zhu M, Li T, Wang J, Shaik N, Zheng X, Ovacik M, Hua F, Jawa V, Boetsch C, Cao Y, Burke J, Datta K, Gadkar K, Upreti V, Betts A (2024) Industry perspective on first-in-human and clinical pharmacology strategies to support clinical development of T-cell engaging bispecific antibodies for cancer therapy. Clin Pharmacol Ther 2024. https://doi.org/10.1002/cpt.3439
50. Saber H, Del Valle P, Ricks TK, Leighton JK (2017) An FDA oncology analysis of CD3 bispecific constructs and first-in-human dose selection. Regul Toxicol Pharmacol 90:144–152. https://doi.org/10.1016/j.yrtph.2017.09.001
51. Liao X, Qi T, Zhou J, Liu C, Cao Y (2024) Optimizing clinical translation of bispecific T-cell engagers through context unification with a quantitative systems pharmacology model. Clin Pharmacol Ther https://doi.org/10.1002/cpt.3302
52. Liu C, Zhou J, Kudlacek S et al (2023) Population dynamics of immunological synapse formation induced by bispecific T cell engagers predict clinical pharmacodynamics and treatment resistance. Elife 12:e83659. https://doi.org/10.7554/eLife.83659
53. Susilo ME, Li CC, Gadkar K et al (2023) Systems-based digital twins to help characterize clinical dose–response and propose predictive biomarkers in a phase I study of bispecific antibody, mosunetuzumab, in NHL. Clin Transl Sci 16:1134–1148
54. Chen KF, Jones HM (2022) PBPK perspective on alternative CYP3A4 inducers for rifampin. CPT Pharmacometrics Syst Pharmacol **11**:1543–1546. https://doi.org/10.1002/psp4.12864
55. Xu Y, Hijazi Y, Wolf A, Wu B, Sun YN, Zhu M (2015) Physiologically based pharmacokinetic model to assess the influence of blinatumomab-mediated cytokine elevations on cytochrome P450 enzyme activity. CPT Pharmacometrics Syst Pharmacol 4(9):507–515. https://doi.org/10.1002/psp4.12003
56. Chen Y, Ma F, Jones N, Deng R, Li C, Li CC (2024) Assessment of CYP3A-mediated drug interaction via cytokine (IL-6) elevation for mosunetuzumab using physiologically-based pharmacokinetic modeling. CPT Pharmacometrics Syst Pharmacol 13(2):234–246. https://doi.org/10.1002/psp4.13073
57. Liu J, Zhu J (2024) Progresses of T-cell-engaging bispecific antibodies in treatment of solid tumors. Int Immunopharmacol 138:112609. https://doi.org/10.1016/j.intimp.2024.112609
58. Xiong F, Janko M, Walker M, Makropoulos D, Weinstock D, Kam M, Hrebien L (2014) Analysis of cytokine release assay data using machine learning approaches. Int Immunopharmacol 22(2):465–479. https://doi.org/10.1016/j.intimp.2014.07.024
59. Mann J, Meshkin H, Zirkle J et al Mechanism-based organization of neural networks to emulate systems biology and pharmacology models. Sci Rep 14:12082. https://doi.org/10.1038/s41598-024-59378-9
60. van de Donk NWCJ, O'Neill C, de Ruijter MEM, Verkleij CPM, Zweegman S (2023) T-cell redirecting bispecific and trispecific antibodies in multiple myeloma beyond BCMA. Curr Opin Oncol 35(6):601–611. https://doi.org/10.1097/CCO.0000000000000983
61. Skokos D, Waite JC, Haber L, Yancopoulos GD et al (2020) A class of costimulatory CD28-bispecific antibodies that enhance the antitumor activity of CD3-bispecific antibodies. Sci Transl Med 12(525):eaaw7888. https://doi.org/10.1126/scitranslmed.aaw7888
62. Carrasco-Padilla C, Hernaiz-Esteban A, Álvarez-Vallina L, Aguilar-Sopeña O, Roda-Navarro P (2022) Bispecific antibody format and the organization of immunological synapses in T cell-redirecting strategies for cancer immunotherapy. Pharmaceutics 15(1):132. https://doi.org/10.3390/pharmaceutics15010132

Part IV

Clinical Applications

Chapter 16

Clinical Development of T-Cell Bispecific Antibodies: Review, Challenges, and Prospective Solutions in Solid Tumor

Chika Ogami and Takahiro Ishiguro

Abstract

T-cell engaging bispecific antibodies (T-BsAbs) have emerged as a promising approach in cancer immunotherapy. This review explores the clinical development of T-BsAbs in solid tumors, highlighting their unique mechanism of action, which involves the simultaneous binding to tumor-associated antigens and the CD3 T-cell receptor, thereby redirecting T cells to tumor cells and activating them in an antigen-dependent manner. Despite their success in hematological malignancies, T-BsAbs face significant challenges in solid tumors, including on- and off-target toxicities, the selection of appropriate tumor-associated antigens, and the complexity of the tumor microenvironment. Safety concerns, such as cytokine release syndrome (CRS) and on-target, off-tumor toxicities, are major obstacles that need to be addressed. Recent advancements offer potential solutions to these challenges. This review provides a comprehensive overview of the current state of T-BsAb development, discussing recent advances, ongoing clinical trials, and emerging strategies to enhance their efficacy and safety in solid tumor treatment.

Keywords T-cell engaging bispecific antibodies (T-BsAbs), Cancer immunotherapy, Solid tumors, Tumor-associated antigens (TAAs), CD3 T-cell receptor, On-target toxicity, Off-target toxicity, Tumor microenvironment, Cytokine release syndrome (CRS)

1 Introduction

T-cell engaging bispecific antibody (T-BsAb) have emerged as a promising approach in cancer immunotherapy, with significant potential for treating solid tumors. While T-BsAbs have shown remarkable success in hematological malignancies, their application to solid tumors presents unique challenges and opportunities. This review focuses on the clinical development of T-BsAbs for solid tumors, exploring the current landscape, key obstacles, and emerging drugs overcoming these hurdles.

T-BsAbs represent a novel class of bispecific antibodies designed to simultaneously bind to tumor-associated antigens (TAAs) and the CD3 T-cell receptor on T cells. This unique structure enables T-BsAbs to redirect T cells to tumor cells and activate them in an

Masahiro Yasunaga (ed.), *Bispecific Antibodies*, Methods in Molecular Biology, vol. 2997,
https://doi.org/10.1007/978-1-0716-5037-0_16,
© The Author(s), under exclusive license to Springer Science+Business Media, LLC, part of Springer Nature 2026

antigen-dependent manner [1]. Specifically, T-BsAbs form an immunological synapse between T cells and tumor cells, promoting the release of cytotoxic proteins (such as perforin and granzymes) from T cells. This mechanism allows for major histocompatibility complex (MHC)-independent tumor cell killing [2].

The clinical development of T-BsAbs has shown significant success, particularly in hematological malignancies [3]. Notable examples include CD19-targeted therapies for acute lymphoblastic leukemia [4], CD20-targeted approaches for B-cell non-Hodgkin lymphoma [5], and BCMA or GPRC5D-targeted treatments for multiple myeloma [6]. While progress in solid tumors has been slower due to challenges such as the lack of tumor-specific target antigens and the complexity of the tumor microenvironment, recent advancements have been made. For instance, a T-BsAb recognizing a gp100 peptide fragment presented on HLA-A2:01 molecules has demonstrated remarkable activity against ocular melanoma [7].

T-BsAbs offer several advantages over checkpoint inhibitors, another prominent class of cancer immunotherapies. While checkpoint inhibitors, such as those targeting CTLA-4 or PD-1/PD-L1, enhance the activation of endogenous T cells, their efficacy depends on the presence of neoantigens presented on tumor cell HLA molecules [8]. In contrast, T-BsAbs can activate T cells polyclonally, regardless of T-cell receptor (TCR) specificity, and redirect them to antigen-expressing tumor cells. This mechanism potentially allows T-BsAbs to be effective against "non-immunogenic" tumors lacking neoantigens [9]. Furthermore, T-BsAbs can utilize a broad range of T cells as effector cells, including both CD8+ and CD4+ T cells [10].

Despite their promise, T-BsAbs face several challenges. Cytokine release syndrome (CRS), caused by the secretion of inflammatory cytokines from activated T cells, is the most frequently observed adverse event [11]. Various resistance mechanisms have also been reported, including downregulation of target antigens by tumor cells [12], activation of apoptosis-resistant signaling pathways [13], and inhibition of antibody activity by factors in the tumor microenvironment [14]. On-target off-tumor toxicity is a particular concern when targeting solid tumors, especially when the target antigen is expressed in normal tissues. For example, the distribution of classical tumor antigens such as EpCAM, EGFR, and HER2 is not strictly tumor-specific, raising concerns about effects on normal tissues. Indeed, the administration of solitomab, an EpCAM-targeted T-BsAb, was limited due to its impact on EpCAM-positive normal tissues, including intestinal epithelium and bile ducts [15].

To overcome these challenges, efforts are underway to develop newT-BsAbs and explore novel combination strategies. Identifying appropriate antigens with high tumor specificity remains a critical challenge, particularly for T-BsAbs targeting solid tumors. These ongoing efforts aim to enhance the depth and durability of T-BsAb efficacy while establishing them as safer and more effective options for cancer treatment.

This review will delve into the current state of T-BsAb development, discussing recent advances, ongoing clinical trials, and emerging strategies to address the challenges associated with this promising immunotherapeutic approach. By examining the potential of T-BsAbs across various cancer types and exploring their integration with other treatment modalities, we aim to provide a comprehensive perspective on the future of T-BsAbs in cancer immunotherapy.

2 Challenges for development of T-BsAb in Solid Tumor

There are four major challenges in the clinical development of T-BsAbs for solid tumors. The first is the occurrence of on- and/or off-target toxicity [16–20]. The second is the selection of TAAs specifically expressed in the target tumor tissues and designing antibody with appropriate affinities [21, 22]. The third is that the efficacy is influenced by the number and state of T cells in the tumor microenvironment [23]. The last is the difficulty in establishing appropriate dosing regimens [24].

3 Occurrence of On- and/or Off-Target Toxicity

The most major on-target toxicity of T-BsAb is CRS, in which the administration of T-BsAbs leads to the activation and degranulation of T cells and tumor cells, resulting in the production of inflammatory cytokines such as interleukin-2, interleukin-6 (IL-6), interferon gamma, and tumor necrosis factor-α, which in turn cause systemic inflammatory reactions [17, 22, 25–28]. CRS is induced by T-BsAbs at a high frequency in solid tumors as well as hematological cancers, and mitigation strategies for CRS have been required for T-BsAb development [17, 18, 29]. Additionally, CRS can also occur as off-target toxicity when the Fc domain of T-BsAbs binds to FcγRs expressed on antigen-presenting cells and others [16, 19, 20]. For example, catumaxomab not only binds to EpCAM on tumor cells and CD3 on T cells but also has the ability to bind to FcγR-positive cells, leading to off-target toxicities such as CRS and hepatotoxicity, thereby limiting its utility for systemic administration due to the lack of a sufficient therapeutic window [19, 20]. Catumaxomab was approved in 2009 by the European Union for intraperitoneal administration for malignant ascites [30], and its development for ovarian and gastric cancers was pursued [31, 32]. However, its development was discontinued due to the occurrence of off-target toxicities caused by FcγR binding and high immunogenicity associated with its non-human

IgG background [20, 33]. Furthermore, it was withdrawn from the market in 2017 for commercial reasons.

To avoid off-target toxicities caused by FcγR binding, the development of antibodies with removed FcγR binding residues has been tried to development. BiTEs, consisting of only two Fv domains that bind to TAA and CD3, are one of the most clinically successful antibody formats among T-BsAbs [16, 21, 34]. Lacking the Fcγ R-binding domain, BiTEs have a smaller molecular size compared to full-length antibodies. Unlike conventional monoclonal antibodies that exhibit a long half-life due to recycling by binding to the neonatal crystallizable fragment receptor (FcRn) [35], BiTEs lack the crystallizable fragment (Fc) domain capable of binding to FcRn, resulting in a short half-life, and frequent or continuous administration is required [17, 36]. While this is a major clinical disadvantage, it may also be unfavorable for antitumor effects in solid tumors [21, 37]. Unlike hematological malignancies where the targets are present in the circulating blood, solid tumors require the transition from the circulating blood to the tumor tissue and the bridging of two cell types. Therefore, achieving efficacy may take longer time than in hematological malignancies, and a short half-life may not provide sufficient antitumor effects [21]. Consequently, the development of BiTEs with extended half-lives, such as those fused with an Fc domain, is underway and being evaluated in clinical trials [38, 39]. Furthermore, not limited to BiTEs, T-BsAbs are being designed and developed to maintain FcRn binding while eliminating FcγR binding [21, 40–43].

4 Selecting Tumor-Associated Antigens and Designing Antibody with Appropriate Affinity

To avoid on-target off-tumor toxicities while achieving efficacy, it is necessary to identify appropriate TAAs. In other words, it is essential to explore TAAs that are expressed on the surface of tumor cells but not on normal tissue cells and design antibodies targeting those TAAs. The recent success of approved agents such as tebentafusp and tarlatamab demonstrates that certain TAAs are suitable targets for T-BsAbs [39, 44]. Furthermore, the success of tebentafusp indicates that T-BsAbs can effectively recognize TAAs with low expression levels. T-BsAbs targeting antigens present in normal tissues, such as CEA or EpCAM, have often exhibited on-target tissue toxicities in preclinical and clinical studies [45–47].

Selecting TAA that do not extremely rapid internalize or shed from the cell surface is also important as well as specificity for tumor cells. The immune activation by T-BsAbs requires simultaneous binding to the TAA expressing on the tumor cell surface and CD3 on T cells. Rapid internalization or shedding of the TAA from the tumor cell surface can lead to a reduction in antitumor effects due to lack of time to bridge two cells and activate T cells [22].

In addition to selecting an appropriate TAA, designing the antibody with appropriate affinity is also crucial. T-BsAbs possess two types of domains that bind to the TAA and CD3, and the binding abilities of each domain to their respective antigens influence efficacy, safety, accumulation of T-BsAb in tumors, target-dependent clearance, and distribution to other tissues. In vitro studies evaluating anti-HER2/CD3 antibodies with different binding affinities to HER2 demonstrated that higher affinity to HER2 was associated with higher cytotoxic activity. However, the study also demonstrated that toxicities were observed with the administration of antibodies with higher HER2 affinity in cynomolgus monkeys [48]. Haber et al. evaluated the in vitro cytotoxic activity against tumor cell lines, as well as the in vivo distribution, serum half-life, and antitumor activity of antibodies targeting multiple TAAs and CD3 with varying CD3 affinities. When PSMA/CD3 or MUC16/CD3 antibodieswere added to cell lines expressing the corresponding TAA in the presence of PSMA or MUC16 inhibitors, the inhibitors had a reduced impact on the cytotoxic activity of antibodies with moderate CD3 affinity compared to those with low CD3 affinity. In other words, when the binding affinity to the TAA is low, antibodies with stronger CD3 binding may be able to maintain antitumor effects. Additionally, the results of antibody distribution studies in mice showed that the MUC16/CD3 antibody with moderate affinity accumulated in MUC16-expressing tumors and the spleen. Furthermore, the accumulation in any tissue was suppressed when anti-CD3 and anti-MUC16 antibodies were administered at saturating concentrations. Regarding splenic accumulation, it was observed that accumulation was suppressed in the case of low CD3 affinity, even without pretreatment with anti-CD3 antibodies [49]. Moreover, in vivo evaluations using HER2/CD3 antibodies with varying CD3 affinities have reported that higher CD3 affinity leads to increased distribution to tissues other than the tumor, affecting the TAA-dependent distribution to tumors [50].

5 Tumor Microenvironment

Similar to cancer immunotherapies with monoclonal antibody, the efficacy of T-BsAbs in solid tumors is influenced by the heterogeneity of the tumor microenvironment across patients, which poses a challenge in development. Preclinical studies have reported that administration of T-BsAbs can convert the tumor microenvironment from a non-inflamed state with few activated T cells to an inflamed state, leading to antitumor activity [51, 52]. However, in clinical trials, off-target or on-target/off-tumor toxicities often occur before sufficient antitumor activity is observed within the tumor

[53]. Evaluating changes in the tumor microenvironment induced by T-BsAb administration in clinical trials requires multiple biopsies, which are invasive, and the assessment is further complicated by tumor heterogeneity and inter-patient variability. Consequently, studies evaluating T-BsAb-induced changes in the tumor microenvironment in clinical trials are limited [20]. In the case of tarlatamab, which received accelerated approval in May 2024, higher clinical efficacy was observed in patients with higher baseline expression levels of the TAA DLL3 [39]. In a Phase 1 trial of cibisatamab targeting CEA/CD3, partial responders (PR) exhibited higher intratumoral expression of CEACAM5 compared to stable disease (SD) or progressive disease (PD) cases [46].

Additionally, cibisatamab administration was suggested to induce a transition from immune-excluded tumors to inflamed tumors by re-localizing CD8+ T cells from the stroma into the tumor. In a Phase 1 trial of pasotuxizumab targeting PSMA/CD3, instead of tumor tissue, changes in the proportion of CD69+CD8+ T cells (activated T cells) among circulating CD8+ T cells were evaluated, and an increase in CD69+CD8+ T cells was observed after pasotuxizumab administration, paralleled by a decrease in circulating tumor cells [17]. Evaluating the relationship between the tumor microenvironment and the efficacy of T-BsAbs in the clinical setting is crucial for informing future development strategies for T-BsAbs in solid tumors, and it is desirable to accumulate insights from various TAA targets and T-BsAb designs.

6 Setting Appropriate Dosing Regimens

Step-up dosing has been considered as a mitigation strategy for CRS induced by T-BsAb administration in both hematological malignancies and solid tumors. This approach leads to increased complexity in the dosing regimen patterns that need to be evaluated, posing a challenge in designing the appropriate dosing regimen [24]. In addition to step-up dosing, subcutaneous administration has also been reported as a useful strategy for mitigating CRS, further increasing the number of regimens that need to be evaluated in clinical trials to identify the optimal dosing regimen. With step-up dosing, it is necessary to determine (1) the appropriate priming dose, (2) the dosing interval for the priming dose (step-up dosing or fractionated step-up dosing), and (3) the number of step-up doses. In clinical trials, dose escalation with flat dosing may be initially conducted to determine the tolerable maximum dose as the priming dose before transitioning to the evaluation of step-up dosing. Alternatively, step-up dosing may be evaluated from the first dose [24, 39, 54]. In the case of tebentafusp, the first-in-human trial employed fixed dosing, and 50 μg was determined as the maximum tolerated dose (MTD). Subsequently, a dose-escalation study using step-up dosing was conducted to determine the recommended Phase 2 dose, which was 36% higher than the target dose with fixed dosing [54]. To support the evaluation of appropriate dosing regimens in solid tumors, the utilization of modeling and simulation(M&S) approaches is being considered [55–57]. Weddell developed a quantitative systems pharmacology (QSP) model that captures the dynamics of cytokines during the clearance phase after T-BsAb administration in solid tumors [56]. Simulations using this model successfully represented the dynamics of cytokines following tebentafusp administration, suggesting the potential utility of QSP models in supporting dose selection.

7 T-BsAb Related Toxicity and Mitigation Strategy

7.1 Cytokine Release Syndrome

As mentioned earlier, the major on-target toxicity associated with T-BsAbs is CRS. Relatively mild symptoms of CRS include flu-like symptoms such as fever, diarrhea, headache, rash, arthralgia, and myalgia. As CRS becomes more severe, hypotension requiring vasopressors, high fever, vascular leak, and coagulation abnormalities may occur, and further progression can lead to multi-organ failure and death [25]. Additionally, in the context of T-BsAb administration for hematological malignancies, neurological symptoms, including immune effector cell-associated neurotoxicity syndrome (ICANS), have been observed [58, 59]. Before the onset of neurological symptoms, a peak in C-reactive protein (CRP) levels is observed, followed by a peak in ferritin levels after symptom onset [60]. The manifestation of ICANS symptoms varies among patients, and the initial symptoms are often nonspecific. Characteristic early symptoms include hesitancy in speech and deterioration in handwriting, along with mild tremors and confusion. As the condition progresses, symptoms such as agitation, seizures, status epilepticus, cerebral edema, and intracranial hemorrhage may occur [61]. Clinical mitigation strategies for CRS include the administration of tocilizumab, steroids, other anti-inflammatory drugs, and step-up dosing.

IL-6 is known as one of the most important cytokines in CRS, and an association between elevated circulating IL-6 levels and the severity of CRS has been reported [11, 62]. Therefore, tocilizumab, an anti-IL-6 receptor antibody, is used for the mitigation of CRS. Tocilizumab binds to both membrane-bound and soluble IL-6R, thereby inhibiting the binding of IL-6 to its receptor and suppressing IL-6 signaling [63]. Since the occurrence of CRS is an on-target toxicity of T-BsAbs, there is a concern that mitigation strategies may attenuate antitumor effects. Kauer et al. reported that the simultaneous addition of dexamethasone with T-BsAbs significantly inhibited T-cell activation and proliferation in vitro [64].

In contrast, no inhibition of T-cell activation and proliferation was observed with tocilizumab in the same evaluation, suggesting that tocilizumab administration can mitigate CRS without compromising the therapeutic effects of T-BsAbs [64]. Tocilizumab is commonly administered to suppress symptoms after the onset of CRS, while steroid administration is recommended for prophylactic use [65]. However, for T-BsAbs, there are reports suggesting the utility of prophylactic tocilizumab administration, such as the suppression of CRS during subsequent T-BsAb doses when tocilizumab is given during the step-up dosing period [66], and the prevention of CRS occurrence and severity when tocilizumab is administered prophylactically before the initial T-BsAb dose [67].

Since tocilizumab is an antibody, it has limited ability to cross the blood–brain barrier [68], suggesting a potential inability to prevent neurotoxicity. In acute leukemia patients receiving CAR-T cell therapy, administration of tocilizumab 2 days after CAR-T cell infusion reduced the incidence of CRS but did not decrease the rate of ICANS [69]. Additionally, as tocilizumab cannot directly address symptoms caused by cytokines other than IL-6, steroid administration is used to suppress neurotoxicity and symptoms that cannot be controlled by tocilizumab [25, 70]. While in vitro studies suggested that dexamethasone administration could attenuate the effects of T-BsAbs, an evaluation in cynomolgus monkeys reported by Kamperschroer et al. in 2024 demonstrated that dexamethasone treatment suppressed CRS symptoms and increased the proportion of CD8 T cells in the blood, suggesting the utility of short-term dexamethasone treatment for CRS mitigation [71]. In a clinical trial of glofitamab in patients with relapsed/refractory diffuse large B-cell lymphoma, the cohort that received mandatory steroid pretreatment before glofitamab administration showed a lower incidence of CRS compared to the cohort without steroid treatment (48% vs. 68%) [72]. Although prospective studies on the appropriate timing and duration of steroid administration in clinical practice are limited, steroid treatment is considered useful as a first-line management strategy for CRS.

If treatment with tocilizumab and steroids is insufficient, other anti-inflammatory drugs such as anakinra or siltuximab may be used [73, 74]. Siltuximab, an anti-IL-6 antibody, has limited clinical reports, but a case report has described the recovery from CRS and ICANS caused by talquetamab following siltuximab administration [75]. In mouse models, both interleukin-1 (IL-1) and IL-6 are involved in CRS induced by CAR-T cells, and administration of anakinra, an IL-1 receptor antagonist, has been shown to mitigate neurotoxicity, suggesting the potential of anakinra for CRS/ICANS suppression [76]. Clinically, while anakinra has been associated with rapid recovery from high-dose-induced CRS/ICANS [77] and fever or cytokine elevation [78], reports are limited, and prospective trials are needed for further verification. Most reports on the suppression of CRS/ICANS by siltuximab and anakinra have been in the context of CAR-T cell therapy, and their efficacy and safety in suppressing CRS/ICANS associated with T-BsAb require further investigation.

Upon administration of T-BsAbs, particularly during the initial dose, high levels of circulating cytokines and CRS occurrence are observed. Therefore, a step-up dosing strategy is implemented as a mitigation approach, where a lower dose than the target dose is administered initially, followed by a gradual increase in dose over subsequent cycles until reaching the target dose [39, 46, 54, 79]. Step-up dosing was initially implemented as a CRS mitigation strategy in the development of T-BsAbs for hematological malignancies, but recent developments in solid tumors have also demonstrated its utility in preventing CRS [39, 54]. Regarding the mechanism of CRS suppression, an in vitro study using an anti-GPC3/CD3 T-BsAb reported that priming treatment with the T-BsAb could reduce cytokine levels without attenuating cytotoxic activity [80]. However, there is no established method to determine the appropriate dose intensity or the number of priming doses in clinical trials, necessitating empirical evaluation in trials or the utilization of M&S to propose regimens [24]. For the CD3/CD20 T-BsAb mosunetuzumab, a mechanistic model incorporating the dynamics of B cells, T cells, lymphoid tissues, and tumors in circulation was developed and simulated to support modifications to the step-fractionated treatment schedule [81].

8 Approved and Promising T-BsAbs for Solid Tumors

This section highlights T-BsAbs that have successfully gained approval for solid tumors, as well as those currently in clinical development that show promise for future success.

8.1 Tebentafusp [82]

Tebentafusp is a gp100/CD3 bispecific antibody designed for the treatment of metastatic uveal melanoma. In the pivotal Phase 3 trial, it demonstrated superior survival compared to the investigator's choice of treatment, with a median overall survival of 22 months versus 16 months and a 6-month progression-free survival rate of 31% versus 19%, respectively. The 1-year overall survival rate was 65%. Based on these results, the FDA approved tebentafusp in April 2022 for HLA-A*02:01-positive adult patients with unresectable or metastatic uveal melanoma. The most common adverse events were CRS (89%), fever (91%), rash and pruritus (83%), fatigue (71%), and chills (69%). As the first T-cell receptor-based bispecific therapy for metastatic uveal melanoma, tebentafusp offers a new treatment option for this rare and aggressive disease.

8.2 Tarlatamab [83]

Tarlatamab (AMG 757) is a DLL3-targeting, half-life-extended BiTE molecule. In the Phase 2 DeLLphi-301 trial, it demonstrated

Table 1
Overview of clinical advances in T-BsAb in the treatment of solid tumors

Tebentafusp-tebn	CD3 × gp100/HLA	Melanoma	Launched	Immunocore	NCT03070392
Tarlatamab	CD3 × DLL3	Lung (small cell)	Registered	Amgen	NCT05060016
Brenetafusp	CD3 × PRAME/HLA	Melanoma	Phase III	Immunocore	NCT06112314
OBT-620	CD3 × DLL3	Lung (neuroendocrine, small cell)	Phase II	Boehringer Ingelheim	NCT05882058
QLS-31905	CD3 × CLDN18.2	Solid tumor	Phase II	Qilu Pharmaceutical	NCT06446388
Ubamatamab	CD3 × MUC16	Fallopian tube ovarian peritoneal	Phase II	Regeneron	NCT06444880
AZD-5863	CD3 × CLDN18.2	Solid tumor	Phase I/II	AstraZeneca	NCT06005493
REGN-4336	CD3 × PSMA	Prostate	Phase I/II	Regeneron	NCT05125016
GEN-1047	CD3 × B7H4	Solid tumor	Phase I/II	Genmab	NCT05180474
CM-350	CD3 × GPC3	Solid tumor	Phase I/II	Chengdu KeyMed Biosciences	NCT05263960
SMET-12	CD3 × EGFR	Solid tumor	Phase I/II	Zhejiang Shimai Pharmaceutical	CTR20231358
MVC-280	CD3 × B7H3	Solid tumor	Phase I/II	Takeda	NCT05220098
MVC-101	CD3 × EGFR	Colorectal head and neck lung (non-small cell)	Phase I/II	Takeda	NCT04844073
Cabotamig	CD3 × CDH17	Gastrointestinal, unspecified	Phase I/II	Arbele	NCT05411133
BNT-142	CD3 × CLDN6	Solid tumor	Phase I/ II	BioNTech	NCT05262530
HPN-328	CD3 × DLL3	Lung (small cell)	Phase I/ II	Merck & Co.	NCT04471727
CMD-011	CD3 × GPC3	Liver	Phase I/II	Zhejiang Shimai Pharmaceutical	CTR20233880
Xaluritamig	CD3 × STEAP1	Prostate	Phase I	Amgen	NCT04221542
ERY974	CD3 × GPC3	Liver	Phase I	Chugai	NCT05022927
Runimotamab	CD3 × HER2	Breast gastrointestinal gastro-oesophageal junction	Phase I	Roche	NCT03448042
EMB-07	CD3 × ROR1	Solid tumor	Phase I	EpimAb Biotherapeutics	NCT05607498
ALPS12	CD3/CD137 × DLL3	Solid tumor	Phase I	Chugai	NCT07107940

a 40% objective response rate and a median duration of response of 9.7 months in extensive-stage small cell lung cancer (ES-SCLC) patients with two or more prior lines of treatment, using a 10 mg every 2 weeks dosing regimen. The median overall survival was 14.3 months. Based on these results, the FDA granted accelerated approval for tarlatamab in adult patients with ES-SCLC who progressed on or after platinum-based chemotherapy. The most common adverse events were CRS (55%), fatigue (51%), and pyrexia (36%). As the first DLL3-targeting BiTE therapy for ES-SCLC, tarlatamab offers a new treatment option for this aggressive disease.

8.3 Xaluritamig [84]

Xaluritamig (AMG 509) is a STEAP1-targeted T-cell engager designed for the treatment of metastatic castration-resistant prostate cancer (mCRPC). In a first-in-human study, it demonstrated encouraging results with a 49% PSA50 response rate and a 24% objective response rate (ORR) across all cohorts, with higher rates (59% PSA50 and 41% ORR) observed at target doses ≥0.75 mg. The maximum tolerated dose was identified as 1.5 mg intravenously weekly via a 3-step dose. The most common treatment-related adverse events were cytokine release syndrome (72%), fatigue (45%), and myalgia (34%). CRS occurred primarily during cycle 1 and improved with premedication and step dosing. As a novel immunotherapy for prostate cancer, xaluritamig shows promising efficacy in early clinical development, supporting its further investigation for this challenging disease.

Table 1 summarizes the clinical advances in T-BsAbs for the treatment of solid tumors. As mentioned above, while the number of T-BsAbs approved for solid tumors is still limited, multiple drug candidates are progressing through clinical development, with several in late-stage development. In the near future, it is anticipated that additional new drugs targeting solid tumors will emerge.

References

1. Baeuerle PA, Reinhardt C (2009) Bispecific T-cell engaging antibodies for cancer therapy. Cancer Res 69(12):4941–4944. https://doi.org/10.1158/0008-5472.CAN-09-0547
2. Kontermann RE, Brinkmann U (2015) Bispecific antibodies. Drug Discov Today 20(7):838–847. https://doi.org/10.1016/j.drudis.2015.02.008
3. Kantarjian H, Stein A, Gokbuget N, Fielding AK, Schuh AC, Ribera JM, Wei A, Dombret H, Foa R, Bassan R, Arslan O, Sanz MA, Bergeron J, Demirkan F, Lech-Maranda E, Rambaldi A, Thomas X, Horst HA, Bruggemann M, Klapper W, Wood BL, Fleishman A, Nagorsen D, Holland C, Zimmerman Z, Topp MS (2017) Blinatumomab versus chemotherapy for advanced acute lymphoblastic leukemia. N Engl J Med 376(9):836–847. https://doi.org/10.1056/NEJMoa1609783
4. Maude SL, Laetsch TW, Buechner J, Rives S, Boyer M, Bittencourt H, Bader P, Verneris MR, Stefanski HE, Myers GD, Qayed M, De Moerloose B, Hiramatsu H, Schlis K, Davis KL, Martin PL, Nemecek ER, Yanik GA, Peters C, Baruchel A, Boissel N, Mechinaud F, Balduzzi A, Krueger J, June CH, Levine BL, Wood P, Taran T, Leung M, Mueller KT, Zhang Y, Sen K, Lebwohl D, Pulsipher MA, Grupp SA (2018) Tisagenlecleucel in children and young adults with B-cell lymphoblastic leukemia. N Engl J Med 378(5):439–448. https://doi.org/10.1056/NEJMoa1709866
5. Dalle S, Reslan L, Besseyre de Horts T, Herveau S, Herting F, Plesa A, Friess T, Umana

P, Klein C, Dumontet C (2011) Preclinical studies on the mechanism of action and the anti-lymphoma activity of the novel anti-CD20 antibody GA101. Mol Cancer Ther 10(1):178–185. https://doi.org/10.1158/1535-7163.MCT-10-0385

6. Cohen AD, Garfall AL, Stadtmauer EA, Melenhorst JJ, Lacey SF, Lancaster E, Vogl DT, Weiss BM, Dengel K, Nelson A, Plesa G, Chen F, Davis MM, Hwang WT, Young RM, Brogdon JL, Isaacs R, Pruteanu-Malinici I, Siegel DL, Levine BL, June CH, Milone MC (2019) B cell maturation antigen-specific CAR T cells are clinically active in multiple myeloma. J Clin Invest 129(6):2210–2221. https://doi.org/10.1172/JCI126397
7. Sadelain M, Riviere I, Riddell S (2017) Therapeutic T cell engineering. Nature 545(7655):423–431. https://doi.org/10.1038/nature22395
8. Ribas A, Wolchok JD (2018) Cancer immunotherapy using checkpoint blockade. Science 359(6382):1350–1355. https://doi.org/10.1126/science.aar4060
9. June CH, O'Connor RS, Kawalekar OU, Ghassemi S, Milone MC (2018) CAR T cell immunotherapy for human cancer. Science 359(6382):1361–1365. https://doi.org/10.1126/science.aar6711
10. Zhang W, Liu L, Su H, Liu Q, Shen J, Dai H, Zheng W, Lu Y, Zhang W, Bei Y, Shen P (2019) Chimeric antigen receptor macrophage therapy for breast tumours mediated by targeting the tumour extracellular matrix. Br J Cancer 121(10):837–845. https://doi.org/10.1038/s41416-019-0578-3
11. Teachey DT, Lacey SF, Shaw PA, Melenhorst JJ, Maude SL, Frey N, Pequignot E, Gonzalez VE, Chen F, Finklestein J, Barrett DM, Weiss SL, Fitzgerald JC, Berg RA, Aplenc R, Callahan C, Rheingold SR, Zheng Z, Rose-John S, White JC, Nazimuddin F, Wertheim G, Levine BL, June CH, Porter DL, Grupp SA (2016) Identification of predictive biomarkers for cytokine release syndrome after chimeric antigen receptor T-cell therapy for acute lymphoblastic leukemia. Cancer Discov 6(6):664–679. https://doi.org/10.1158/2159-8290.CD-16-0040
12. Maus MV, Grupp SA, Porter DL, June CH (2014) Antibody-modified T cells: CARs take the front seat for hematologic malignancies. Blood 123(17):2625–2635. https://doi.org/10.1182/blood-2013-11-492231
13. Long AH, Haso WM, Shern JF, Wanhainen KM, Murgai M, Ingaramo M, Smith JP, Walker AJ, Kohler ME, Venkateshwara VR, Kaplan RN, Patterson GH, Fry TJ, Orentas RJ, Mackall CL (2015) 4-1BB costimulation ameliorates T cell exhaustion induced by tonic signaling of chimeric antigen receptors. Nat Med 21(6):581–590. https://doi.org/10.1038/nm.3838
14. Chmielewski M, Abken H (2015) TRUCKs: the fourth generation of CARs. Expert Opin Biol Ther 15(8):1145–1154. https://doi.org/10.1517/14712598.2015.1046430
15. Amann M, D'Argouges S, Lorenczewski G, Brischwein K, Kischel R, Lutterbuese R, Mangold S, Rau D, Volkland J, Pflanz S, Raum T, Munz M, Kufer P, Schlereth B, Baeuerle PA, Friedrich M (2009) Antitumor activity of an EpCAM/CD3-bispecific BiTE antibody during long-term treatment of mice in the absence of T-cell anergy and sustained cytokine release. J Immunother 32(5):452–464. https://doi.org/10.1097/CJI.0b013e3181a1c097
16. Baeuerle PA, Wesche H (2022) T-cell-engaging antibodies for the treatment of solid tumors: challenges and opportunities. Curr Opin Oncol 34(5):552–558. https://doi.org/10.1097/CCO.0000000000000869
17. Hummel HD, Kufer P, Grullich C, Seggewiss-Bernhardt R, Deschler-Baier B, Chatterjee M, Goebeler ME, Miller K, de Santis M, Loidl W, Dittrich C, Buck A, Lapa C, Thurner A, Wittemer-Rump S, Koca G, Boix O, Docke WD, Finnern R, Kusi H, Ajavon-Hartmann A, Stienen S, Sayehli CM, Polat B, Bargou RC (2021) Pasotuxizumab, a BiTE((R)) immune therapy for castration-resistant prostate cancer: phase I, dose-escalation study findings. Immunotherapy 13(2):125–141. https://doi.org/10.2217/imt-2020-0256
18. Lim EA, Schweizer MT, Chi KN, Aggarwal R, Agarwal N, Gulley J, Attiyeh E, Greger J, Wu S, Jaiprasart P, Loffredo J, Bandyopadhyay N, Xie H, Hansen AR (2023) Phase 1 study of safety and preliminary clinical activity of JNJ-63898081, a PSMA and CD3 bispecific antibody, for metastatic castration-resistant prostate cancer. Clin Genitourin Cancer 21(3):366–375. https://doi.org/10.1016/j.clgc.2023.02.010
19. Mau-Sorensen M, Dittrich C, Dienstmann R, Lassen U, Buchler W, Martinius H, Tabernero J (2015) A phase I trial of intravenous catumaxomab: a bispecific monoclonal antibody targeting EpCAM and the T cell coreceptor CD3. Cancer Chemother Pharmacol 75(5):1065–1073. https://doi.org/10.1007/s00280-015-2728-5
20. Borlak J, Langer F, Spanel R, Schondorfer G, Dittrich C (2016) Immune-mediated liver injury of the cancer therapeutic antibody catumaxomab targeting EpCAM, CD3 and Fcgamma receptors. Oncotarget 7(19):28059–28074. https://doi.org/10.18632/oncotarget.8574

21. Kamakura D, Asano R, Yasunaga M (2021) T cell bispecific antibodies: an antibody-based delivery system for inducing antitumor immunity. Pharmaceuticals (Basel) 14(11). https://doi.org/10.3390/ph14111172
22. van de Donk N, Zweegman S (2023) T-cell-engaging bispecific antibodies in cancer. Lancet 402(10396):142–158. https://doi.org/10.1016/S0140-6736(23)00521-4
23. Middelburg J, Kemper K, Engelberts P, Labrijn AF, Schuurman J, van Hall T (2021) Overcoming challenges for CD3-bispecific antibody therapy in solid tumors. Cancers (Basel) 13(2). https://doi.org/10.3390/cancers13020287
24. Ball K, Dovedi SJ, Vajjah P, Phipps A (2023) Strategies for clinical dose optimization of T cell-engaging therapies in oncology. MAbs 15(1):2181016. https://doi.org/10.1080/19420862.2023.2181016
25. Shimabukuro-Vornhagen A, Godel P, Subklewe M, Stemmler HJ, Schlosser HA, Schlaak M, Kochanek M, Boll B, von Bergwelt-Baildon MS (2018) Cytokine release syndrome. J Immunother Cancer 6(1):56. https://doi.org/10.1186/s40425-018-0343-9
26. Budde LE, Sehn LH, Matasar M, Schuster SJ, Assouline S, Giri P, Kuruvilla J, Canales M, Dietrich S, Fay K, Ku M, Nastoupil L, Cheah CY, Wei MC, Yin S, Li CC, Huang H, Kwan A, Penuel E, Bartlett NL (2022) Safety and efficacy of mosunetuzumab, a bispecific antibody, in patients with relapsed or refractory follicular lymphoma: a single-arm, multicentre, phase 2 study. Lancet Oncol 23(8):1055–1065. https://doi.org/10.1016/S1470-2045(22)00335-7
27. Goto H, Ogawa C, Iida H, Horibe K, Oh I, Takada S, Maeda Y, Minami H, Nakashima Y, Morris JD, Kormany W, Chen Y, Miyamoto T (2022) Safety and efficacy of blinatumomab in Japanese adult and pediatric patients with relapsed/refractory B-cell precursor acute lymphoblastic leukemia: final results from an expansion cohort. Acta Haematol 145(6):592–602. https://doi.org/10.1159/000525835
28. Hutchings M, Morschhauser F, Iacoboni G, Carlo-Stella C, Offner FC, Sureda A, Salles G, Martinez-Lopez J, Crump M, Thomas DN, Morcos PN, Ferlini C, Broske AE, Belousov A, Bacac M, Dimier N, Carlile DJ, Lundberg L, Perez-Callejo D, Umana P, Moore T, Weisser M, Dickinson MJ (2021) Glofitamab, a novel, bivalent CD20-targeting T-cell-engaging bispecific antibody, induces durable complete remissions in relapsed or refractory B-cell lymphoma: a phase I trial. J Clin Oncol 39(18):1959–1970. https://doi.org/10.1200/JCO.20.03175
29. Dorff T, Horvath LG, Autio K, Bernard-Tessier A, Rettig MB, Machiels JP, Bilen MA, Lolkema MP, Adra N, Rottey S, Greil R, Matsubara N, Tan DSW, Wong A, Uemura H, Lemech C, Meran J, Yu Y, Minocha M, McComb M, Penny HL, Gupta V, Hu X, Jurida G, Kouros-Mehr H, Janat-Amsbury MM, Eggert T, Tran B (2024) A phase I study of acapatamab, a half-life extended, PSMA-targeting bispecific T-cell engager for metastatic castration-resistant prostate cancer. Clin Cancer Res 30(8):1488–1500. https://doi.org/10.1158/1078-0432.CCR-23-2978
30. Seimetz D, Lindhofer H, Bokemeyer C (2010) Development and approval of the trifunctional antibody catumaxomab (anti-EpCAM x anti-CD3) as a targeted cancer immunotherapy. Cancer Treat Rev 36(6):458–467. https://doi.org/10.1016/j.ctrv.2010.03.001
31. Knodler M, Korfer J, Kunzmann V, Trojan J, Daum S, Schenk M, Kullmann F, Schroll S, Behringer D, Stahl M, Al-Batran SE, Hacker U, Ibach S, Lindhofer H, Lordick F (2018) Randomised phase II trial to investigate catumaxomab (anti-EpCAM x anti-CD3) for treatment of peritoneal carcinomatosis in patients with gastric cancer. Br J Cancer 119(3):296–302. https://doi.org/10.1038/s41416-018-0150-6
32. Sehouli J, Reinthaller A, Marth C, Reimer D, Reimer T, Stummvoll W, Angleitner-Boubenizek L, Brandt B, Chekerov R (2014) Intra- and postoperative catumaxomab in patients with epithelial ovarian cancer: safety and two-year efficacy results from a multicentre, single-arm, phase II study. Br J Cancer 111(8):1519–1525. https://doi.org/10.1038/bjc.2014.443
33. Ruf P, Kluge M, Jager M, Burges A, Volovat C, Heiss MM, Hess J, Wimberger P, Brandt B, Lindhofer H (2010) Pharmacokinetics, immunogenicity and bioactivity of the therapeutic antibody catumaxomab intraperitoneally administered to cancer patients. Br J Clin Pharmacol 69(6):617–625. https://doi.org/10.1111/j.1365-2125.2010.03635.x
34. Einsele H, Borghaei H, Orlowski RZ, Subklewe M, Roboz GJ, Zugmaier G, Kufer P, Iskander K, Kantarjian HM (2020) The BiTE (bispecific T-cell engager) platform: development and future potential of a targeted immuno-oncology therapy across tumor types. Cancer 126(14):3192–3201. https://doi.org/10.1002/cncr.32909
35. Ramdani Y, Lamamy J, Watier H, Gouilleux-Gruart V (2022) Monoclonal antibody engineering and design to modulate FcRn activities: a comprehensive review. Int J Mol Sci 23(17). https://doi.org/10.3390/ijms23179604

36. Sanford M (2015) Blinatumomab: first global approval. Drugs 75(3):321–327. https://doi.org/10.1007/s40265-015-0356-3
37. Goebeler ME, Bargou RC (2020) T cell-engaging therapies – BiTEs and beyond. Nat Rev Clin Oncol 17(7):418–434. https://doi.org/10.1038/s41571-020-0347-5
38. Giffin MJ, Cooke K, Lobenhofer EK, Estrada J, Zhan J, Deegen P, Thomas M, Murawsky CM, Werner J, Liu S, Lee F, Homann O, Friedrich M, Pearson JT, Raum T, Yang Y, Caenepeel S, Stevens J, Beltran PJ, Canon J, Coxon A, Bailis JM, Hughes PE (2021) AMG 757, a half-life extended, DLL3-targeted bispecific T-cell engager, shows high potency and sensitivity in preclinical models of small-cell lung cancer. Clin Cancer Res 27(5):1526–1537. https://doi.org/10.1158/1078-0432.CCR-20-2845
39. Paz-Ares L, Champiat S, Lai WV, Izumi H, Govindan R, Boyer M, Hummel HD, Borghaei H, Johnson ML, Steeghs N, Blackhall F, Dowlati A, Reguart N, Yoshida T, He K, Gadgeel SM, Felip E, Zhang Y, Pati A, Minocha M, Mukherjee S, Goldrick A, Nagorsen D, Hashemi Sadraei N, Owonikoko TK (2023) Tarlatamab, a first-in-class DLL3-targeted bispecific T-cell engager, in recurrent small-cell lung cancer: an open-label, phase I study. J Clin Oncol 41(16):2893–2903. https://doi.org/10.1200/JCO.22.02823
40. Ishiguro T, Sano Y, Komatsu SI, Kamata-Sakurai M, Kaneko A, Kinoshita Y, Shiraiwa H, Azuma Y, Tsunenari T, Kayukawa Y, Sonobe Y, Ono N, Sakata K, Fujii T, Miyazaki Y, Noguchi M, Endo M, Harada A, Frings W, Fujii E, Nanba E, Narita A, Sakamoto A, Wakabayashi T, Konishi H, Segawa H, Igawa T, Tsushima T, Mutoh H, Nishito Y, Takahashi M, Stewart L, ElGabry E, Kawabe Y, Ishigai M, Chiba S, Aoki M, Hattori K, Nezu J (2017) An anti-glypican 3/CD3 bispecific T cell-redirecting antibody for treatment of solid tumors. Sci Transl Med 9(410). https://doi.org/10.1126/scitranslmed.aal4291
41. Lehmann S, Perera R, Grimm HP, Sam J, Colombetti S, Fauti T, Fahrni L, Schaller T, Freimoser-Grundschober A, Zielonka J, Stoma S, Rudin M, Klein C, Umana P, Gerdes C, Bacac M (2016) In vivo fluorescence imaging of the activity of CEA TCB, a novel T-cell bispecific antibody, reveals highly specific tumor targeting and fast induction of T-cell-mediated tumor killing. Clin Cancer Res 22(17):4417–4427. https://doi.org/10.1158/1078-0432.CCR-15-2622
42. Root AR, Guntas G, Katragadda M, Apgar JR, Narula J, Chang CS, Hanscom S, McKenna M, Wade J, Meade C, Ma W, Guo Y, Liu Y, Duan W, Hendershot C, King AC, Zhang Y, Sousa E, Tam A, Benard S, Yang H, Kelleher K, Jin F, Piche-Nicholas N, Keating SE, Narciandi F, Lawrence-Henderson R, Arai M, Stochaj WR, Svenson K, Mosyak L, Lam K, Francis C, Marquette K, Wroblewska L, Zhu HL, Sheehan AD, LaVallie ER, D'Antona AM, Betts A, King L, Rosfjord E, Cunningham O, Lin L, Sapra P, Tchistiakova L, Mathur D, Bloom L (2021) Discovery and optimization of a novel anti-GUCY2c x CD3 bispecific antibody for the treatment of solid tumors. MAbs 13(1):1850395. https://doi.org/10.1080/19420862.2020.1850395
43. Yu S, Zhang J, Yan Y, Yao X, Fang L, Xiong H, Liu Y, Chu Q, Zhou P, Wu K (2019) A novel asymmetrical anti-HER2/CD3 bispecific antibody exhibits potent cytotoxicity for HER2-positive tumor cells. J Exp Clin Cancer Res 38(1):355. https://doi.org/10.1186/s13046-019-1354-1
44. First Drug Approved for Rare Eye Cancer (2022) Cancer Discov 12(4):879–880. https://doi.org/10.1158/2159-8290.CD-NB2022-0009
45. Kebenko M, Goebeler ME, Wolf M, Hasenburg A, Seggewiss-Bernhardt R, Ritter B, Rautenberg B, Atanackovic D, Kratzer A, Rottman JB, Friedrich M, Vieser E, Elm S, Patzak I, Wessiepe D, Stienen S, Fiedler W (2018) A multicenter phase 1 study of solitomab (MT110, AMG 110), a bispecific EpCAM/CD3 T-cell engager (BiTE(R)) antibody construct, in patients with refractory solid tumors. Oncoimmunology 7(8):e1450710. https://doi.org/10.1080/2162402X.2018.1450710
46. Segal NH, Melero I, Moreno V, Steeghs N, Marabelle A, Rohrberg K, Rodriguez-Ruiz ME, Eder JP, Eng C, Manji GA, Waterkamp D, Leutgeb B, Bouseida S, Flinn N, Das Thakur M, Elze MC, Koeppen H, Jamois C, Martin-Facklam M, Lieu CH, Calvo E, Paz-Ares L, Tabernero J, Argiles G (2024) CEA-CD3 bispecific antibody cibisatamab with or without atezolizumab in patients with CEA-positive solid tumours: results of two multi-institutional Phase 1 trials. Nat Commun 15(1), 4091. https://doi.org/10.1038/s41467-024-48479-8
47. Lutterbuese R, Raum T, Kischel R, Hoffmann P, Mangold S, Rattel B, Friedrich M, Thomas O, Lorenczewski G, Rau D, Schaller E, Herrmann I, Wolf A, Urbig T, Baeuerle PA, Kufer P (2010) T cell-engaging BiTE antibodies specific for EGFR potently eliminate KRAS- and BRAF-mutated colorectal cancer cells. Proc Natl Acad Sci U S A 107(28):12605–12610. https://doi.org/10.1073/pnas.1000976107

48. Staflin K, Zuch de Zafra CL, Schutt LK, Clark V, Zhong F, Hristopoulos M, Clark R, Li J, Mathieu M, Chen X, Johnston J, Low J, Ybarra R, Slaga D, Yang J, Ovacik M, Dybdal NO, Totpal K, Junttila MR, Ellerman D, Lee G, Dennis MS, Prell R, Junttila TT (2020) Target arm affinities determine preclinical efficacy and safety of anti-HER2/CD3 bispecific antibody. JCI Insight 5(7). https://doi.org/10.1172/jci.insight.133757
49. Haber L, Olson K, Kelly MP, Crawford A, DiLillo DJ, Tavare R, Ullman E, Mao S, Canova L, Sineshchekova O, Finney J, Pawashe A, Patel S, McKay R, Rizvi S, Damko E, Chiu D, Vazzana K, Ram P, Mohrs K, D'Orvilliers A, Xiao J, Makonnen S, Hickey C, Arnold C, Giurleo J, Chen YP, Thwaites C, Dudgeon D, Bray K, Rafique A, Huang T, Delfino F, Hermann A, Kirshner JR, Retter MW, Babb R, MacDonald D, Chen G, Olson WC, Thurston G, Davis S, Lin JC, Smith E (2021) Generation of T-cell-redirecting bispecific antibodies with differentiated profiles of cytokine release and biodistribution by CD3 affinity tuning. Sci Rep 11(1):14397. https://doi.org/10.1038/s41598-021-93842-0
50. Mandikian D, Takahashi N, Lo AA, Li J, Eastham-Anderson J, Slaga D, Ho J, Hristopoulos M, Clark R, Totpal K, Lin K, Joseph SB, Dennis MS, Prabhu S, Junttila TT, Boswell CA (2018) Relative target affinities of T-cell–dependent bispecific antibodies determine biodistribution in a solid tumor mouse model. Mol Cancer Ther 17(4):776–785. https://doi.org/10.1158/1535-7163.Mct-17-0657
51. Bacac M, Fauti T, Sam J, Colombetti S, Weinzierl T, Ouaret D, Bodmer W, Lehmann S, Hofer T, Hosse RJ, Moessner E, Ast O, Bruenker P, Grau-Richards S, Schaller T, Seidl A, Gerdes C, Perro M, Nicolini V, Steinhoff N, Dudal S, Neumann S, von Hirschheydt T, Jaeger C, Saro J, Karanikas V, Klein C, Umana P (2016) A novel carcinoembryonic antigen T-cell bispecific antibody (CEA TCB) for the treatment of solid tumors. Clin Cancer Res 22(13):3286–3297. https://doi.org/10.1158/1078-0432.CCR-15-1696
52. Dang K, Castello G, Clarke SC, Li Y, Balasubramani A, Boudreau A, Davison L, Harris KE, Pham D, Sankaran P, Ugamraj HS, Deng R, Kwek S, Starzinski A, Iyer S, van Schooten W, Schellenberger U, Sun W, Trinklein ND, Buelow R, Buelow B, Fong L, Dalvi P (2021) Attenuating CD3 affinity in a PSMAxCD3 bispecific antibody enables killing of prostate tumor cells with reduced cytokine release. J Immunother Cancer 9(6). https://doi.org/10.1136/jitc-2021-002488
53. Wei J, Yang Y, Wang G, Liu M (2022) Current landscape and future directions of bispecific antibodies in cancer immunotherapy. Front Immunol 13:1035276. https://doi.org/10.3389/fimmu.2022.1035276
54. Carvajal RD, Nathan P, Sacco JJ, Orloff M, Hernandez-Aya LF, Yang J, Luke JJ, Butler MO, Stanhope S, Collins L, McAlpine C, Holland C, Abdullah SE, Sato T (2022) Phase I study of safety, tolerability, and efficacy of tebentafusp using a step-up dosing regimen and expansion in patients with metastatic uveal melanoma. J Clin Oncol 40(17):1939–1948. https://doi.org/10.1200/JCO.21.01805
55. Ma H, Wang H, Sove RJ, Jafarnejad M, Tsai CH, Wang J, Giragossian C, Popel AS (2020) A quantitative systems pharmacology model of T cell engager applied to solid tumor. AAPS J 22(4):85. https://doi.org/10.1208/s12248-020-00450-3
56. Weddell J (2023) Mechanistically modeling peripheral cytokine dynamics following bispecific dosing in solid tumors. CPT Pharmacometrics Syst Pharmacol 12(11):1726–1737. https://doi.org/10.1002/psp4.12928
57. Betts A, van der Graaf PH (2020) Mechanistic quantitative pharmacology strategies for the early clinical development of bispecific antibodies in oncology. Clin Pharmacol Ther 108(3):528–541. https://doi.org/10.1002/cpt.1961
58. Ludwig H, Terpos E, van de Donk N, Mateos MV, Moreau P, Dimopoulos MA, Delforge M, Rodriguez-Otero P, San-Miguel J, Yong K, Gay F, Einsele H, Mina R, Caers J, Driessen C, Musto P, Zweegman S, Engelhardt M, Cook G, Weisel K, Broijl A, Beksac M, Bila J, Schjesvold F, Cavo M, Hajek R, Touzeau C, Boccadoro M, Sonneveld P (2023) Prevention and management of adverse events during treatment with bispecific antibodies and CAR T cells in multiple myeloma: a consensus report of the European Myeloma Network. Lancet Oncol 24(6):e255–e269. https://doi.org/10.1016/S1470-2045(23)00159-6
59. Balendran S, Tam C, Ku M (2023) T-cell engaging antibodies in diffuse large B cell lymphoma-an update. J Clin Med 12(21). https://doi.org/10.3390/jcm12216737
60. Karschnia P, Jordan JT, Forst DA, Arrillaga-Romany IC, Batchelor TT, Baehring JM, Clement NF, Gonzalez Castro LN, Herlopian A, Maus MV, Schwaiblmair MH, Soumerai JD, Takvorian RW, Hochberg EP, Barnes JA, Abramson JS, Frigault MJ, Dietrich J (2019) Clinical presentation, management, and biomarkers of neurotoxicity after adoptive immunotherapy with CAR T cells. Blood 133(20):2212–2221. https://doi.org/10.1182/blood-2018-12-893396

61. Rees JH (2022) Management of immune effector cell-associated neurotoxicity syndrome (ICANS). In: Kroger N, Gribben J, Chabannon C, Yakoub-Agha I, Einsele H (eds) The EBMT/EHA CAR-T cell handbook. Cham (CH), pp 141–145. https://doi.org/10.1007/978-3-030-94353-0_27
62. Wei Z, Xu J, Zhao C, Zhang M, Xu N, Kang L, Lou X, Yu L, Feng W (2023) Prediction of severe CRS and determination of biomarkers in B cell-acute lymphoblastic leukemia treated with CAR-T cells. Front Immunol 14:1273507. https://doi.org/10.3389/fimmu.2023.1273507
63. Tanaka T, Narazaki M, Kishimoto T (2014) IL-6 in inflammation, immunity, and disease. Cold Spring Harb Perspect Biol 6(10):a016295. https://doi.org/10.1101/cshperspect.a016295
64. Kauer J, Horner S, Osburg L, Muller S, Marklin M, Heitmann JS, Zekri L, Rammensee HG, Salih HR, Jung G (2020) Tocilizumab, but not dexamethasone, prevents CRS without affecting antitumor activity of bispecific antibodies. J Immunother Cancer 8(1). https://doi.org/10.1136/jitc-2020-000621
65. Geraud A, Hueso T, Laparra A, Bige N, Ouali K, Cauquil C, Stoclin A, Danlos FX, Hollebecque A, Ribrag V, Gazzah A, Goldschmidt V, Baldini C, Suzzoni S, Bahleda R, Besse B, Barlesi F, Olivier L, Massard C, Marabelle A, Castilla-Llorente C, Champiat S, Michot JM (2024) Reactions and adverse events induced by T-cell engagers as anti-cancer immunotherapies, a comprehensive review. Eur J Cancer 205:114075. https://doi.org/10.1016/j.ejca.2024.114075
66. Martin TG, Mateos MV, Nooka A, Banerjee A, Kobos R, Pei L, Qi M, Verona R, Doyle M, Smit J, Sun W, Trancucci D, Uhlar C, van de Donk N, Rodriguez C (2023) Detailed overview of incidence and management of cytokine release syndrome observed with teclistamab in the MajesTEC-1 study of patients with relapsed/refractory multiple myeloma. Cancer 129(13):2035–2046. https://doi.org/10.1002/cncr.34756
67. Maria-Victoria M, Bahlis NJB, Andrew S, Kaedbey RK, Paula R, Harrison SH, Wong CW, Goodman GG, Rin N, Voleak C, Cooper JC, Suzanne T (2023) P946: Tocilizumab pre-treatment significantly reduces the incidence of cytokine release syndrome in patients with relapsed/refractory multiple myeloma (RRMM) who receive cevostamab. HemaSphere 7:e75458a75458. https://doi.org/10.1097/01.HS9.0000970688.75458.a8
68. Nellan A, McCully CML, Cruz Garcia R, Jayaprakash N, Widemann BC, Lee DW, Warren KE (2018) Improved CNS exposure to tocilizumab after cerebrospinal fluid compared to intravenous administration in Rhesus macaques. Blood 132(6):662–666. https://doi.org/10.1182/blood-2018-05-846428
69. Sheth VS, Gauthier J (2021) Taming the beast: CRS and ICANS after CAR T-cell therapy for ALL. Bone Marrow Transplant 56(3):552–566. https://doi.org/10.1038/s41409-020-01134-4
70. Lee DW, Santomasso BD, Locke FL, Ghobadi A, Turtle CJ, Brudno JN, Maus MV, Park JH, Mead E, Pavletic S, Go WY, Eldjerou L, Gardner RA, Frey N, Curran KJ, Peggs K, Pasquini M, DiPersio JF, van den Brink MRM, Komanduri KV, Grupp SA, Neelapu SS (2019) ASTCT consensus grading for cytokine release syndrome and neurologic toxicity associated with immune effector cells. Biol Blood Marrow Transplant 25(4):625–638. https://doi.org/10.1016/j.bbmt.2018.12.758
71. Kamperschroer C, Guffroy M, Shen A, Dokmanovich M, Stubbs M, O'Donnell LM (2024) Nonclinical investigation of cytokine mitigation strategies for T-cell-engaging bispecifics in the cynomolgus macaque. J Immunother 47(5):160–171. https://doi.org/10.1097/CJI.0000000000000512
72. Dickinson MJ, Carlo-Stella C, Morschhauser F, Bachy E, Corradini P, Iacoboni G, Khan C, Wrobel T, Offner F, Trneny M, Wu SJ, Cartron G, Hertzberg M, Sureda A, Perez-Callejo D, Lundberg L, Relf J, Dixon M, Clark E, Humphrey K, Hutchings M (2022) Glofitamab for relapsed or refractory diffuse large B-cell lymphoma. N Engl J Med 387(24):2220–2231. https://doi.org/10.1056/NEJMoa2206913
73. Maus MV, Alexander S, Bishop MR, Brudno JN, Callahan C, Davila ML, Diamonte C, Dietrich J, Fitzgerald JC, Frigault MJ, Fry TJ, Holter-Chakrabarty JL, Komanduri KV, Lee DW, Locke FL, Maude SL, McCarthy PL, Mead E, Neelapu SS, Neilan TG, Santomasso BD, Shpall EJ, Teachey DT, Turtle CJ, Whitehead T, Grupp SA (2020) Society for Immunotherapy of Cancer (SITC) clinical practice guideline on immune effector cell-related adverse events. J Immunother Cancer 8(2). https://doi.org/10.1136/jitc-2020-001511
74. Yakoub-Agha I, Chabannon C, Bader P, Basak GW, Bonig H, Ciceri F, Corbacioglu S, Duarte RF, Einsele H, Hudecek M, Kersten MJ, Kohl U, Kuball J, Mielke S, Mohty M, Murray J, Nagler A, Robinson S, Saccardi R, Sanchez-Guijo F, Snowden JA, Srour M, Styczynski J, Urbano-Ispizua A, Hayden PJ, Kroger N (2020) Management of adults and children undergoing chimeric antigen receptor T-cell therapy: best

practice recommendations of the European Society for Blood and Marrow Transplantation (EBMT) and the Joint Accreditation Committee of ISCT and EBMT (JACIE. Haematologica 105(2):297–316. https://doi.org/10.3324/haematol.2019.229781

75. Lipe BC, Renaud T (2023) Siltuximab as a primary treatment for cytokine release syndrome in a patient receiving a bispecific antibody in a clinical trial setting. J Oncol Pharm Pract 29(4):1006–1010. https://doi.org/10.1177/10781552221140320
76. Norelli M, Camisa B, Barbiera G, Falcone L, Purevdorj A, Genua M, Sanvito F, Ponzoni M, Doglioni C, Cristofori P, Traversari C, Bordignon C, Ciceri F, Ostuni R, Bonini C, Casucci M, Bondanza A (2018) Monocyte-derived IL-1 and IL-6 are differentially required for cytokine-release syndrome and neurotoxicity due to CAR T cells. Nat Med 24(6):739–748. https://doi.org/10.1038/s41591-018-0036-4
77. Wehrli M, Gallagher K, Chen YB, Leick MB, McAfee SL, El-Jawahri AR, DeFilipp Z, Horick N, O'Donnell P, Spitzer T, Dey B, Cook D, Trailor M, Lindell K, Maus MV, Frigault MJ (2022) Single-center experience using anakinra for steroid-refractory immune effector cell-associated neurotoxicity syndrome (ICANS). J Immunother Cancer 10(1). https://doi.org/10.1136/jitc-2021-003847
78. Gazeau N, Liang EC, Wu QV, Voutsinas JM, Barba P, Iacoboni G, Kwon M, Ortega JLR, Lopez-Corral L, Hernani R, Ortiz-Maldonado V, Martinez-Cibrian N, Martinez AP, Maziarz RT, Williamson S, Nemecek ER, Shadman M, Cowan AJ, Green DJ, Kimble E, Hirayama AV, Maloney DG, Turtle CJ, Gauthier J (2023) Anakinra for refractory cytokine release syndrome or immune effector cell-associated neurotoxicity syndrome after chimeric antigen receptor T cell therapy. Transplant Cell Ther 29(7):430–437. https://doi.org/10.1016/j.jtct.2023.04.001
79. Salvaris R, Ong J, Gregory GP (2021) Bispecific antibodies: a review of development, clinical efficacy and toxicity in B-cell lymphomas. J Pers Med 11(5). https://doi.org/10.3390/jpm11050355
80. Iwata Y, Narushima Y, Harada A, Mishima M (2022) Priming treatment with T-cell redirecting bispecific antibody ERY974 reduced cytokine induction without losing cytotoxic activity in vitro by changing the chromatin state in T cells. Toxicol Appl Pharmacol 441:115986. https://doi.org/10.1016/j.taap.2022.115986
81. Hosseini I, Gadkar K, Stefanich E, Li CC, Sun LL, Chu YW, Ramanujan S (2020) Mitigating the risk of cytokine release syndrome in a Phase I trial of CD20/CD3 bispecific antibody mosunetuzumab in NHL: impact of translational system modeling. NPJ Syst Biol Appl 6(1):28. https://doi.org/10.1038/s41540-020-00145-7
82. Dhillon S (2022) Tebentafusp: first approval. Drugs 82(6):703–710. https://doi.org/10.1007/s40265-022-01704-4
83. Dhillon S (2024) Tarlatamab: first approval. Drugs 84(8):995–1003. https://doi.org/10.1007/s40265-024-02070-z
84. Kelly WK, Danila DC, Lin CC, Lee JL, Matsubara N, Ward PJ, Armstrong AJ, Pook D, Kim M, Dorff TB, Fischer S, Lin YC, Horvath LG, Sumey C, Yang Z, Jurida G, Smith KM, Connarn JN, Penny HL, Stieglmaier J, Appleman LJ (2024) Xaluritamig, a STEAP1 x CD3 XmAb 2+1 immune therapy for metastatic castration-resistant prostate. Cancer: results from dose exploration in a first-in-human study. Cancer Discov 14(1):76–89. https://doi.org/10.1158/2159-8290.CD-23-0964

Check for updates

Chapter 17

Clinical Development of Bispecific Antibodies: Review, Advances, and Challenges in Hematological Malignancies

Anri Saitoh

Abstract

Precision therapies such as monoclonal antibodies (mAbs), immune-checkpoint inhibitors (ICIs), and chimeric antigen receptor T (CAR T)-cell therapies have revolutionized the treatment of cancer. Of these, bispecific antibodies (bsAbs) are gaining attention for their ability to simultaneously target two distinct molecular antigens, aimed at engaging T cells to circumvent cancer immune evasion mechanisms, resulting in tumor cytolysis. Several bsAbs have now been approved in the United States, Europe, and Japan for the treatment of hematological malignancies, most of which are CD3-redirecting bsAbs, although other immune-activating mechanisms are also being explored. This chapter reviews bsAbs in clinical development for hematological malignancies, their basic structure and mechanism of action, and efficacy and safety results of the most advanced bsAbs under clinical investigation.

Keywords Antibodies, B-cell malignancies, Bispecific antibodies, Bispecific T-cell engagers, Hematological malignancies, Lymphoma, Leukemia, Multiple myeloma

1 Introduction

Hematological malignancies cause disruption of normal hematopoietic function and have the potential to affect any system or organ in the body [1]. Pharmacological treatment has historically consisted of chemotherapy and radiotherapy, and combination chemotherapy in particular has been a fundamental approach for both pediatric and adult patients with hematologic malignancies. However, these approaches lack specificity and therefore have broad effects, some of which may be deleterious. Chemotherapy agents target not only neoplastic cells but also rapidly dividing healthy cells, resulting in significant toxicity and adverse effects, including hair loss, nausea, mucositis, and thrombocytopenia [2]. Radiotherapy can damage surrounding tissue and lead to permanent sequelae (e.g., scarring and tissue fibrosis) [3]. Additionally, many cancers develop resistance to chemotherapy and radiotherapy over time [4, 5].

Masahiro Yasunaga (ed.), *Bispecific Antibodies*, Methods in Molecular Biology, vol. 2997,
https://doi.org/10.1007/978-1-0716-5037-0_17,

© The Author(s), under exclusive license to Springer Science+Business Media, LLC, part of Springer Nature 2026

The development of precision therapies that selectively target molecules or pathways involved in the proliferation, survival, and spread of malignant cells over the past two decades has marked a significant shift in the treatment of cancer, resulting in improved outcomes and reduced off-target effects compared with traditional modalities [6, 7]. In addition to monoclonal antibodies (mAbs)—engineered proteins that target a single molecule on cancer cells—immune-checkpoint inhibitors (ICIs), chimeric antigen receptor T (CAR T)-cell therapies, and other agents that utilize strategies to circumvent cancer immune evasion mechanisms have been developed and approved in recent years for hematological malignancies [8].

For hematological malignancies, immunotherapeutic approaches, such as redirecting T cells to target cancer cells, have shown particular promise, and several bsAbs are now approved for this indication. T-cell-redirecting bsAbs are designed to engage immune cells and surface antigens or molecules that are overexpressed on cancer cells simultaneously, resulting in T cell lysis of the tumor cell. In this way, T cell-redirecting bsAbs provide a way to overcome major histocompatibility complex class I (MHC-I) downregulation by cancer cells. Overcoming tumor immune evasion and reactivating the immune system to kill cancer cells is a key mechanism of action of bsAbs.

2 Development of Bispecific Antibodies in Hematological Malignancies

In 2014, blinatumomab (Blincyto®), an anti-CD19 × anti-CD3 bispecific T-cell engager (BiTE®) antibody, was the first bsAb approved for hematological malignancies by the Food and Drug Administration (FDA) in the setting of relapsed/refractory B-cell acute lymphoblastic leukemia (ALL). BiTEs are a type of bsAb that simultaneously bind to an antigen on tumor cells and a CD3 on T cells to induce tumor lysis. Since the approval of blinatumomab, there has been a dramatic increase in the number of new clinical trials registered evaluating novel bsAb drugs in various indications, representing an annual increase of 20.4% [9]. Today, almost 300 bsAbs are in early stage clinical development, all of which are T-cell-redirecting bsAbs, with approximately 75% of these trials conducted in solid tumors, and 25% in hematological malignancies (Fig. 1). Of the early stage clinical trials for hematological malignancies, 28% are for leukemia, 40% are for lymphoma, 25% are for multiple myeloma (MM), and 7% are for nonspecified hematological malignancies at the time of writing (Search made March 2024).

To date, eight T-cell-redirecting bsAbs have been approved by the FDA, seven of which have been approved for the treatment of hematological malignancies. In 2022, mosunetuzumab (Lunsumio™), which targets CD20 × CD3, and teclistamab (Tecvayli™), which targets B-cell maturation antigen (BCMA) × CD3, were approved for follicular lymphoma (FL) and MM, respectively. In 2023, elranatamab

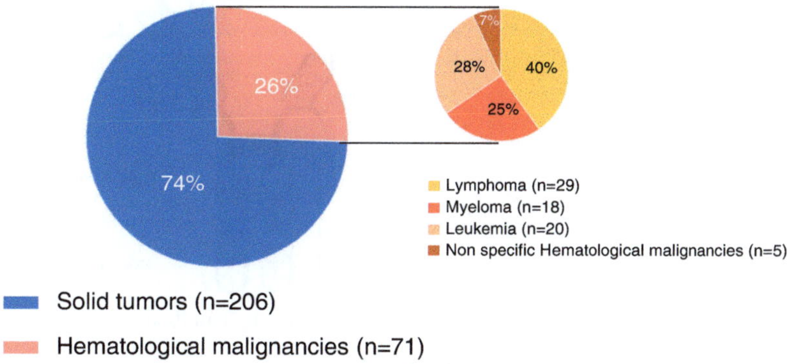

Fig. 1 Clinical trials of bsAbs in early stage development for solid tumors and hematological malignancies

(Elrexfio™), which targets BCMA × CD3, and talquetamab (Talvey™), which targets G protein-coupled receptor class C group 5 member D (GPRC5D) and CD3, were approved for the treatment of MM. In the same year, epcoritamab (Epkinly™) and glofitamab (Columvi®), which target CD20 × CD3, were also approved for the treatment of diffuse large B-cell lymphoma (DLBCL).

The development of bsAbs in hematological malignancies has been stimulated by the need for innovative and effective therapeutic strategies in the context of these complex and heterogeneous diseases.

3 Basic Structure of Antibodies and Development of Bispecific Antibodies

Natural bivalent IgG antibodies are Y-shaped proteins with four polypeptide chains comprising two identical light chains and heavy chains (Fig. 2a). They are produced by B cells and most innate immune cells to identify and neutralize antigens in the body, such as pathogens [10]. The variable region is responsible for antigen binding (Fab) and the constant region (Fc) interacts with other elements of the immune system to promote removal of the antigen and determines the functional properties of the antibody. The most well-known Fc-mediated antibody effector functions are antibody—dependent cell-mediated cytotoxicity (ADCC), antibody-dependent cellular phagocytosis (ADCP), and complement-dependent cytotoxicity (CDC) [10]. The variable region of the heavy and light chains pair in each arm to generate two identical binding sites. The flexible hinge region connects the Fab and Fc regions, allowing for movement and effective antigen binding (Fig. 2) [11, 12].

In recent decades, therapeutic mAbs have become the most widely used and approved immunotherapy in clinical practice for various malignancies. Monoclonal antibodies have been successful

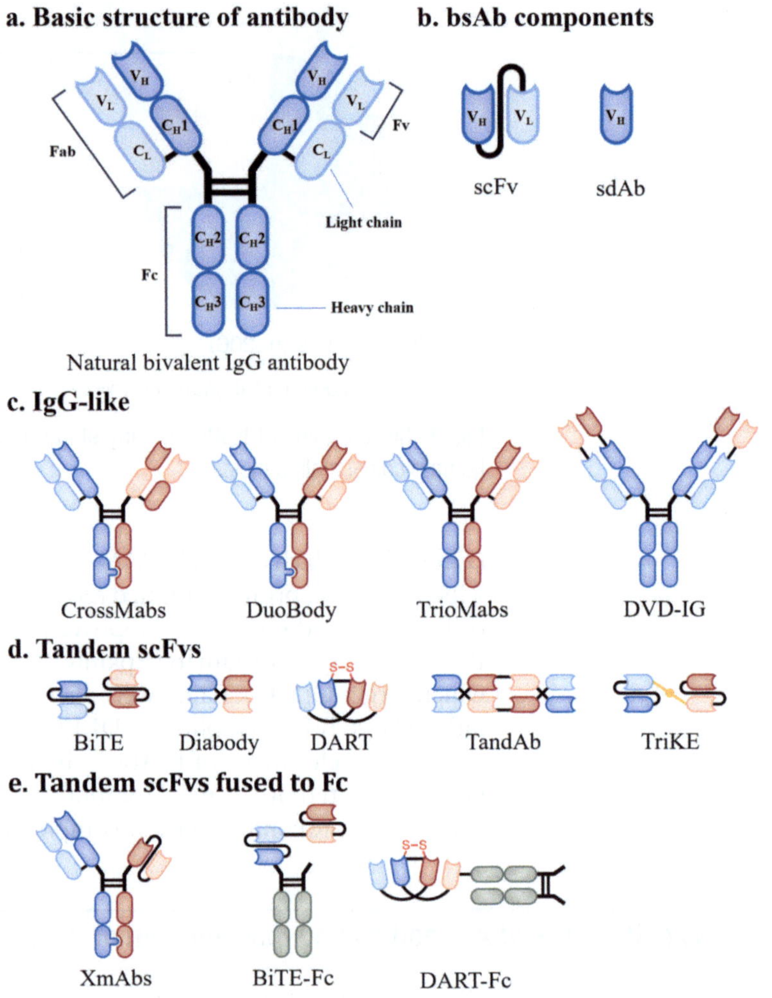

Fig. 2 Bispecific antibodies: structure and formats. Basic structure of immunoglobulin G (IgG) and a schematic representation of several representative bispecific antibodies (bsAbs) developing especially in the field of hematological malignancies: (a) Basic structure of antibody, (b) Components of bsAb, (c) IgG-like bsAbs, (d) Non-IgG-like bsAbs - Tandem scFvs, (e) Non-IgG-like bsAbs - Tandem scFvs fused to Fc

as cancer therapeutics by targeting surface antigens overexpressed or uniquely expressed on tumor cells. However, significant limitations also exist with mAb therapy, including low tumor penetration and retention rate, with studies showing that ~20% of the administered dose interacts with the surface proteins of tumors [13]. For inducing a robust response, an intact immune system is needed. Furthermore, the suboptimal response to treatment also comes from extensive cross-talk among some signaling pathways in cancer cells and nearby cells, which contribute to disease relapse [13]. As such, bispecific antibodies have been developed that bind to two distinct antigens or 2 different antigens with a 2 + 1 approach (i.e., two epitopes for a tumor antigen), aimed at overcoming some of the limitations associated with natural mAbs

3.1 Role of the Fc Domain in Bispecific Antibodies

BsAbs may be broadly divided into two categories based on the presence or absence of an Fc domain: IgG-like subtypes (containing an Fc region) and non-IgG-like subtypes (missing an Fc region) [14]. **IgG-like bsAbs** are constructed to resemble traditional IgG antibodies and typically consist of two antigen-binding Fab regions derived from two different monoclonal antibodies, each with specificity toward a particular antigen (Fig. 2c). Fab and Fc portions are joined by a flexible linker, allowing independent binding to their respective targets.

Bispecific antibodies that include an Fc region may be further subdivided into those with a structure resembling an IgG molecule and those that contain an appended or modified Ig-like structure. In standard bispecific antibodies that have the IgG structure, significant efforts have been made to engineer Fc regions of conventional monospecific antibodies to reduce Fc effector functions. If Fc-mediated function was intact for T-cell-redirecting bsAbs, cross-linking with CD3 T cells and NK cells would potentially occur without binding to the tumor antigen, which could lead to systemic T-cell activation, NK cell depletion, and severe cytokine release syndrome (CRS). Catumaxomab, the first approved bsAb, possesses normal Fc-mediated functions and is associated with severe cytokine-related AEs. Hence, recently approved T-cell-redirecting bsAbs, such as the BiTE blinatumomab, lack an Fc region and therefore do not mediate Fc-dependent effector functions. Later and more recent drugs using BiTE technology have an Fc-like region but also have a mutation for silencing effector function to avoid toxicity while retaining a longer half-life [15].

CrossMabs retain the general structure and size of a conventional IgG but employ knobs-in-holes technology to create a "knob" or "hole" on each heavy chain to promote Fc heterodimerization. Correct association of the light chains and their cognate heavy chains is achieved by exchange of heavy- and light-chain domains within the Fab of one half of the bsAb. This "crossover" retains the antigen-binding affinity but makes the two arms sufficiently different that light-chain mispairing is not possible. This approach allows generation of various bsAb formats, including bi (1+1), tri (2+1), and tetra (2+2) antibodies as well as non-Fc tandem Fab-based antibodies [16].

Another way to generate bsAbs is by Fab-arm exchange. Human IgG4 molecules naturally form bsAbs in vivo through a process called Fab-arm exchange, in which half molecules recombine with half molecules from other IgG4, which results in randomly paired bsAbs that are monovalent for a respective antigen [17]. Controlled Fab-arm exchange combines two parental mAbs with a matched point mutation (F405L and K409R) in the CH3 domains, which are then purified and recombined following a reduction/oxidation reaction to form a bsAb with a very high yield (typically >95%) [18]. Developed by Genmab, controlled Fab-arm exchange is used to generate bsAbs belonging to the **DuoBody®** platform.

TrioMabs are trifunctional IgG-like bsAbs produced using quadroma technology. They are constructed from two half antibodies, each with one light and one heavy chain that originate from parental rat IgG2b and mouse IgG2a isotypes, as well as an additional binding site in the Fc region, resulting in a trivalent structure. Heavy and light chain mispairings can occur with IgG-like bsAbs made from two distinct polypeptides, leading to ineffective antibodies or unwanted homodimers.

Dual Variable Domain (DVD)-Ig CrossMab formats are constructed when four variable domains from one antibody are connected to the N terminus of the heavy and light chains of a second antibody via flexible naturally occurring linkers to create a tetravalent IgG-like antibody. As they are made up of two identical light-heavy chain pairs, these bsAbs avoid the potential for undesired chain pairing and streamline manufacturing and purification procedures [19].

Various other technologies have also been developed, allowing the generation of IgG-based bsAbs such as **cross-over dual variable (CODV)-Ig, dual action FABs (DABs),** and **dutadabs**.

3.2 Characteristics of Non-IgG-like bsAb Format

Depending on the desired mechanism of action, some bsAbs may be designed without an Fc region entirely. **Non-IgG-like bsAbs** are comprised of only two antigen-binding Fab fragments, multivalent single-chain variable fragments (scFvs), or fusion proteins that incorporate the variable chains of two mAbs. Due to the lack of an Fc domain, these non-IgG-like bsAbs are smaller (<65 kDa) and provide faster biodistribution and strong tissue penetration. T-cell receptor (TCR) bsAbs are designed to bind to a tumor-associated antigen (TAA) and a T-cell receptor to simultaneously recruit, activate, and stimulate the patient's own T cells to attack the tumor cells. These bsAbs include formats such as **BiTE, diabodies,** and **dual-affinity retargeting (DART) antibodies** (Fig. 2d).

BiTE™ antibody constructs consist of two scFvs tandemly arranged on a short single polypeptide linker [12]. One fragment binds to a specific tumor-associated antigen (TAA) and the other binds to a T-cell-specific receptor, often CD3, thereby inducing target cell-dependent T-cell activation and proliferation, resulting in apoptosis of the bound tumor cell. A competing format to the tandem scFv format of the T-cell-redirecting bsAbs are **diabodies**, which consist of the variable regions of the heavy (VH) and light (VL) chains covalently linked to each other with a short linker to form a dimer. However, research suggests that the Fv contact's instability impacts overall flexibility, and the diabodies are too rigid to crystallize. **DART antibodies** are based on the diabody format but are formed by linking VH and VL sequences with other antibody VL and VH sequences, respectively, on two independent polypeptide chains, with a cysteine residue added at the end of each heavy chain to form a C-terminal disulfide bond to improve stability [20]. Other diabody formats include **tandem diabodies (TandAbs)**, which are tetravalent antibodies that have been constructed by tandemly fusing two scFvs into a single polypeptide chain and have two binding sites for each antigen to maintain the avidity of a bivalent antibody and possess a molecular weight exceeding the renal clearance threshold [21]. **Bispecific (BiKEs)** and **trispecific killer cell engagers (TriKEs)** consist of either two (BiKE) or three (TriKE) variable antigen regions, one of which activates natural killer cells by binding to interleukin (IL)-16. For TriKEs, the addition of an IL-15 crosslinker increases survival and proliferation of natural killer cells (NKs) and enhances the immune response [22].

The size of fragment-based bsAbs and their lack of an Fc region mean that they have a shorter serum half-life and undergo more rapid renal clearance. As such, bi- and multi-specific bsAb formats that are fused to an IgG Fc region have been developed to extend their half-life and improve therapeutic efficacy, such as **XmAbs, BiTE-Fc**, and **DART-Fc antibodies** (Fig. 2e).

4 Main Mechanisms of Action of Bispecific Antibodies

The advantage of bsAbs lies in their ability to bind to two different antigens or epitopes on the same antigen, as well as the variety of formats that can be constructed to suit the desired mechanisms of action. BsAbs may exert their therapeutic effects via three main mechanisms of action, including (1) activation and redirection of immune cells toward tumor cells, (2) co-stimulation and enhancement of T-cell-receptor mediated T-cell activation, and (3) restoration and enhancement of T-cell activity with immune checkpoint blockade (Fig. 3).

4.1 Activation and Redirection of Immune Cells Toward Tumor Cells (e.g., T-Cell Engagers)

Cytotoxic T cells typically recognize and remove abnormal cells during routine cellular immune surveillance by bridging with them via the T-cell receptor major histocompatibility peptide complex. However, malignant cells can evade this step by reducing their expression of the human leukocyte antigen (HLA) peptide and secreting various immune suppression substances, to create a hostile tumor

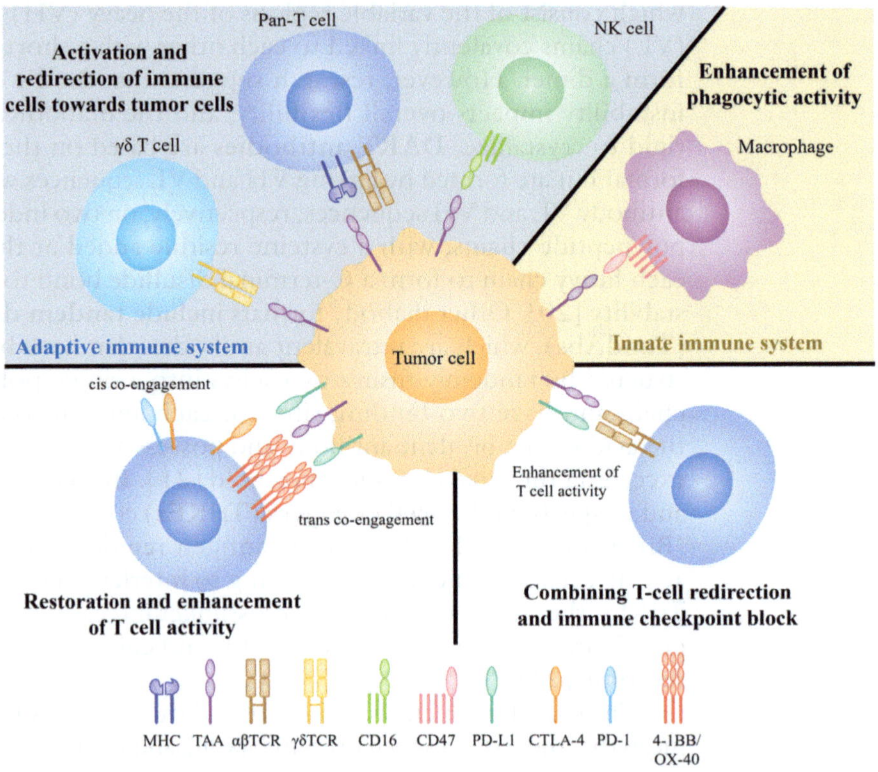

Fig. 3 Mechanisms of action of bispecific antibodies in cancer immunity. BsAbs may exert their therapeutic effects via three main mechanisms of action, including (1) activation and redirection of immune cells toward tumor cells, (2) co-stimulation and enhancement of T-cell-receptor-mediated T cell activation, and (3) restoration and enhancement of T cell activity with immune checkpoint blockade

microenvironment (TME) for infiltrating T cells. Immune cell-engaging bsAbs have been designed that simultaneously bind to antigens present on both immune cells and tumor cells, and thus are able to redirect immune effector cells to tumor cells to create an artificial synapse and induce apoptosis (Fig. 3). For example, T cells expressing CD3, CD4, CD8, and NK cells expressing primarily CD16 are frequent targets of bsAbs. Redirection of T cells toward extracellular tumor-associated antigens (TAA) or intracellular neoantigens may occur via pan-T cell engagers or subset-specific T-cell engagers (e.g., γδ T cells). An illustrative example of this mechanism is the development of the BiTE antibody, blinatumomab. The unique mechanism of action of blinatumomab involves its ability to bind simultaneously to CD19 on malignant B cells and CD3 on T cells in ALL. By doing so, blinatumomab brings T cells in close proximity to leukemia cells, facilitating the formation of an immunological synapse and activation of T-cells that lead to perforin granzyme secretion, which facilitates tumor cell lysis. This redirection of T-cell cytotoxicity is a powerful strategy in treating certain types of hematologic cancers, including

ALL. Factors affecting the potency of T-cell engagers include CD3 binding affinity and epitope, which play a crucial role in T-cell activation and cell killing potency (Fig. 4a) and require an optimal balance between cell killing and cytokine release. As shown in Fig. 4a, TCR-CD3 clustering is believed to play a role in T-cell activation and monovalent CD3 binding is preferred to avoid peripheral activation in the absence of target cell binding [23]. Epitope and affinity are the primary mechanisms for tuning the CD3 interaction. The properties of the TAA and the TAA binding are also relevant factors to consider when generating engager antibodies, with the density of specific peptide-major histocompatibility complex (pMHC) molecules impacting the degree of T-cell activation (Fig. 4b). Intrinsic properties of the TAA, including size, surface density, and on-target off-tumor distribution inform T cell engager design in terms of epitope and apparent affinity, with shorter immune synapse distance increasing potency. Other factors that can influence the formation of the immune synapse are also important and include TAA molecules like LFA-1 or format (size, geometry, and valence), which are involved in cell–cell distance. However, mechanisms such as antigen shedding can inhibit the formation of the immune synapse itself (Fig. 4c) [24].

Most of the focus to date has been on immune cell redirection in the adaptive immune system. However, there is also a mechanism of action that utilizes the innate immune response and involves the engagement of macrophages, which may be exploited by bsAbs. Phagocytosis plays a vital role in maintaining homeostasis [25]. BsAbs that link macrophages to phagocytic targets could be valuable for restoring tissue homeostasis without triggering inflammation [26]. In

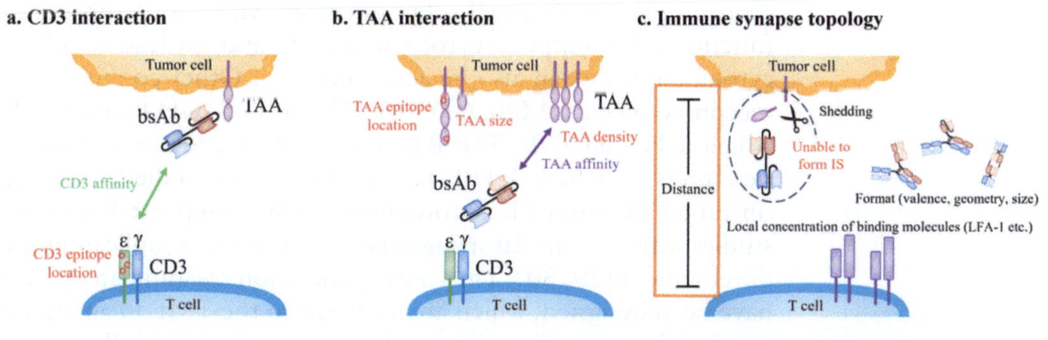

Fig. 4 Factors affecting the effectiveness of T-cell engagers in cancer immunotherapy. (**a**) The potency of T-cell engagers is influenced by CD3 binding affinity and epitope, which are essential for T-cell activation and cytotoxic effectiveness. (**b**) The properties of tumor-associated antigens (TAA) and its binding is important when creating engager antibodies, as the density of specific peptide-major histocompatibility complex (pMHC) molecules affects T-cell activation levels. (**c**) Additionally, intrinsic properties of TAA, such as size, surface density, and on-target off-tumor distribution, are vital for T-cell engager design, with a shorter immune synapse distance contributing to increased potency. Factors such as TAA molecules (e.g., LFA-1) and format can affect immune synapse formation, while antigen shedding may inhibit synapse establishment. This figure was created by the author, conceptually inspired by [Chen RP et al., Expert Opin Biol Ther, 2022] 23, and modified with original design and interpretation

hematological malignancies, some ongoing clinical studies are testing agents that target "don't eat me" signal molecules (such as CD47/SIRPα) expressed on macrophage and tumor cells markers [27].

4.2 Co-Stimulation and Enhancement of T-Cell-Receptor Mediated T-Cell Activation

Naïve T cells require at least two signals for activation, both of which are provided by an antigen-presenting cell (usually a dendritic cell). Signal 1 is generated after the interaction of MHC-peptide complexes binding to T-cell receptors, while signal 2 is generated via B7 costimulatory proteins on the antigen-presenting cell binding to counterreceptors on the T cell, typically CD28 on the T cell surface [25]. Co-stimulation significantly potentiates the immune response to antigenic stimuli, with various co-stimulatory receptors utilized for cancer immunotherapy. As mentioned, most of the immune co-stimulation targets to date focus on members of the B7-CD28 and TNF receptor (TNFR) superfamily [26]. CD28 is one of the first identified co-stimulatory molecules expressed on the surface of T cells and is the main co-stimulatory receptor target in the B7-CD28 family [26]. It contributes to lowering the threshold necessary for T cell receptor-mediated T cell activation, resulting in enhanced T-cell proliferation, cytokine production and secretion, and cell survival. Findings from preclinical studies have shown that CD28-costimulation is required for CD8 T cell rescue and lung cancer patients who responded to PD-1 therapy have higher levels of CD28+T cells [27]. An anti-HER2, anti-CD3, and anti-CD28 tri-specific antibody have recently been developed by Sanofi that can target, stimulate, and prolong the lifespan of T cells and stimulated regression of breast cancers in a humanized mouse model [28].

To date, CD3 has been the main target for binding interaction with T cells as it enables a potent and robust polyclonal T-cell response, although T-cell subpopulations with counterproductive functions that suppress cytotoxic activity and/or lead to adverse cytokine release can also be activated [23]. Other co-stimulatory receptors, such as OX40, CD40, CD27, 4-1BB, GITR, and CD30, which all belong to the TNFR family, represent promising therapeutic targets, of which 4-1BB has been the most studied. However, on-target off-tumor hepatotoxicity has been reported in clinical studies with co-stimulatory agonistic antibodies, complicating their development [29, 30]. Thus, next-generation co-stimulatory bsAbs have been designed, which are localized to the TME in an attempt to minimize systemic toxicity with systemic effector T-cell co-stimulation. Cinrebafusp alfa (PRS-343) is a bispecific fusion protein that combines a 4-1BB-targeting Anticalin protein to a variant of trastuzumab that has a mutation-modified IgG4 isotype [31]. This modification removes the risk of ADCC and non-tumor-target activation of 4-1BB-positive lymphocytes, facilitating HER2+ tumor-localized targeting of T cells, and has demonstrated higher efficacy and reduced peripheral toxicity compared with current monospecific approaches [31]. Recently, the results of a phase 1 dose-escalation study were published, which evaluated the safety and efficacy of glofitamab, a

CD20 × CD3 T-cell engaging bsAb with englumafusp alfa, a CD19 × 4-1BBL costimulatory bispecific antibody-like fusion protein in R/R B-cell non-Hodgkin's lymphoma (NHL) [32]. Across all doses investigated, englumafusp alfa in combination with glofitamab shows promising anti-tumor activity in patients with R/R aNH, with CR rates and best overall response rates of 57% and 67%, respectively. Combination therapy had a comparable safety and CRS profile to glofitamab monotherapy. Pharmacodynamically, expansion of activated and memory CD8+ T cells was observed, with limited expansion of terminally differentiated PD1+ CD8+ T cells in blood, in association with the deepening of the molecular response over time.

4.3 Restoration and Enhancement of T-Cell Activity with Immune Checkpoint Blockade

T cells are exposed to persistent antigen stimulation in the TME, resulting in the expression of multiple immune checkpoint receptors on the surface of T cells. Although these inhibitory receptors maintain self-tolerance under normal conditions, their sustained expression leads to T-cell exhaustion and an immunosuppressive TME. To date, several inhibitory checkpoints have been identified and may be potential targets in hematological malignancies including the programmed cell death 1 (PD)-1/PD-L1 axis [33], cytotoxic T-lymphocyte associated protein 4 (CTLA-4) [34], lymphocyte activation gene 3 (LAG-3) [35], T-cell immunoglobulin domain and mucin domain 3 (TIM-3) [36], and T-cell immunoreceptor with Ig and ITIM domains (TIGIT) [37]. Although inhibitory checkpoint blockade can restore effector functions to exhausted tumor-infiltrating T cells and has shown remarkable efficacy across a broad range of cancers, most patients either do not respond (primary resistance) or develop resistance after initial tumor regression (acquired resistance) [38]. A main mechanism exploited by cancer cells to evade blockade with single-target ICIs and undergo immune escape is by switching signaling pathways or activating intracellular signals. Secondly, the success of ICI therapy is largely dependent on the pre-existing immune response, with limited effects on T-cell priming and proliferation [39]. The use of bsAbs to bind to two or more signaling receptors or their ligand simultaneously (*trans* co-engagement) or two molecules on the same cell membrane (*cis* co-engagement) could, therefore, reduce tumor cell escape, stimulate the immune response, and improve therapeutic efficacy. The use of bsAbs in conjunction with PD-1 inhibitors are also under clinical investigation in the treatment of MM (NCT05338775).

Similarly, bsAbs have been designed that combine an immune co-stimulation target and immune checkpoint blockade. ATG-101 (also known as YN-051) is a next-generation, first-in-class tetravalent PD-L1 × 4-1BB bsAb GEN1046 (DuoBody-PD-L1×4-1BB) under development by Antengene in patients with advanced solid tumors and B-cell lymphomas. ATG-101 binds to PD-L1 and 4-1BB simultaneously, with a greater affinity for PD-L1, and potently activates 4-1BB positive T cells when crosslinked with PD-L1 positive cells. A mutation on the CH2 domain of the Fc region eliminates the binding capacity to most FcγRs but preserves the ability to bind to FcRn. Strong anti-tumor efficacy has been demonstrated in a variety of tumor models, including those that were resistant to or experienced progression on anti-PD-L1 therapy, turning "cold" tumors "hot." The safety profile was manageable with no evidence of hepatotoxicity in non-human primates and a low risk of CRS. Nevertheless, the landscape for co-stimulatory bsAb therapeutics is still in its infancy, and more research on their effectiveness and safety is still needed.

5 Clinical Efficacy and Safety of Bispecific Antibodies for Hematological Malignancies

Several bsAbs have been approved for hematological malignancies, all of which are CD3 T-cell engagers (Fig. 5). Numerous others are in early stage clinical trials evaluating a combination of various targets (Fig. 5). Tables 1 and 2 summarize the main clinical efficacy and safety results, respectively, of various bsAbs in advanced development for hematological malignancies.

5.1 Acute Lymphoblastic Leukemia

CD19, a critical mediator of B-cell signaling, is expressed on almost all cells of the B-lymphocyte lineage and is maintained during the neoplastic transition of precursor B cells in B-cell ALL (B-ALL), making it an attractive therapeutic target [40]. Anti-CD19 CAR

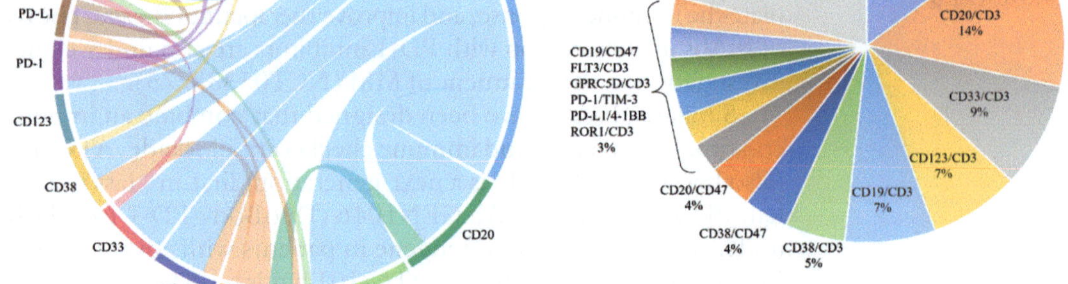

Fig. 5 Bispecific antibodies in clinical development for hematological malignancies. (**a**) Target combinations, (**b**) percentage of bispecific antibody targets in hematological malignancies

Table 1

Clinical efficacy of main bispecific antibodies in advanced clinical development for hematological malignancies

Agent	Indication	NCT	Phase	N	PLOT, median	TCR	ORR	Other	References
Blinatumomab (CD19 × CD3)	R/R B-cell precursor ALL	TOWER NCT020[illegible]3167	3	405	NS	NS	NS	Blinatumomab vs. chemotherapy Median OS:7.7 mo (95% CI, 5.6–9.6) versus 4.0 mo (95% CI, 2.9–5.3)HR for death: 0.71 (95% CI: 0.55–0.93; $p = 0.01$) Estimated 6-mo survival: 54% vs. 39% CRR: 34% versus 16%; $p < 0.001$) CR, CRh, or CRi: 44% versus 25%; $p <$ 0.0016-mo EFS: 31% versus 12%	Kantarjian et al. [44]
	MRD + R/R B-cell precursor ALL	BLAST NCT01207388	2	116	NS	NS	NS	Complete MRD response: 78% 18-mo RFS: 54% mOS: 36.5 mo MRD responders versus nonresponders:RFS: 23.6 versus 5.7 mo OS: 38.9 versus 12.5 mo	Gokbuget et al. [96]
	R/R PH+ B-cell ALL	ALCANTARA NCT02000427	2	45	NS	NS	NS	CR/CRh (≤2 cycles): 35.6% mRFS: 6.8 mo mOS: 9.0 mo	Martinelli et al. [97]

(continued)

Table 1
(continued)

Agent	Indication	NCT	Phase	N	PLOT, median	TCR	ORR	Other	References
Flotetuzumab (CD123 × CD3)	R/R AML	VOYAGE NCT02152956	1/2	88	4	NS	13.6%	CR/CRh rate: 11.7% RP2D CR/CRh rate and ORR (RP2D): 18% and 24%	Uy et al. [51]
Vibecotamab (CD123 × CD3)	R/R AML	NCT02730312	1	120	NS	NS	9.0%	9.0% (10/111 pts) achieved an overall response of morphologic leukemia-free state or better	Short et al. [98]
AMG330 (CD33 × CD3)	R/R AML	NCT02520427	1	55	NS	NS	19.0%	CR: 7.1% CRih: 9.5% Morphologic leukemia-free state: 2.4%	Ravandi et al. [99]
AZD7789 (PD-1 × TIM-3)	R/R HL	NCT05216835	1/2	39	NS	NS	NS	On-treatment scan results: SD: 36.8% (7/19) μPR: 10.5% (2/19) PD: 57.9% (11/19) NE: 52.6% (1/19) Target lesion shrinkage observed in 8 pts.	Mei et al. [100]
A-319 (CD19 × CD3)	R/R B-NHL	NCT04056975	1	21	NS	NS	NS	CR: 4.8% PR: 14.3% SD: 33.3%	Liu et al. [101]
EX103 (CD20 × CD3)	R/R B-NHL	NCT06021678	1	13	5	NS	67.7%	DCR: 84.6%	Qi et al. [102]
GB261 (CD20 × CD3)	R/R B-NHL, CLL	NCT04923048	1/2	47	3	NS	73%	CRR: 45.5% ORR/CRR3 mg: 100% each 10 mg: 56% and 22%;30 mg: 67% and 33%. mTTR: 1.3 mo	Song et al. [103]

Mosunetuzumab (CD20 × CD3)	R/R FL	NCT02500407	2	90	3	NS	60.0%	CRR significantly higher than historical control ($p < 0.0001$). ORR: 80.0%	Budde et al. [58]
TNB-486 (CD19 × CD3)	R/R FL	NCT04594642	1	17	NS	NS	91%	6-mo PFS: 91%	Nair et al. [104]
Glofitamab (RG6026) (CD20 × CD3)	R/R DLBCL	NP30179NCT03075696	2	155	3	NS	NS	CRR: 39.0% 12-mo PFS: 37%	Dickenson |et al. [105
Epcoritamab (CD20 × CD3)	R/R DLBCL	EPCORE NHL-1NCT03625037	1/2	157	3	NS	63.1%	CR rate: 38.9%	Thieblemont et al. [61]
Odronextamab (CD20 × CD3)	R/R DLBCL	ELM-2NCT03888105	2	127	2	NS	52%	CR rate: 31% DoR:10.2 mo	Ayyappan et al. [106]
Teclistamab (BCMA × CD3)	R/R MM	MajesTEC-1NCT03145181 NCT04557098	1/2	165 (ph 1: 40; ph 2: 125)	5	77.6%	63.0%	≥CR: 46.1% MRD-negativity rate in MRD evaluable pts: 85.7% mDOR: 24.0 mo mPFS: 11.4 mo	Garfall et al. [72]
Elranatamab (BCMA × CD3)	R/R MM	MagnetisMM-3 (NCT04649359)	2	123	5	96.7%	61.0%	≥CR: 37.4% mDOR: NR (among responders) mPFS: 17.2 mo Median	Tomasson et al. [73]
Linvoseltamab (REGN5458) (BCMA × CD3)	R/R MM	LINKER MM-1 (NCT03761108)	1/2	221	NS	82.1 (200 mg); 93.3 (50 mg)	70.9% (200 mg); 48.1% (50 mg)	mDOR: 29.4 (200 mg)	Bumma et al. [74]

(continued)

Table 1
(continued)

Agent	Indication	NCT	Phase	*N*	PLOT, median	TCR	ORR	Other	References
REGN5459 (BCMA × CD3)	R/R MM	NCT04083534	1/2	43	5	NS	67%	≥VGPR: 58%m TTR: 0.8 mom DOR: NR12-mo DOR: 66.7%	Suvannasankha et al. [107]
ABBV-383 (TNB-383B) (BCMA × CD3)	R/R MM	NCT03933735	1	124	5	82%	57%	≥VGPR: 43% MRD-negative rate: 73%6-mo DOR rate 79.9% 12-mo DOR rate: 79.9% mPFS: 10.4 mo	D'Souza et al. [75]
Talquetamab (GPRC5D × CD3)	R/R MM	MonumenTAL-1NCT03399799	1/2	232	6	79%	74.1% (405 ug dose); 69.5% (800 ug dose)	Median time to response: 1.2 mo (405 μg dose) and 1.3 mo (800 μg dose) mDOR: 9.5 mo (405 μg dose) and 17.5 mo (800 μg dose) mPFS: 7.5 mo (405 μg dose) and 11.2 mo (800 μg dose)	Rasche et al. [76]
AMG420 (CD38 × CD3)	R/R MM	NCT02514239	1	42	NS	NS	31%	ORR at MTD of 400 μg/d: 70%	Topp et al. [109]
Cevostamab (FcRH5 × CD3)	R/R MM	GO39775NCT03275103	1	16	6	81.3%	62.5%	sCR: 43.8%, CR: 18.8%; VGPR: 31.3%; PR: 6.3%	Lesokhin et al. [110]

ALL, acute lymphoblastic leukemia; AML, acute myeloid leukemia; CR, complete response; DLBCL, diffuse large B cell lymphoma; DOR, duration of response; FL, follicular lymphoma; HL, Hodgkin's lymphoma; mDOR, median duration of response; MM, multiple myeloma; MRD, minimal residual disease; MTD, maximum tolerated dose; mTTR, median time to response; NHL, non-Hodgkin's lymphoma; NR, not reported; NS, not specified; ORR, objective response rate; OS, overall survival; PFS, progression-free survival; PR, partial response; R/R, relapse/refractory; sCR, stringent complete response; TCR, triple-class refractory; VGPR, very good partial response

Table 2

Clinical safety of main bispecific antibodies in advanced clinical development for hematological malignancies

Agent	Indication	NCT	Phase	*N*	Safety	References
Blinatumomab (CD19 × CD3)	R/R B-cell precursor ALL	TOWER NCT02013167	3	405	Blinatumomab vs. chemotherapy AEs: 99% each Grade ≥3 AEs: 87% versus 92% SAEs: 62% versus 45% Fatal AEs: sepsis (3% vs. 4%), septic shock (2% vs. 0%), multiorgan failure (1% vs. 0%), respiratory failure (<1% vs. 2%), and bacteremia (0% vs. 2%)	Kantarjian et al. [44]
	MRD + R/R B-cell precursor ALL	BLAST NCT01207388	2	166	AEs: Cycle 1: 100%; Cycle 2: 85%; Cycle 3: 79%; Cycle 4: 75%. Gr 3: 33%; Gr 4: 27% CRS: 3% Neurological events: 53%	Gokbuget et al. [96]
	R/R PH+ B-cell ALL	ALACANTARA NCT02000427	2	45	TEAEs (gr ≥3): 100% (84%)Treatment-related AEs (gr ≥3): 91% (44%) CRS: 9% (gr ≥3: 0) Neurologic events: 62% Fatal AEs: 11%	Martinelli et al. [97]
Flotetuzumab (CD123 × CD3)	R/R AML	VOYAGE NCT02152956	1/2	88	IRR/CRS: 96% (gr ≥3: 8.0%)Nausea: 28% Peripheral edema: 30% (gr ≥3: 2.0%)Pyrexia: 22% Diarrhea: 22%	Uy et al. [51]
Vibecotamab (CD123 × CD3)	R/R AML	NCT02730312	1	120	DLT: 13.3% CRS: 59.2%	Short et al. [98]
AMG330 (CD33 × CD3)	R/R AML	NCT02520427	1	55	TEAEs: 100% AMG 330-related AEs: 89% (49/55) CRS: 67% (gr ≥3: 13%) Nausea: 20%	Ravandi et al. [99]
AZD7789 (PD-1 × TIM-3)	R/R HL	NCT05216835	1/2	39	TEAEs: 82% (gr ≥3: 23%). The most common TEAE was increased blood creatinine (18%); no TEAE led to discontinuation. TRAEs: 41%; the most common was asthenia (8%). No DLTs or gr ≥3 TRAEs.	Mei et al. [100]

(continued)

Table 2 (continued)

Agent	Indication	NCT	Phase	*N*	Safety	References
EX103 (CD20 × CD3)	R/R B-NHL	NCT06021678	1	13	No DLTs observed MTD: NR The most common AEs: CRS, pyrexia, neutropenia, thrombocytopenia All CRS gr 1–2, no gr ≥ 3 No ICANS or other clinically significant neurologic AEs	Qi et al. [102]
A-319 (CD19 × CD3)	R/R B-NHL	NCT04056975	1	21	AEs: 85.7% TRAEs: 57.1% CRS: 23.8% (gr 1–2: 100%) ICANS: 28.6%Fever: 66.7% Decreased immunoglobulin: 23.8%	Liu et al. [101]
GB261 (CD20 × CD3)	R/R B-NHL, CLL	NCT04923048	1/2	47	TEAEs: 95.7% TRAEs: 85.1%.The most common TEAEs were COVID-19 (40.4%; gr 1–2: 27.6%; grade ≥3: 12.8%) and neutropenia (31.9%; gr 1–2: 14.9%; grade ≥ 3: 17.0%). AE-related discontinuation and death were reported in 2 patients, who were due to COVID-19 pneumonia.CRS: 12.8% (gr 1: 8.5%; gr 2: 4.3%; gr 3: 0%)ICANS: 0%	Song et al. [103]
Mosunetuzumab (CD20 × CD3)	R/R FL	NCT02500407	2	90	CRS: 44% (gr 1: 26%; gr 2: 17%)Most common Gr 3/4 AEs: Neutropenia/neutrophil count decreased 27%; hypophosphataemia 17%; hyperglycemia 8%; anemia 8%. SAEs: 47%	Budde et al. [58]
TNB-486 (CD19 × CD3)	R/R FL	NCT04594642	1	17	CRS: 71% (gr 1: 59%; gr 2: 12%)	Nair et al. [104]
Glofitamab (RG6026) (CD20 × CD3)	R/R DLBCL	NP30179 NCT03075696	2	155	Discontinuation due to AEs: 9% CRS: 63% (gr ≥3: 4%) Grade ≥3 AEs: 62% Gr ≥3 neurological events: 3%	Dickenson et al. [105]
Epcoritamab (CD20 × CD3)	R/R DLBCL	EPCORE NHL-1 NCT03625037	1/2	157	CRS: 49.7% (gr 1–2: 47.1%; gr 3: 2.5%) Pyrexia: 23.6% Fatigue: 22.9% ICANS: 6.4%	Thieblemont et al. [61]

Odronextamab (CD20 × CD3)	R/R DLBCL	ELM-2 NCT03888105	2	141	TRAEs leading to: Treatment interruption/ delay: 53% Discontinuation: 10% CRS: 55% (gr 1/2: 98%) Anemia: 43% Pyrexia: 42%	Ayyappan et al. [106]
Teclistamab (BCMA × CD3)	R/R MM	MajesTEC-1 NCT03145181 NCT04557098	1/2	165	Neutropenia: 71.5% (gr 3/4: 65.5%) Anemia: 55.2% (gr 3/4: 37.6%) Thrombocytopenia: 41.8% (gr 3/4: 23.0%) Infections: 78.8% (gr 3/4: 55.2%) CRS: 72.1% (gr 3/4: 0.6%)	Costa et al. [95]
Elranatamab (BCMA × CD3)	R/R MM	MagnetisMM-3 (NCT04649359)	2	123	Neutropenia: 49.6% (gr 3/4: 49.6%);Anemia: 48.8% (gr 3/4: 37.4%);Thrombocytopenia: 31.7% (gr 3/4: 23.6%); Lymphopenia: 26.8% (gr 3/4: 25.2%) Infections: 70.7% (gr 3/4: 42.3%; fatal: 6.5%) CRS: 57.7% (gr 3/4:0%) Neurotoxicity: 4.9% (gr 3/4: 0%)	Tomasson et al. [73]
Linvoseltamab (REGN5458) (BCMA × CD3)	R/R MM	LINKER MM-1 (NCT03761108)	1/2	221	TEAEs: 100% (gr ≥3: 73.5%) [200 mg group] 98.1% (gr ≥3: 72.1%) [50 mg group] CRS: 46.2% (gr 3: 1%) [200 mg group] 54.8% (gr 3: 2%) [50 mg group] Fatigue: 33.3% (gr ≥3: 0%) [200 mg group] 29.8% (gr ≥ 3: 0%) [50 mg group] Anemia: 38.5% (gr ≥ 3: 30.8%) [200 mg group] 42.3% (gr ≥3: 37.5%) [50 mg group]Gr ≥3 ICANS: 7.7% [200 mg group] Infections: 74.4% (gr ≥3: 33.3%) [200 mg group]	Bumma et al. [74]
ABBV-383 (TNB-383B) (BCMA × CD3)	R/R MM	NCT03933735	1	124	TEAEs: 98% (gr ≥3: 72%). The most common (≥25% pts/ gr ≥3) hematologic TEAEs were neutropenia (37%/34%) and anemia (29%/16%). The most common (≥25% pts/ gr ≥3) nonhematologic were CRS (57%), fatigue (30%), nausea (29%), and diarrhea (27%). Hypogammaglobulinemia: 14%.	D'Souza et al. [75]

(continued)

Table 2
(continued)

Agent	Indication	NCT	Phase	*N*	Safety	References
REGN5459 (BCMA × CD3)	R/R MM	NCT04083534	1/2	43	TEAEs: 100% of pts (gr ≥3: 74%) Common (≥ 30% of pts) TEAEs:CRS: 54%; fatigue: 44%; neutropenia: 37%; anemia: 35%; cough: 30%; diarrhea: 30%.Common (≥15% of pts) gr ≥ 3 TEAEs:Neutropenia: 37%; anemia: 26%; lymphopenia: 23%; thrombocytopenia: 19%; hypertension: 16%. Gr 1, 2, and 3 CRS: 47%, 2%, and 5%, respectively No Gr 4–5 CRS.ICANS: 2% (Gr 2). Serious TEAEs and TEAEs leading to tx discontinuation: 63% and 16%, respectively. Infections: 61% (Gr ≥ 3: 37%). Two deaths (COVID pneumonia and COVID infection)	Suvannasankha et al. [107]
Talquetamab (GPRC5D × CD3)	R/R MM	MonumenTAL-1NCT03399799	1	232	CRS: 77% (405 μg dose); 80% (800 μg dose) Skin-related AEs: 67% (405 μg dose); 70% (800 μg dose) Dysgeusia: 63% (405 μg dose); 57% (800 μg dose)	Chari et al. [108]
AMG420 (CD38 × CD3)	R/R MM	NCT02514239	1	42	Serious AEs: 48% Infections: 33.3%; polyneuropathy: 4.8% Treatment-related SAEs: gr 3 polyneuropathies: 4.8%; gr 3 edema: 2.4%.No gr ≥3 CNS toxicities	Topp et al. [109]
Cevostamab (FcRH5 × CD3)	R/R MM	GO39775NCT03275103	1	16	N/A	Lesokhin et al. [110]

AE, adverse event; ALL, acute lymphoblastic leukemia; AML, acute myeloid leukemia; CNS, central nervous system; CRS, cytokine release syndrome; DLBCL, diffuse large B cell lymphoma; DLT, dose-limiting toxicities; FL, follicular lymphoma; Gr, grade; HL, Hodgkin's lymphoma; ICANS, Immune effector cell-associated neurotoxicity syndrome; IRR, infusion-related reaction; MM, multiple myeloma; N/A, not applicable; NHL, non-Hodgkin's lymphoma; R/R, relapse/refractory; SAE, serious adverse event; TEAE, treatment-emergent adverse event; tx, treatment

T-cell therapies represent a major breakthrough in the treatment of B-ALL; however, relapse is common, with loss or reduction in CD19 surface expression increasingly recognized as a major cause of treatment failure [41]. Blinatumomab, a BiTE that targets CD19 and CD3 on T cells, initially gained accelerated approval in 2014 for Philadelphia chromosome-negative relapsed or refractory (r/r) B-cell precursor ALL in adults [42], supported by data from a clinical study conducted in 189 adults, in which 43% of patients achieved a complete response (CR) of their r/r B-cell ALL [43]. Full approval was granted in 2018 for both adult and pediatric patients with B-cell ALL in first or second complete remission with minimal residual disease (MRD) ≥0.1%, based largely on the results of the phase 3 TOWER study conducted in 405 patients with relapsed/refractory B-ALL. In that study, treatment with blinatumomab was associated with an overall survival (OS) advantage (median OS: 7.7 months vs. 4.0 months) and an improved CR rate (34% vs. 16%) compared with standard-of-care chemotherapy [44]. Although initially investigated for Philadelphia chromosome-negative disease, the phase 2 ALCANTARA study also demonstrated activity in patients with Philadelphia chromosome (Ph+) B cell ALL, including patients with prior tyrosine kinase exposure and T315I mutation [45]. In addition to r/r B-cell ALL, blinatumomab has now also been approved by the FDA for the treatment of MRD-positive B-cell ALL [46]. The basis of this approval was a phase 2 study in 116 patients with MRD-positive B-ALL in complete hematological remission, in which 78% achieved a complete MRD response and OS (38.9 vs. 12.5 months; $p = 0.002$) and relapse-free disease (23.6 vs. 5.7 months; $p = 0.002$) were significantly longer in MRD responders versus nonresponders, respectively [47]. The safety profile of blinatumomab is generally favorable, with primary toxicities including infection, hematologic toxicity, and neurotoxicity [44, 47]. Compared with other bsAbs, the incidence of cytokine release syndrome (CRS) is lower, but the incidence of neurotoxicity is higher, although the clinical symptoms are mostly transient and resolve quickly following dose interruption and initiation of supportive treatment [48].

5.2 Acute Myeloid Leukemia

The development of bsAbs for the treatment of AML has to date been limited by the (1) lack of suitable target antigens that are ubiquitously expressed on malignant AML cells and (2) overlap of these antigens with those on hematopoietic cells, increasing the risk of hematological toxicities and cytopenias. Currently, the most frequent targets for AML are CD33, CD123, CLL-1, and FLT-3. In contrast with CD33, which is also expressed on normal hematopoietic stem/progenitor cells and thus may potentially impair normal hematopoiesis [49], CD123, is highly expressed in AML and relatively less expressed by hematopoietic stem and progenitor cells, making it a more attractive therapeutic target. Nevertheless, several CD33 × CD3 bsAbs have been evaluated in phase 1/2 clinical trials for the treatment of AML, including JNJ-67561244, AMV564, and AMG-330, although the latter has been discontinued due to safety issues [50]. Flotetuzumab, a bispecific DART antibody-based molecule targeting CD123 × CD3, has demonstrated encouraging anti-leukemic activity and acceptable safety in a multicenter, open-label, phase 1/2 study in 88 adults with relapsed/refractory AML [51]. However, in 2022, MacroGenics announced that they were discontinuing development in favor of their next-generation drug, MGD024, which has been designed to have a longer half-life and reduced likelihood of CRS [52].

Vibecotamab (formally XmAb14045), another bsAb targeting CD123 × CD3, achieved an overall response of morphologic leukemia-free state or better with an overall response rate (ORR) of 9.0% in a phase 1 study in 120 patients with r/r AML, and is currently being assessed in a phase 2 study [53].

The incidence of CRS was high (59.2%) but was mostly grade ≤2 and manageable with premedication. Identification of a recommended phase 2 dose comprised three step-up doses in week 1 was also noted to reduce CRS. Other CD123 × CD3 bsAbs, including APVO436, SAR440334, and JNJ-63709178, are currently being evaluated in ongoing phase 2 clinical studies after showing promising anti-leukemic activity in preclinical and phase 1 studies [54–56].

5.3 Non-Hodgkin's Lymphoma

Non-Hodgkin's lymphoma (NHL) comprises a diverse group of lymphoproliferative neoplasms, the most common being indolent FL and DLBCL. Although the prognosis for B-cell NHL has vastly improved following the introduction of the anti-CD20 monoclonal antibody rituximab, a proportion of patients experience relapse or are refractory to treatment [57].

Currently, several CD20 × CD3 bsAbs, including mosunetuzumab [58], odronextamab [59], glofitamab [60], and epcoritamab [61], have demonstrated substantial efficacy in both indolent and aggressive NHL subtypes in phase 1 and 2 clinical trials, which has resulted in the accelerated approval of mosunetuzumab, glofitamab, epcoritamab, and fast-track designation of odronextamab by the FDA. In patients with relapsed/refractory FL, treatment with mosunetuzumab and odronextamab has led to CR rates of 60 and 75%, respectively [58, 62].

In patients with DLBCL, positive CR rates of 37–39% have similarly been achieved with glofitamab [60], epcoritamab [61], and odronextamab [62]. However, durability of responses was superior with glofitamab, with 70% of patients still in CR after 18 months versus 48% with odronextamab [60, 62]. Regarding AEs, low-grade CRS is the most common AE associated with CD20 × CD3 bsAbs [63], although grade ≥3 events are rare [64]. CRS events typically occur during the first treatment cycle, and their seriousness can be minimized with step-up dosing and administration of premedication and appropriate supportive treatment [60, 62, 65]. Other AEs, such as pyrexia, neutropenia, anemia, and electrolyte disturbances, are mostly transient and not clinically significant. Notably, the frequency and severity of immune effector cell-associated neurotoxicity syndrome (ICANS) are lower with anti-CD20 × anti-CD3 bsAbs than with anti-CD20 CAR-T therapy [63, 66].

CD19-targeting bsAbs, such as blinatumomab, duvortuxizumab, and TNB-486, have also been investigated in NHL, but were discontinued due to neurotoxicity concerns [67–69]. Treatment with TNB-486, a human CD19 × CD3 × T cell-redirecting bsAb, induced a high CR rate of 91% in patients with relapsed/refractory FL in a phase 1 trial and had a lower incidence of ICANS and CRS than blinatumomab, likely due to its anti-CD3 moiety, which binds with low affinity to CD3 on T cells [67, 68, 70].

5.4 Multiple Myeloma

BCMA is selectively expressed on malignant plasma cells, making it an attractive therapeutic target for bsAbs in MM. In 2022, following initial conditional approval in the EU, the FDA granted accelerated approval for teclistamab (JNJ-64007957), a humanized IgG4-proline, alanine, alanine (IgG4-PAA) bsAb targeting BCMA × CD3, for the treatment of relapsed-refractory MM (RRMM) [71]. In the phase 1/2 MajesTEC-1 trial conducted in 165 patients at the Recommended Phase II Dose (RP2D) with triple-class exposed MM, treatment with teclistamab resulted in an ORR of 63.0% and a CR or better rate of 46.1% at a median follow-up of 30.4 months [72]. The median duration of response (DOR) and progression-free survival (PFS) was 24.0 months and 11.4 months, respectively [72]. Common AEs (any grade; grade 3–4) included CRS, the majority of which were grade 1–2 in severity, cytopenias (neutropenia [71.5%; 65.5%], anemia [55.2%; 37.6%], thrombocytopenia [41.8%; 23.0%]), and infections (78.8%; 55.2%). In 2023, elranatamab, an IgG2A bsAb also targeting BCMA × CD3, was granted accelerated approval by the FDA following the results of the phase 2 Magnetis MM-3 trial, which reported an ORR of 61.0% and a CR or better rate of 37.4% in 123 patients with triple class-refractory MM at median follow-up of 28.4 months (by reverse Kaplan–Meier)[73]. Common AEs (any grade; grade 3–4) included CRS (57.7%; 0%), neutropenia (49.6%; 49.6%), anemia (48.8%; 37.4%), and infections (70.7%; 42.3%) [73]. In 2024, the updated data from the phase 2 LINKER-MM1 study evaluating the use of the anti-BCMA T cell-redirecting bsAb linvoseltamab (REGN5458) in patients with triple-class refractory MM was published [74]. At data cutoff, the ORR was 70.9% in patients receiving the higher (200 mg) dose and 48.1% for those on the lower (50 mg) dose [74]. The median DOR for 200 mg patients was 29.4 months and the probability of maintaining response at 12 months was 80.9%. The safety profile was generally tolerable, with CRS (46% [200 mg], 54.8% [50 mg]) and fatigue (33.3% [200 mg],

29.8% [50 mg]) the most common AEs. ABBV-383, an anti-BCMA T cell-redirecting bsAb, has the advantage of not requiring step-up dosing, achieved an objective response rate of 57% and a VGPR or better in 43% of 124 RRMM patients in a phase 1 study [75].

Talquetamab (JNJ-64407564) is a first-in-class, bispecific IgG4 antibody that binds to both GPRC5D and CD3 receptors, mediating T-cell-activated lysis of GPRC5D-positive MM cells. In the phase 1/2 MonumenTAL-1 study, an ORR of 74.1% was achieved in patients receiving the 0.4 mg/kg subcutaneous (SC) once weekly (QW) dose and 69.5% in patients receiving the 0.8 mg/kg SC twice weekly (Q2W) dose; the proportion of patients achieving a very good partial response (VGPR) or better was 59.4% and 59.1%, and the median time to first confirmed response (range) was 1.2 (0.2–10.9) and 1.3 (0.2–4.9) months, respectively [76]. The most common AEs were CRS (77%–80%; grade 3: 0–3%) and cytopenias, including neutropenia (36–67%; grade 3/4: 23–53%), which were reversible and mostly occurred during step-up and cycles 1–2. Infections occurred in 34–47% (grade 3/4: 7–9%) of patients [77]. Taste changes in the form of dysgeusia and dry mouth are also common, occurring in approximately 60% and 30–57% of patients, respectively [78]. Other bsAbs currently in development for the treatment of MM include cevostamab and AMG 424, which target Fc receptor homolog 5 (FcRH5) and CD38, respectively [79].

6 Future Perspectives and Research Directions

Over the past two decades, the field of oncology has been transformed with the development of targeted therapies that selectively target molecules or pathways involved in cancer immunology, such as CAR T-cell therapies, ICIs, and bsAbs. Several CAR-T-cell therapies have been granted approval for the treatment of various hematological cancers, most of which target CD19 or BCMA [80]. BsAbs, and in particular T-cell-redirecting bsAbs, have transformed the treatment landscape for advanced hematological malignancies, with several agents now approved and numerous others demonstrating promising efficacy and safety in early stage clinical trials. However, the use of targeted therapies in oncology is complicated by several challenges that need to be overcome in order to facilitate broader uptake and more streamlined and safer usage in clinical practice.

scFv-induced CAR aggregation on the surface of T cells has been shown to result in premature and antigen-independent CAR T-cell exhaustion, which has been recognized as a major cause of nonresponse and relapse after CAR T-cell therapy [81]. Advancements in genetic engineering have similarly led to the development of next-generation nanobody-based bivalent CAR-T cell therapies with the VHH target-binding domain [82] (Fig. 6) that are designed

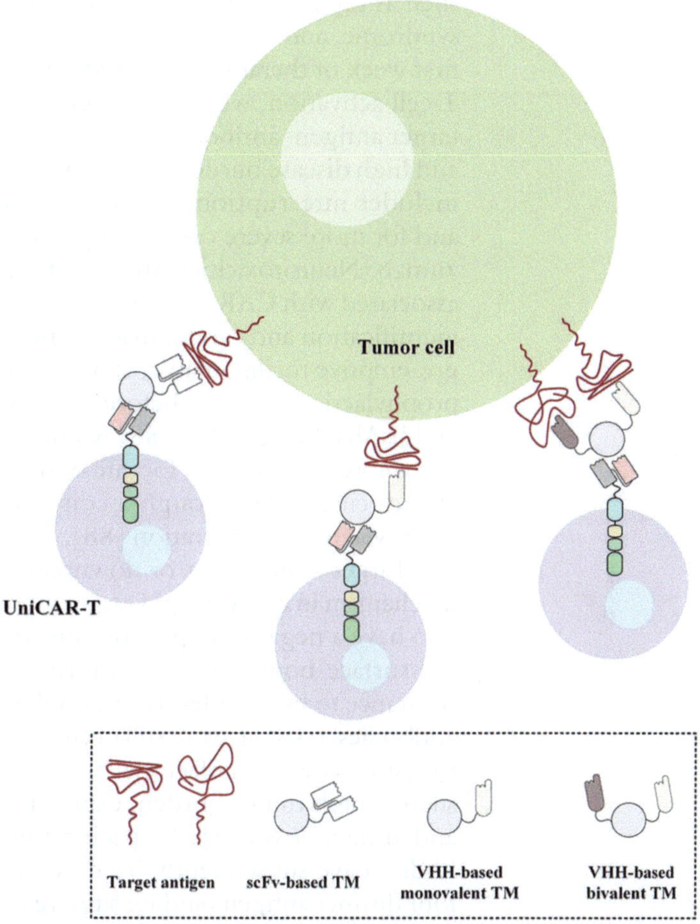

Fig. 6 Next-generation nanobody-based bivalent CAR-T with the VHH target-binding domain. Redirection of UniCAR-expressing T cells using scFv-based, monovalent VHH-based, and bivalent VHH-based targeting modules targeting a particular target antigen. This figure is adapted from Pouya Safarzadeh Kozani et al. [82], licensed under CC BY. © 2022, Pouya Safarzadeh Kozani et al. CAR, chimeric antigen receptor; scFv, single-chain variable fragment; TM, target molecule; VHH, variable heavy domain of heavy chain

to replicate the effectiveness of conventional CAR-T-cell therapies without the tendency to aggregate on the surface of T cells due to their monomeric structure [82, 83].

CRS is a potentially life-threatening AE that frequently occurs in patients receiving CAR T-cell therapies (30–100% [grade ≥3: 6%]) [84] and, to a lesser extent, T-cell-engaging bsAbs (70% [grade ≥3: 0.2%]) [85, 86]. Activation of effector immune cells leads to hypersecretion of several pro-inflammatory cytokines such as C-reactive protein (CRP), ferritin, IL-6, interferon (INF)-γ, tumor necrosis factor (TNF)-α, and IL-2, which may lead to life-threatening complications such as capillary leak, vasodilatory shock, acute respiratory distress syndrome, and end-organ failure. Onset typically occurs during the first week of therapy and is predominantly due to the magnitude of T-cell activation, which is influenced by several factors such as the target antigen, antibody construct, epitope binding affinities, dose, and high disease burden and/or antigen load. Management of CRS includes interruption the infusion of the drug and antipyretics, and for more severe cases, vasopressors, corticosteroids, and tocilizumab. Neurotoxicities, such as ICANS are also a concerning AE associated with CAR T-cell therapies and bsAbs that require prompt identification and careful management. Recently, strategies include pre-emptive tocilizumab or steroids to prevent severe symptoms, or prophylactic treatment before any toxicity appears both in CAR-T and bsAbs therapies [87, 88]. Clinical trials are currently underway to explore the potential for safe outpatient bsAbs treatment, aiming at reducing CRS through a combination of step-up dosing and prophylactic administration [88].

Target antigen loss or downregulation is a common resistance mechanism in cancer; rapid target internalization and shedding may also have a negative impact on tumor cell elimination because only cell surface-bound bsAbs can form an immune synapse. Although resistance to bsAbs is less common than with traditional monoclonal antibodies, resistance can still occur. Mechanisms that may contribute to resistance are multifactorial but may include tumor-related factors, such as high tumor burden, expression of T-cell inhibitory ligands, and antigen loss [89]. Trispecific antibodies and tetraspecific antibodies have subsequently been developed, which target three and four distinct antigen binding sites, respectively, aimed at overcoming these issues [90]. Antibody-based proteolysis-targeting chimeras (PROTACs) called AbTACs have also been designed, which are fully recombinant bispecific antibodies that bind to a degrader moiety such as membrane-bound E3 ligase and a surface antigen (TAA), resulting in the internalization of a cell surface target and proteasomal degradation [91, 92]. Furthermore, the results of the first Phase 1b study evaluating two bsAbs in combination, the anti-BCMA bsAb teclistamab with the anti-GPRC5D bsAb talquetamab, have recently been published in patients with RRMM aimed at strengthening efficacy outcomes in this difficult-to-treat population, particularly in patients with unmet needs such as those with extramedullary disease, supporting further evaluation of the combination [93].

7 Conclusion

BsAbs represent the next generation of therapeutic antibodies affording clinicians with the ability to implement more sophisticated targeting strategies with the aim of achieving superior therapeutic efficacy for the treatment of hematological malignancies. As the number of bsAbs with new and unique formats entering clinical trials continues to expand, it is anticipated that these agents will feature increasingly prominently in the treatment of hematological malignancies. Regulatory agencies have recently revised their guidance in order to keep up with the pace of development for these molecules, with the latest FDA guidance document published in May 2021 outlining various aspects that should be taken into consideration during the development of bsAbs, including design strategy, preclinical studies, quality control, drug metabolism, and toxicity [94]. Most of the bsAbs research in hematological malignancies to date has focused on the relapsed/refractory disease setting. Recently, data on treatment outcomes with bsAbs in patients with earlier-stage disease have also emerged, and research is progressing [95]. More recently, next-generation targeted therapies have been developed, such as bivalent structures using VHH in CAR T-cell constructs and multi-specific antibodies. Along with the development of these multi-target CAR-T/bsAbs, clinical trials are also progressing employing them in combination with currently available drugs and approaches that activate endogenous immunity. The success of developing various approaches may possibly lead to an expansion of treatment options for patients with hematological malignancies in the future.

Acknowledgments

I wish to thank Zhu Yan (Angeline) from Medical Affairs at Johnson & Johnson Asia Pacific, and members of the Hematology group, Oncology department of Medical Affairs at Johnson & Johnson, especially Yusuke Izumi, Ryo Yano, and Maiko Murota for their support in obtaining internal approval. I also wish to thank Colleen Kane (J&J Innovative Medicine US) and Luis Ralat (J&J World Headquarters US) for their insightful comments and suggestions, which greatly improved this manuscript.

I also would like to thank Dr. Hiroshi Fujiwara (Mie University) for his invaluable advice regarding figure illustrations.

Medical writing assistance was provided by Jordana Campbell, BSc, CMPP of inScience Communications, Springer Healthcare. This medical writing assistance was funded by Johnson & Johnson.

Finally, Dr. Kohei Tsumoto (University of Tokyo) is acknowledged for his critical feedback on this review article.

References

1. Karagianni P, Giannouli S, Voulgarelis M (2021) From the (epi)genome to metabolism and vice versa; examples from hematologic malignancy. Int J Mol Sci 22(12):6321
2. Amjad MT, Chidharla A, Kasi A (2024) Cancer chemotherapy. StatPearls [Internet]. Available from: https://www.ncbi.nlm.nih.gov/books/NBK564367/. Date Accessed: 21 Nov 2024

3. Wang K, Tepper JE (2021) Radiation therapy-associated toxicity: etiology, management, and prevention. CA Cancer J Clin 71(5):437–454
4. Alfarouk KO, Stock CM, Taylor S et al (2015) Resistance to cancer chemotherapy: failure in drug response from ADME to P-gp. Cancer Cell Int 15:71. https://doi.org/10.1186/s12935-015-0221-1
5. Wu Y, Song Y, Wang R, Wang T (2023) Molecular mechanisms of tumor resistance to radiotherapy. Mol Canc 22(1):96
6. Zhong L, Li Y, Xiong L et al (2021) Small molecules in targeted cancer therapy: advances, challenges, and future perspectives. Signal Transduct Target Ther 6(1):201
7. Zhou Z, Li M (2022) Targeted therapies for cancer. BMC Med 20(1):90
8. Sochacka-Ćwikła A, Mączyński M, Regiec A (2021) FDA-approved drugs for hematological malignancies—the last decade review. Cancers (Basel) 14(1):87. https://doi.org/10.3390/cancers14010087
9. Wang S, Chen K, Lei Q et al (2021) The state of the art of bispecific antibodies for treating human malignancies. EMBO Mol Med 13(9):e14291
10. van Erp EA, Luytjes W, Ferwerda G, van Kasteren PB (2019) Fc-mediated antibody effector functions during respiratory syncytial virus infection and disease. Front Immunol 10:548. https://doi.org/10.3389/fimmu.2019.00548
11. Janeway CA Jr, Travers P, Walport M, et al (2001) The structure of a typical antibody molecule. Immunobiology: the immune system in health and disease. 5th edn. Garland Science, New York, https://www.ncbi.nlm.nih.gov/books/NBK27144/
12. Suurs FV, Lub-de Hooge MN, de Vries EGE, de Groot DJA (2019) A review of bispecific antibodies and antibody constructs in oncology and clinical challenges. Pharmacol Ther 201:103–119
13. Wang Q, Chen Y, Park J et al (2019) Design and production of bispecific antibodies. Antibodies (Basel) 8(3):43
14. Duell J, Lammers PE, Djuretic I et al (2019) Bispecific antibodies in the treatment of hematologic malignancies. Clin Pharmacol Ther 106(4):781–791
15. Lejeune M, Köse MC, Duray E et al (2020) Bispecific, T-cell-recruiting antibodies in B-cell malignancies. Front Immunol 7(11):762
16. Klein C, Schaefer W, Regula JT (2016) The use of CrossMAb technology for the generation of bi- and multispecific antibodies. Mabs 8(6):1010–1020
17. Yanakieva D, Pekar L, Evers A et al (2022) Beyond bispecificity: controlled Fab arm exchange for the generation of antibodies with multiple specificities. Mabs 14(1):2018960
18. Paul S, Connor J, Nesspor T et al (2016) An efficient process of generating bispecific antibodies via controlled Fab-arm exchange using culture supernatants. Protein Expr Purif 121:133–140
19. Li H, Er Saw P, Song E (2020) Challenges and strategies for next-generation bispecific antibody-based antitumor therapeutics. Cell Mol Immunol 17(5):451–461
20. Moshaver B, Wouters RF, Kelder A et al (2019) Relationship between CD34/CD38 and side population (SP) defined leukemia stem cell compartments in acute myeloid leukemia. Leuk Res 81:27–34
21. Kang J, Sun T, Zhang Y (2022) Immunotherapeutic progress and application of bispecific antibody in cancer. Front Immunol 13:1020003
22. Allen C, Zeidan AM, Bewersdorf JP (2021) BiTEs, DARTS, BiKEs and TriKEs—are antibody based therapies changing the future treatment of AML?. Life 11(6):465
23. Chen RP, Shinoda K, Rampuria P et al (2022) Bispecific antibodies for immune cell retargeting against cancer. Expert Opin Biol Ther 22(8):965–982
24. Roda-Navarro P, Álvarez-Vallina L (2019) Understanding the spatial topology of artificial immunological synapses assembled in T cell-redirecting strategies: a major issue in cancer immunotherapy. Front Cell Dev Biol 7:370
25. Bretscher PA (1999) A two-step, two-signal model for the primary activation of precursor helper T cells. Proc Natl Acad Sci U S A 96(1):185–190
26. Wei J, Yang Y, Wang G, Liu M (2022) Current landscape and future directions of bispecific antibodies in cancer immunotherapy. Front Immunol 13:1035276
27. Kamphorst AO, Wieland A, Nasti T et al (2017) Rescue of exhausted CD8 T cells by PD-1-targeted therapies is CD28-dependent. Science 355(6332):1423–1427
28. Seung E, Xing Z, Wu L et al (2022) A trispecific antibody targeting HER2 and T cells inhibits breast cancer growth via CD4 cells. Nature 603(7900):328–334
29. Peper-Gabriel JK, Pavlidou M, Pattarini L et al (2022) The PD-L1/4-1BB bispecific antibody-anticalin fusion protein PRS-344/S095012 elicits strong T-cell stimulation in a tumor-localized manner. Clin Cancer Res 28(15):3387–3399

30. Jeong S, Park E, Kim HD et al (2021) Novel anti-4-1BB×PD-L1 bispecific antibody augments anti-tumor immunity through tumor-directed T-cell activation and checkpoint blockade. J Immunother Cancer 9(7):e002428. https://doi.org/10.1136/jitc-2021-002428
31. Hinner MJ, Aiba RSB, Jaquin TJ et al (2019) Tumor-localized costimulatory T-cell engagement by the 4-1BB/HER2 bispecific antibody-anticalin fusion PRS-343. Clin Cancer Res 25(19):5878–5889
32. Hutchings M, Dickinson MJ, Gritti G et al (2024) Englumafusp Alfa (CD19-4-1BBL) Combined with Glofitamab Is Safe and Efficacious in Patients with r/r B-NHL: Extended Follow up Analysis of the Dose-Escalation Part of Phase 1 Trial BP41072. Blood 144(Supplement 1):990.
33. Karczmarczyk A, Korpysz M, Bilska S et al (2022) Programmed cell death-1 and its ligands as targets for therapy of multiple myeloma patients. Cancer Manag Res 14:1267–1281
34. Sadeghi M, Khodakarami A, Ahmadi A et al (2022) The prognostic and therapeutic potentials of CTLA-4 in hematological malignancies. Expert Opin Ther Targets 26(12):1057–1071
35. Shapiro M, Herishanu Y, Katz BZ et al (2017) Lymphocyte activation gene 3: a novel therapeutic target in chronic lymphocytic leukemia. Haematologica 102(5):874–882
36. Tan J, Tan H, Li Y (2023) Targeting TIM-3 for hematological malignancy: latest updates from the 2022 ASH annual meeting. Exp Hematol Oncol 12(1):62
37. Jin S, Zhang Y, Zhou F et al (2022) TIGIT: A promising target to overcome the barrier of immunotherapy in hematological malignancies. Front Oncol 12:1091782
38. Seto T, Sam D, Pan M (2019) Mechanisms of primary and secondary resistance to immune checkpoint inhibitors in cancer. Med Sci (Basel) 7(2):14. https://doi.org/10.3390/medsci7020014
39. Zhang T, Lin Y, Gao Q (2023) Bispecific antibodies targeting immunomodulatory checkpoints for cancer therapy. Cancer Biol Med 20(3):181–195
40. Katz BZ, Herishanu Y (2014) Therapeutic targeting of CD19 in hematological malignancies: past, present, future and beyond. Leuk Lymphoma 55(5):999–1006
41. Aparicio-Pérez C, Carmona M, Benabdellah K, Herrera C (2023) Failure of ALL recognition by CAR T cells: a review of CD 19-negative relapses after anti-CD 19 CAR-T treatment in B-ALL. Front Immunol 14:1165870
42. Jen EY, Xu Q, Schetter A et al (2019) FDA approval: blinatumomab for patients with B-cell precursor acute lymphoblastic leukemia in morphologic remission with minimal residual disease. Clin Cancer Res 25(2):473–477
43. Topp MS, Gökbuget N, Stein AS et al (2015) Safety and activity of blinatumomab for adult patients with relapsed or refractory B-precursor acute lymphoblastic leukaemia: a multicentre, single-arm, phase 2 study. Lancet Oncol 16(1):57–66
44. Kantarjian H, Stein A, Gökbuget N et al (2017) Blinatumomab versus chemotherapy for advanced acute lymphoblastic leukemia. N Engl J Med 376(9):836–847
45. Martinelli G, Boissel N, Chevallier P et al (2017) Complete hematologic and molecular response in adult patients with relapsed/refractory Philadelphia chromosome-positive B-precursor acute lymphoblastic leukemia following treatment with blinatumomab: results from a phase II, single-arm, multicenter study. J Clin Oncol 35(16):1795–1802
46. Litzow MR, Sun Z, Mattison RJ et al (2024) Blinatumomab for MRD-Negative Acute Lymphoblastic Leukemia in Adults. N Engl J Med 391(4):320–333.
47. Gökbuget N, Dombret H, Bonifacio M et al (2018) Blinatumomab for minimal residual disease in adults with B-cell precursor acute lymphoblastic leukemia. Blood 131(14):1522–1531
48. Jain T, Litzow MR (2020) Management of toxicities associated with novel immunotherapy agents in acute lymphoblastic leukemia. Ther Adv Hematol 11:2040620719899897
49. Pizzitola I, Anjos-Afonso F, Rouault-Pierre K et al (2014) Chimeric antigen receptors against CD33/CD123 antigens efficiently target primary acute myeloid leukemia cells in vivo. Leukemia 28(8):1596–1605
50. Tian C, Chen Z (2023) Immune therapy: a new therapy for acute myeloid leukemia. Blood Sci 5(1):15–24
51. Uy GL, Aldoss I, Foster MC et al (2021) Flotetuzumab as salvage immunotherapy for refractory acute myeloid leukemia. Blood 137(6):751–762
52. Winer ES, Maris M, Sharma MR et al (2022) A phase 1, first-in-human, dose-escalation study of MGD024, a CD123 x CD3 bispecific Dart® molecule, in patients with relapsed or refractory CD123-positive (+) hematologic malignancies. Blood 140(Supplement 1):11753–11754
53. Ravandi F, Bashey A, Foran J et al (2023) Phase 1 study of vibecotamab identifies an optimized dose for treatment of relapsed/refractory acute myeloid leukemia. Blood Adv 7(21):6492–6505
54. Watts J, Maris M, Lin TL et al (2022) Updated results from a phase 1 study of

APVO436, a novel bispecific anti-CD123 x anti-CD3 Adaptir™ molecule, in relapsed/refractory acute myeloid leukemia and myelodysplastic syndrome. *Blood* 140(Supplement 1):6204–6205
55. Uckun FM, Lin TL, Mims AS et al (2021) A clinical phase 1B study of the CD3xCD123 bispecific antibody APVO436 in patients with relapsed/refractory acute myeloid leukemia or myelodysplastic syndrome. Cancers 13(16):4113
56. Boyiadzis M, Desai P, Daskalakis N et al (2023) First-in-human study of JNJ-63709178, a CD123/CD3 targeting antibody, in relapsed/refractory acute myeloid leukemia. Clin Transl Sci 16(3):429–435
57. Dhillon S (2017) Obinutuzumab: a review in rituximab-refractory or -relapsed follicular lymphoma. Target Oncol 12(2):255–262
58. Budde LE, Sehn LH, Matasar M et al (2022) Safety and efficacy of mosunetuzumab, a bispecific antibody, in patients with relapsed or refractory follicular lymphoma: a single-arm, multicentre, phase 2 study. Lancet Oncol 23(8):1055–1065
59. Bannerji R, Arnason JE, Advani RH et al (2022) Odronextamab, a human CD20×CD3 bispecific antibody in patients with CD20-positive B-cell malignancies (ELM-1): results from the relapsed or refractory non-Hodgkin lymphoma cohort in a single-arm, multicentre, phase 1 trial. Lancet Haematol 9(5):e327–e339
60. Dickinson MJ, Carlo-Stella C, Morschhauser F et al (2022) Glofitamab for relapsed or refractory diffuse large B-cell lymphoma. N Engl J Med 387(24):2220–2231
61. Thieblemont C, Phillips T, Ghesquieres H et al (2023) Epcoritamab, a novel, subcutaneous CD3xCD20 bispecific T-cell–engaging antibody, in relapsed or refractory large B-cell lymphoma: dose expansion in a phase I/II trial. J Clin Oncol 41(12):2238–2247
62. Kim TM, Taszner M, Cho S-G et al (2022) Odronextamab in patients with relapsed/refractory (R/R) follicular lymphoma (FL) grade 1-3a: Results from a prespecified analysis of the pivotal phase II study ELM-2. Blood 140(Supplement 1):2280–2282
63. Shao R, Li Z, Xin H et al (2023) Biomarkers as targets for CAR-T/NK cell therapy in AML. Biomarker Res 11(1):65
64. Salvaris R, Ong J, Gregory GP (2021) Bispecific antibodies: a review of development, clinical efficacy and toxicity in B-cell lymphomas. J Pers Med 11:5
65. Bartlett NL, Sehn LH, Matasar MJ et al (2022) Mosunetuzumab monotherapy demonstrates durable efficacy with a manageable safety profile in patients with relapsed/refractory follicular lymphoma who received ≥2 prior therapies: Updated results from a pivotal phase II study. Blood 140(Supplement 1):1467–1470
66. Bachy E, Le Gouill S, Di Blasi R et al (2022) A real-world comparison of tisagenlecleucel and axicabtagene ciloleucel CAR T cells in relapsed or refractory diffuse large B cell lymphoma. Nat Med 28(10):2145–2154
67. Goebeler ME, Knop S, Viardot A et al (2016) Bispecific T-cell engager (BiTE) antibody construct blinatumomab for the treatment of patients with relapsed/refractory non-Hodgkin lymphoma: final results from a phase I study. J Clin Oncol 34(10):1104–1111
68. Viardot A, Goebeler ME, Hess G et al (2016) Phase 2 study of the bispecific T-cell engager (BiTE) antibody blinatumomab in relapsed/refractory diffuse large B-cell lymphoma. Blood 127(11):1410–1416
69. Topp M, Dlugosz-Danecka M, Skotnicki AB et al (2023) Safety of AFM11 in the treatment of patients with B-cell malignancies: findings from two phase 1 studies. Trials 24(1):4
70. Jacobs R, Nair R, Cho SG et al (2023) S224: High complete response rate with TNB-486, a novel CD19xCD3 T-cell engager, in relapsed/refractory follicular lymphoma: Interim results from an ongoing phase 1 study. Hemasphere 7(Supplement):e60614fc
71. Kang C (2022) Teclistamab: first approval. Drugs 82(16):1613–1619
72. Garfall AL, Nooka AK, Donk NWCJvd et al (2024) Long-term follow-up from the phase 1/2 MajesTEC-1 trial of teclistamab in patients with relapsed/refractory multiple myeloma. J Clin Oncol 42(Supplement 16):7540–7540
73. Tomasson MH, Iida S, Niesvizky R et al (2024) Long-term survival and safety of elranatamab in patients with relapsed or refractory multiple myeloma: Update from the MagnetisMM-3 study. Hemasphere 8(7):e136
74. Bumma N, Richter JR, Jagannath S et al (2024) Linvoseltamab for treatment of relapsed/refractory multiple myeloma. J Clin Oncol 42:2702–2712
75. D'Souza A, Shah N, Rodriguez C et al (2022) A phase I first-in-human study of ABBV-383, a B-cell maturation antigen × CD3 bispecific T-cell redirecting antibody, in patients with relapsed/refractory multiple myeloma. J Clin Oncol 40(31):3576–3586
76. Rasche L, Schinke C, Touzeau C et al (2023) MM-492 long-term efficacy and safety results from the phase 1/2 MonumenTAL-1 study of talquetamab, a GPRC5D×CD3 bispecific antibody, in patients with relapsed/refractory multiple myeloma (RRMM) Clin Lymphoma

Myeloma. 2024; 24: S561-S562efractory multiple myeloma (RRMM). J Clin Oncol 41:8036–8036
77. Minnema MC, Krishnan AY, Berdeja JG et al (2022) Efficacy and safety of talquetamab, a G protein-coupled receptor family C group 5 member D x CD3 bispecific antibody, in patients with relapsed/refractory multiple myeloma (RRMM): Updated results from MonumenTAL-1. J Clin Oncol 40(Supplement 16):8015–8015
78. Laheij AMGA, van de Donk NWCJ (2023) Characterization of dysgeusia and xerostomia in patients with multiple myeloma treated with the T-cell redirecting GPRC5D bispecific antibody talquetamab. Support Care Cancer 32(1):20
79. Cho SF, Yeh TJ, Anderson KC, Tai YT (2022) Bispecific antibodies in multiple myeloma treatment: a journey in progress. Front Oncol 12:1032775
80. Wang C, Wang J, Che S, Zhao H (2023) CAR-T cell therapy for hematological malignancies: history, status and promise. Heliyon 9(11):e21776
81. Kouro T, Himuro H, Sasada T (2022) Exhaustion of CAR T cells: potential causes and solutions. J Transl Med 20(1):239
82. Safarzadeh Kozani P, Naseri A, Mirarefin SMJ et al (2022) Nanobody-based CAR-T cells for cancer immunotherapy. Biomarker Res 10(1):24
83. Long AH, Haso WM, Shern JF et al (2015) 4-1BB costimulation ameliorates T cell exhaustion induced by tonic signaling of chimeric antigen receptors. Nat Med 21(6):581–590
84. Frey N, Porter D (2019) Cytokine release syndrome with chimeric antigen receptor T cell therapy. Biol Blood Marrow Transpl 25(4):e123–e127
85. Sharma S, Wang D, Lon H-K et al (2022) A systematic meta-analysis of cytokine release syndrome incidence in B-cell maturation antigen-targeting chimeric antigen receptor T-cell therapy and bispecific antibodies for patients with relapsed and/or refractory multiple myeloma. Blood 140(Supplement 1): 10036–10038
86. Usmani SZ, Garfall AL, van de Donk NWCJ et al (2021) Teclistamab, a B-cell maturation antigen × CD3 bispecific antibody, in patients with relapsed or refractory multiple myeloma (MajesTEC-1): a multicentre, open-label, single-arm, phase 1 study. Lancet 398(10301):665–674
87. Jain MD, Smith M, Shah NN (2023) How I treat refractory CRS and ICANS after CAR T-cell therapy. Blood 141(20):2430–2442
88. Korst C, Groen K, Bosman PWC et al (2024) Prophylactic tocilizumab reduces the incidence of cytokine release syndrome in relapsed/refractory myeloma patients treated with teclistamab: Implications for outpatient step-up dosing. Hemasphere 8(7):e132
89. van de Donk NWCJ, Chari A, Mateos MV (2024) Mechanisms of resistance against T-cell engaging bispecific antibodies in multiple myeloma: implications for novel treatment strategies. Lancet Haematol 11(9):e693–e707
90. Yao Y, Hu Y, Wang F (2023) Trispecific antibodies for cancer immunotherapy. Immunol 169(4):389–399
91. Cotton AD, Nguyen DP, Gramespacher JA et al (2021) Development of antibody-based PROTACs for the degradation of the cell-surface immune checkpoint protein PD-L1. J Am Chem Soc 143(2):593–598
92. Klein C, Brinkmann U, Reichert JM, Kontermann RE (2024) The present and future of bispecific antibodies for cancer therapy. Nat Rev Drug Discov 23(4):301–319
93. Cohen YC, Magen H, Gatt ME et al (2025) Talquetamab plus teclistamab in relapsed or refractory multiple myeloma. N Engl J Med 392(2):138–149
94. Sun Y, Yu X, Wang X et al (2023) Bispecific antibodies in cancer therapy: target selection and regulatory requirements. Acta Pharmaceutica Sin B 13(9):3583–3597
95. Costa LJ, van de Donk NWCJ, Rosinol L et al (2024) Long-term follow-up from the phase 1/2 MajesTEC-1 trial of teclistamab in patients with relapsed/refractory multiple myeloma: subgroup analysis by lines of therapies. Hematol Transfus Cell Ther 46(S4):S515
96. Gökbuget Hervé N, Massimiliano D, Albrecht B et al (2018) Blinatumomab for minimal residual disease in adults with B-cell precursor acute lymphoblastic leukemia. Abstract Blood 131(14):1522–1531. https://doi.org/10.1182/blood-2017-08-798322
97. Martinelli G, Boissel N, Chevallier P et al (2021) Long-term follow-up of blinatumomab in patients with relapsed/refractory Philadelphia chromosome–positive B-cell precursor acute lymphoblastic leukaemia: Final analysis of ALCANTARA study. Eur J Cancer 146:107–114. https://doi.org/10.1016/j.ejca.2020.12.022
98. Short NJ, Bachireddy P, Huang X et al (2023) A phase II study of vibecotamab a CD3-CD123 bispecific T-cell engaging antibody for MRD-positive AML and MDS after hypomethylating agent failure. J Clin Oncol 41(16_suppl):TPS7076-TPS7076.https://doi.org/10.1200/JCO.2023.41.16_suppl.TPS7076
99. Ravandi F, Walter RB, Subklewe M et al (2020) Updated results from phase I dose-escalation

study of AMG 330 a bispecific T-cell engager molecule in patients with relapsed/refractory acute myeloid leukemia (R/R AML). J Clin Oncol 38(15_suppl):7508-7508. https://doi.org/10.1200/JCO.2020.38.15_suppl.7508

100. Mei MG, Corazzelli G, Morschhauser F et al (2023) Safety and Preliminary Efficacy of Sabestomig (AZD7789) an Anti-PD-1 and Anti-TIM-3 Bispecific Antibody in Patients with Relapsed or Refractory Classical Hodgkin Lymphoma Previously Treated with Anti-PD-(L)1 Therapy. Blood 142(Supplement 1): 4433-4433. https://doi.org/10.1182/blood-2023-180926
101. Liu WP, Jiang B, Shuang YR et al (2023) Safety and Efficacy of a-319 (a CD3xCD19 T cell engager) in Patients with Relapsed or Refractory B-Cell Non-Hodgkin Lymphoma. Blood 142(Supplement 1):6158-6158. https://doi.org/10.1182/blood-2023-188071
102. Qi J, Sun M, Zhou K et al (2023) EX103 a Novel CD20xCD3 Bispecific Antibody Showed a Favorable Safety and Potent Single-Agent Antitumor Activity in Heavily Pretreated Relapsed/Refractory Patients with B-Cell Non-Hodgkin Lymphoma. Blood 142(Supplement 1):6230-6230. https://doi.org/10.1182/blood-2023-173486
103. Song Y, Li Z, Li L et al (2023) GB261 an Fc-Function Enabled and CD3 Affinity De-Tuned CD20/CD3 Bispecific Antibody Demonstrated a Highly Advantageous Safety/Efficacy Balance in an Ongoing First-in-Human Dose-Escalation Study in Patients with Relapsed/Refractory Non-Hodgkin Lymphoma. Blood 142(Supplement 1):1719-1719. https://doi.org/10.1182/blood-2023-188444
104. Nair R, Jacobs R, Cho S-G et al (2023) High complete response rate with TNB-486 in relapsed/refractory follicular lymphoma: Interim results from an ongoing phase 1 study. J Clin Oncol 142(16_suppl):7524-7524. https://doi.org/10.1200/JCO.2023.41.16_suppl.7524
105. Dickinson MJ, Carlo-Stella C, Morschhauser F et al (2022) Glofitamab for Relapsed or Refractory Diffuse Large B-Cell Lymphoma. N Engl J Med 387(24):2220-2231. https://doi.org/10.1056/NEJMoa2206913
106. Ayyappan S, Kim WS, Kim TM et al (2023) Final Analysis of the Phase 2 ELM-2 Study: Odronextamab in Patients with Relapsed/Refractory (R/R) Diffuse Large B-Cell Lymphoma (DLBCL). Blood 142(Supplement 1):436-436. https://doi.org/10.1182/blood-2023-179818
107. Suvannasankha A, Kapoor P, Pianko MJ et al (2023) Abstract CT013: Safety and efficacy from the phase 1/2 first-in-human study of REGN5459 a BCMA×CD3 bispecific antibody with low CD3 affinity in patients with relapsed/refractory multiple myeloma. Cancer Res 83(8_Supplement):CT013-CT013. https://doi.org/10.1158/1538-7445.AM2023-CT013
108. Chari A, Minnema MC, Berdeja JG et al (2022) Talquetamab a T-Cell–Redirecting GPRC5D Bispecific Antibody for Multiple Myeloma. N Engl J Med 387(24):2232-2244. https://doi.org/10.1056/NEJMoa2204591
109. Topp MS, Duell J, Zugmaier G et al (2020) Anti–B-Cell Maturation Antigen BiTE Molecule AMG 420 Induces Responses in Multiple Myeloma. J Clin Oncol 38(8):775-783. https://doi.org/10.1200/JCO.19.02657
110. Lesokhin AM, Richter J, Trudel S et al (2022) Enduring Responses after 1-Year Fixed-Duration Cevostamab Therapy in Patients with Relapsed/Refractory Multiple Myeloma: Early Experience from a Phase I Study. Blood 140(Supplement 1):4415-4417. https://doi.org/10.1182/blood-2022-157547

Subject Index

Masahiro Yasunaga (ed.), *Bispecific Antibodies*, Methods in Molecular Biology, vol. 2997,
https://doi.org/10.1007/978-1-0716-5037-0

The Author(s), under exclusive license to Springer Science+Business Media, LLC, part of Springer Nature 2026